A Phenomenological Revision of E. E. Harris's Dialectical Holism

Fig. 1 Errol E. Harris at Northwestern University, early 1970s. (Photo recreated with permission from Nigel Harris)

James Schofield

A Phenomenological Revision of E. E. Harris's Dialectical Holism

palgrave
macmillan

James Schofield
Plymouth State University
Plymouth, NH, USA

ISBN 978-3-030-65028-5 ISBN 978-3-030-65029-2 (eBook)
https://doi.org/10.1007/978-3-030-65029-2

This Palgrave Macmillan imprint is published by the registered company Springer Nature Switzerland AG.
The registered company address is: Gewerbestrasse 11, 6330 Cham, Switzerland

Dedicated to all those beautiful souls who reflected, inspired, challenged, and knew me in Christchurch, New Zealand.

Contents

List of Figures

PART I

An Introduction to Dialectical Holism

CHAPTER 1

The Need for a New Metaphysics

1.1 The Historical Development of Dialectical Holism

In the winter of 2007, at Plymouth State University, NH, I took a seat in what was to be my first course in philosophy. As it happened, Dr. David Haight was my professor and the course text was Harris's *Cosmos and Anthropos* (1991). After subsequent readings, it was this very text—which I now believe had no business being in an introductory philosophy course—that was to provide a road map for my own approach to philosophy and an essential inspiration for the present work.

Harris grew up in apartheid South Africa and was educated at Rhodes and Oxford. Years later, Harris taught philosophy at Northeastern University, where he and Haight had become friends. Harris's career was poised in a unique span of history that permitted him to witness first-hand the rise and fall of logical positivism, the Second World War, and the quantum revolution. His direct experiences with these and many other key developments in philosophy and science contributed tremendously to his arguments in philosophy of science, ethics, and epistemology, and helped to solidify his overarching worldview. By framing specific theories across the natural sciences within an account of phenomenology and process philosophy, Harris produced a metaphysics that he believed to be our best bet of addressing some

J. Schofield, *A Phenomenological Revision of E. E. Harris's Dialectical Holism*,
https://doi.org/10.1007/978-3-030-65029-2_1

of our most long-standing philosophical problems. I aim to assess this system in light of recent developments in the very disciplines to which he appealed.

My claim is that by outlining and assessing Harris's system in the face of contemporary research there will emerge a form of mutual enlightenment: Harris's metaphysics reveals as yet unnoticed connections between and implications for a number of sympathetic theories, from psychology to cosmology on the one hand, while, on the other, recent research in these fields may also update the empirical content of Harris's original system. I maintain that the resulting metaphysics frames a transdisciplinary paradigm shift and provides a viable solution to the hard problem of consciousness.

Leaving aside monographs and handbooks devoted to the subjects, recent textbooks in metaphysics have largely excluded any mention of process ontology; while those textbooks devoted to the philosophy of science or mind have made little if any mention of *phenomenology*. What is distinctive of Harris's system is the effort to unite a transdisciplinary approach to consciousness with a metaphysics of process and a phenomenological methodology. Importantly, as will be a central theme in the chapters to follow, this approach led Harris to maintain that to be successful, philosophical theories of consciousness, cosmology, and everything in between had to be contextually examined in light of one another. This rendered Harris among the few system-builders who survived a mass extinction of his kind of philosophy by the hands of the empiricists in the early twentieth century. Nevertheless, various veins of research today—from 4E cognitive science to Bohmian mechanics and emergence—appear to at least in part and unwittingly be developing many of the same arguments as Harris had been advancing since 1965. What is especially noteworthy is that proponents of these mutually supportive camps do not appear to be fully aware of their would-be supporters working outside of their home discipline. For these reasons, illuminating Harris's efforts to establish such a transdisciplinary process philosophy is well warranted today.

To understand the metaphysics that Harris developed, and which I claim can be identified in numerous present-day theories, it is important to first become familiar with the traditions that have been influential in its formation. Roughly, but chronologically speaking, these traditions consist of the following four camps:

1. *Process ontology*—developed by ancient Greek philosophers and reignited in the early twentieth century by A.N. Whitehead (1861–1947),

these views hold that what exists (reality or being) is ultimately constituted by some kind of process(es).

2. *Objective* idealism—concerning the works of G.W.F Hegel (1770–1831) and his followers, this thesis argued that mind (Idea) is fundamental in nature and that its genesis aims at ***maximally coherent self-reflection*** (Absolute).
3. *Phenomenology*—first systematized by Edmund Husserl (1859–1938) and later supported by gestalt and pragmatic psychologies, it aims to produce a rigorous method for revealing the necessary conditions and qualities of experience.
4. *Systems theory*—K.L. von Bertalanffy (1901–1972) first proposed this theory about the nature of complex systems, which has since been widely adopted as a framework for understanding organization and emergence across many of the special sciences.

A straightforward contention to follow is that if these camps do in fact reflect the core principles of Harris's metaphysics, then some viable version of each must be consistent with one another. While I endorse this contention, I also go significantly further. I argue that Harris's metaphysics provides a means of linking current research and arguments from each of the above camps and that doing so will be both scientifically and philosophically fruitful. If this is so, then "dialectical holism" will have survived Harris's original formation.

Broadly construed, Harris's system fits neatly within the tradition of process philosophy, which is based on the notion that ontology or Being is fundamentally dynamic and that this should be the central focus in any philosophical examination of natural phenomena. Traditional philosophies involving neo-empirical theories, by contrast, tend to sidestep or merely ignore the primary questions that process philosophers consider of utmost importance:

(a) What is the role of mind in our experience of reality as becoming?
(b) Are there different types of becoming?
(c) If so, how ought we categorize the respective modes of becoming?
(d) Is there a common and fundamental (evolutionary) form to all modes of becoming?
(e) Is there a purpose or telos to the process(es) of nature?

Because the methods of traditional philosophy capture a static view of reality via formal systems such as predicate logic, process philosophers have had to develop their own concepts and methods to support their arguments. Proponents of this approach have consequently contributed immensely to theoretical philosophy, both in Europe and in the US. Throughout history numerous key figures have dealt with the above questions and posited answers that have been particularly influential for Harris's system. In what follows I provide a brief overview of these key historical theses and gesture towards how Harris has both utilized and critically responded to each.

1.1.1 *Ancient Philosophy*

In the Western tradition, Greek philosopher Heraclitus of Ephesus (born ca. 560 BCE) is recognized as the founder of process metaphysics and has been famously credited with the saying, "it is not possible to step twice into the same river."[1] He believed that ultimate reality is the transformation of a cosmic fire, which produces apparently stable forms of matter such as sea and earth only to consume them again in an ongoing cycle of creation and disintegration. While fire is postulated as an underlying, all pervasive cosmic order that is creative and self-moving (what he called *logos*), the changes produced by fire were believed to occur in regulated ways. Heraclitus articulated three seminal insights in the history of Western process philosophy: (1) he considered process or dynamicity as an explanatory feature, rather than just a feature of nature to be explained; (2) he argued that processes form organizational units and occur in a quantitatively measurable fashion; and (3) he contrasted dynamic transitions or alterations with dynamic permanence, that is, he differentiated between two forms of dynamicity (Kirk 1951).

For the later Stoics, the logos was still considered a cosmic fire, dynamic in all existence, but also as divine and *rationally directed*. By calling God "cosmic fire", the Stoics reaffirmed their *ontological* monism, the view that what exists is a *singular* whole. In this vein, God is not over and above matter, but is rather the purest and subtlest aspect of matter. For the stoics, our nearest approach to this subtlest nature is in our reason, which is a spark of the divine fire within the human being (Tsanoff 2015, 112). This conception of "reason" was an important anticipation of what would later become the guiding theme of objective idealism, which had tremendous influence on Harris's work. A remaining challenge for the Stoics and objective idealists alike, is that of reconciling conceptions of teleology (or purpose) and causal determinism in nature.

Aristotle (384–33 BCE) provided a characterization of matter that gave rise to the basic thesis of "substance metaphysics", which further blossomed into a plethora of analytic accounts that have traditionally been opposed to process philosophy. Nevertheless, many of Aristotle's conceptualizations also helped lay the foundation for and increased the sophistication of process thought today. One notable contribution in this regard includes his classification of changes (*Kinêsis*) as including generation, destruction, alteration, and locomotion. This classification rests on a sophisticated doctrine of "potential" versus "actual" features attributed to some continuous substances that could come into and go out of existence in predictable ways. His approach to such prediction largely depended upon an understanding of purpose in nature and formal order in its constituents, that is, the *scala naturae*.

In his *Categories*, Aristotle claims that *form* (*eidos*) makes something the type of thing it is by permitting and sometimes facilitating the changes it undergoes. Thus, he takes the form of an entity in question to be the "principle" of that being's unity, not the summation of its parts. Aristotle maintained that temporal order depends upon a prior order that holds between stages of a change. He claims that "change is the actuality of that which potentially is" (Physics, Bk 3, Ch. 1). By this Aristotle appears to mean that any change depends upon something existing prior to the change that holds the potential to be in the end state after the change (Coope 2009, 43). Hence, he argues that it is the aim of a thing and the stages of that aim that both account for a being's organization at any given stage and give rise to time itself.

Aristotle (1957) distinguished four types of causation, the first two being associated with a scientific conception of material interactions, whereas three and four have led to a great deal of debate in philosophy of science today:

- *Material*—influences by a material composition that bring about a particular resulting phenomenon (e.g. the clumpiness of clay).
- *Efficient*—influences by one object upon another in causal chains that result in a particular phenomenon (e.g. a potter moulding the clay).
- *Formal*—the organizational force that brings about a particular phenomenon (e.g. the form of a statue constraining its parts).

- *Final*—the purpose (or *telos*), for which some cause is done (e.g. the intention of the potter to form the clay into a statue).

Such an account of causation was deeply opposed to the atomistic views of Aristotle's time and was to become crucially important for Harris's later arguments against the reductive empirical theories of the early twentieth century. From early in his philosophical career at Rhodes University (1925) under the guidance of A.R. Lord, Harris was first subjected to these and many other ancient theories (Harris 2015, ff. 36). As evidenced by his detailed analyses of ancient and early modern theories in his *Fundamentals of Philosophy* (1969), Harris took to heart the long-standing traditional problems concerning reason, ontology, and the nature of causation. Consequently, he was ever mindful to contextualize his solutions within the historical debates surrounding these problems.

Along with most other process philosophers today, Harris faced the challenge of establishing a consistency between accounts of formal and final causation on the one hand and the theories of the natural sciences on the other, which typically recognize only efficient causation. In the current literature, avoiding reductivism typically involves appeals to *emergence*: the material generation of an irreducible property, relation, process, or function that arises at the level of a system-as-a-whole. As will be a recurring theme below (particularly Chap. 5), Harris provides extensive arguments from emergence for formal and final causation that are essential to his metaphysics.

1.1.2 Spinoza

Though not a process ontologist himself, Dutch philosopher Baruch Spinoza (1632–1677) provided systemic accounts of nature, mind, and knowledge that have been invaluable for Harris's system. Spinoza's philosophy signified a radical break from Rene Descartes's (1596–1650) widely accepted view that mind and matter constitute two separate substances in nature. Central to Spinoza's philosophy was the thesis that "substance" is both sufficient in itself (ontologically independent) and conceived through itself (conceptually independent) (Spinoza 2002, ff. 217). A central aim of Spinoza's system is that of distinguishing between our knowledge and the nature of substance. For Spinoza, this singular Substance is identified with God or the *infinite idea of* nature, while the diversity of phenomena in nature and our conceptualization of it is analysed in terms of *modes* and *attributes* of this Substance.

In his Ethics, Spinoza proposed that "mode" means the affections or modification of Substance that appear to occur through something other than Substance and by which it is conceived. Modes are thus ontologically and conceptually dependent upon Substance and are instantiated by all the ways things appear in the world. Spinoza introduced the idea of an "attribute" as that which the intellect perceives of Substance and constitutes its essence. Hence, attributes are posited as the properties that compose the singular Substance of nature. Though he claimed that there were infinite attributes, Spinoza maintained that only material extension and thought are available to the intellect. While attributes and their Substance depend upon one another, modes conceptually and ontologically depend upon their attribute. Importantly, while modes may be considered contingent insofar as their essences neither involve nor exclude their existence, those that do exist are necessitated by the essential attributes of Substance. Though he rejected teleology, Spinoza argues from analogy with geometric figures that this is because everything that is possible is made actual by the whole (E IP17; 33).

Harris published three texts on Spinoza's philosophy: *Salvation from Despair* (1974), *Spinoza's Philosophy an Outline* (1992), and *The Substance of Spinoza* (1995). Although Harris agreed with a qualified version of Spinoza's pantheistic thesis and found many of his arguments relevant for contemporary moral and political debates, my interest is only with Harris's appeals to Spinoza's understanding of the relation between nature and mind. For Spinoza, the more perfection or reality a substance has, the more attributes it has, which meant that contrary to Descartes, a substance can have any quantity of attributes according to its degree of perfection, so it is possible for the thinking substance to be extended. As will be discussed at length in Chap. 6, Harris has argued: "Spinoza's theory of body-mind relation is in effect an identity theory, and not, as is usually held, a theory of parallelism. Body and mind are one thing viewed in two alternative ways: as a mode of extension or as a mode of thought" (1973, 82). He claims that Spinoza conception of modes and aspects provides a framework for a non-reductive account of mind, which may be identified with the dynamic body as a whole (the relation being that of matter to form).

For Spinoza, the conatus (or self-preserving aim) of mind is that of acquiring adequate ideas in an effort to free oneself from the illusory passions. Though total completion of this task is not possible for any one life because only Substance is determined to act by itself alone (E ID7), individuals can purportedly achieve a significant increase in freedom by aiming at truth. Harris claims this consists in elucidating "the comprehensive and systematic coherence of the whole" (1995, 5). Harris's sustained argument has been that by identifying the idea and extension of nature with nature itself, Spinoza anticipates theories advanced by Hegel and Collingwood. The key difference being that the latter philosophers recognized that "Substance" must be qualified as a particular kind of process; one which links a coherence *theory of knowledge* and *dialectical evolution*:

> Further development of the implications of the nature of Substance reveals that it differentiates itself into a scale of forms—infinite attributes, infinite modes, finite things that follow from it of necessity in infinite ways (modes) [...] all governed and determined by the same principle of order inherent in the attribute. (1995, 13)

1.1.3 Hegel

German philosophers such as G.W.F. Hegel (1770–1831) and F.W.J. Schelling (1775–1854) attempted to establish a view of nature in which mind was fundamental. Growing out of Immanuel Kant's (1724–1804) tradition of *transcendental* idealism, these later idealists were at pains to overcome the distinction advanced in his *Critique of Pure Reason* between the *phenomenal* realm of conditioned experience and the unknowable *noumenal* realm of things in themselves (1998). These post-Kantian idealists dealt with Kant's challenge by developing phenomenological methods that focused upon the knowable appearance of the world and our capacities for reflective reasoning. This approach helped inspire the later phenomenological tradition and was invaluable in the development of Harris's system.

Hegel (1977) maintained that understanding reality required that we understand the genesis of consciousness as a process within nature, which he called the dialectic. This he considered a gradual self-awareness of the whole of being, that is, an evolution towards Absolute (*Begriffe*). This is described as a "concretization" because as thought becomes increasingly self-aware, it more fully manifests the process and conditions of its own becoming. This process was later articulated within the following triadic structure:

1. *Thesis*—presuming a realism about one's object, the moment of its particularity (multiplicity, abstractness, and externality) explicitly becomes the object of thought's immanent self-reflection. In so doing, what was initially considered an adequate representation of infinite thought is revealed to be an incomplete, finite object.
2. *Antithesis*—a realism about thought's object is suspended, which results in both self and idea being sceptically divided from their object (the world) via negation. Consequently, the *concept of* being *as a whole* is polarized.
3. *Synthesis*—finite determinations are understood as phases within a total movement. This movement is conceptualized as being governed by a principle of order that reveals the interdependence and unity of prior oppositions (e.g. self-object).

For Hegel, the inner movement of reality is the process of God coming to self-realization in and through the evolution of the universe and living beings. More specifically, Hegel argued that when fully understood, reality is the *Concept* being thought by God as manifested in a person's comprehension of this process as an enactment of wisdom. Just as humans continually correct their concepts of reality through a dialectical process, so God is said to become more fully manifested through the dialectical process of becoming. Ultimately, Hegel maintained that recognizing the dialectic within one's own mind permits a translation from the structure of our ideas (logic) to the essential features of nature (ontology), thereby identifying the former in the latter.

Though nearly every part of his system was in some way influenced by Hegel's work, Harris published two texts specifically devoted to his philosophy: *An Interpretation of the Logic of Hegel* (1983) and *The Spirit of Hegel* (1993). In the former (1983) Harris summarizes dialectical logic thus:

> True explanation, or proof, in dialectical logic, is nothing more nor less than the tracing out of the development of the Concept (or whole) itself—its self-specification. It is the process of self-explication of the system of the real, and explanation is seeing, or finding the proper place of each phase and item in the light of the whole and its principle of order. (171)

Crucial for later discussions, Harris's assertion that the function of "explication" is that of *clarifying the relations among phases* will take on increasing significance as we work towards his solution to the problem of consciousness. In his latter (1993) text, Harris expounded upon the essential insight from Hegel's line of reasoning that had guided much of his professional career, maintaining that dialectical evolution could be paired with modern sciences to provide a (monistic and process ontological) solution to the mind-body problem:

> The soul, and subsequently the mind, are presented as successive phases of integrating, idealizing activity, the effect of which is the attainment of progressively higher degrees of organization and diversified unity. What is unified and idealized is primarily the organic process of the body, for body and mind are not two substances in external relation, but one in which different dialectical phases of concrete organization are manifested. (173)

Harris's greatest development and criticism of Hegel has certainly been an update of compatible scientific theories, but Harris remained committed to the idea that Hegel's core thesis provided the best way to systematize the natural sciences and solve long-standing metaphysical problems. To this end, Harris spent much of his career in dialogue with physicists, biologists, and psychologists to work out the most empirically secure interpretation of ***dialectical evolution***, broadly construed. While Harris's resulting conception of evolution may provide a unique means of naturalizing mind, contra Harris, I argue that this reasoning does not entail objective idealism. Rather, I contend that if guided by pragmatist reservations and contemporary phenomenological insights, following Harris's own reasoning opens the way towards a much more fruitful research programme than either he or Hegel could clearly foresee.

1.1.4 Pragmatism

The formation of pragmatism developed through conversations among a small group of American philosophers in Cambridge. Chief among them were Charles S. Peirce (1839–1914), William James (1842–1910), and later John Dewey (1859–1952). While their respective visions of the philosophy differed in pronounced ways, I claim that at least the following currents of pragmatism are consistent with one another and may provide crucial developments within Harris's system.

A scientist and mathematical genius, Peirce was so influenced by the development of modern biology and the German idealists that he posited a triadic evolutionary process as primary to both phenomenology and nature:

- *Firstness*—signifies the qualitative immediacy characteristic of anything *in itself.*
- *Secondness*—highlights an irreducible relational opposition.
- *Thirdness*—references the network of *connections* in and through which any experience acquires its defining properties.

However, Peirce rejected Hegel's grand conclusion about our capacity to apprehend the whole of nature via Reason and proposed the pragmatist maxim: we should consider what practical effects we can conceive the object of our belief to have and then our conception of those effects is the whole of our conception of the object. A realist about science, he maintained that we work at our best when engaging in a cooperative community to overcome common problems. By extension, he claimed that truth is to be understood as the result of an endless investigation (1931, 5.565); that which is fated to be ultimately agreed upon by all investigators (1931, 5.407). Thus, he supported *fallibilism*, which holds that no questions are unanswerable, no answers are absolutely true, no formulations are final, and no level of examination is ultimate (1931, ff. CP 1.141). By implication, inquiry is never complete and so we ought to question the *habits* of thought and action, so as to keep our *inquiry* moving forward.

A philosopher and psychologist, William James developed the pragmatist epistemology, in which truth was not abstract, objective, or pre-established in rationalist principles, but a context-dependent aspect of human cooperation. He maintained that "we carve out everything", just as we identify constellations to serve human purposes. For James, neither do we create our truths out of nothing, nor is truth independent of human thought. He embraced the humanistic principle that we cannot weed out the human contribution and his metaphysics of process was endorsed on the basis that reality unfolds in and through human participation. Consequently, James maintained that there is no real distinction between unknown reality and the knowing consciousness; nor between objective matter and subjective mind. Reality, he claimed, is ultimately "pure experience", which is *neutral* between *subjective* phenomena and *objective* material events (1976).

Counter his analytic contemporaries, psychologist John Dewey developed his theories of mind and nature upon a Hegelian metaphysic. Specifically, it was Hegel's organicism, dynamism, and interrelated conceptions of reality that Dewey found so appealing for their ability to fuse subject and object (Dewey 1981, 6–7). Dewey drew upon conceptions of a reflex arc developed in laboratory experiments to reject both theses of mind-as-substance and contemporary scientific projects that reduced mind to brain states. He anticipated modern dynamic field theory by maintaining that mind is a range of dynamic interactions between organism and world. Taking life to be fundamental, he proposed the term "situation" to denote the ongoing feedback loops of development that exist between organism and environment, which he believed to constitute the very becoming of reality. For Dewey however, this view of the mind was framed as a realist, naturalistic, non-reductive, and emergentist process metaphysics.

Following these original founders, pragmatists have been deeply critical of metaphysics insofar as it is a pursuit of insight into reality itself. More recently, Putnam (2002) has been credited for making clear what metaphysical claims are being rejected: (1) there is a definite class of objects and properties that exist mind- and discourse-independently; (2) the world can, in principle, be truly and completely described by means of a single true theory, from a "God's-Eye View;" and (3) truth is to be understood as a non-epistemic relation of *correspondence* between propositions, sentences, and so on (i.e. the ultimate truth-bearers) and the non-linguistic items of the world itself. Instead, pragmatists have endorsed metaphysics as an inquiry into the fundamental, historically changing, and reinterpretable features of the *human world*. Following Kant, Pihlstrom (2011) claims pragmatism posits that the world emerges in and through our inquiry and world-categorizing practices. Metaphysics is thus reconceived as an examination of the basic features of a humanly categorized reality and the practice-embedded conditions necessary for us to be able to experience an objectively structured world.

Thus, process philosophy has been consistently endorsed by pragmatists since its original formation and has historically served as a unique bridge between process thought and phenomenology. "Temporality is, therefore, a fundamental feature of the world we live in, even if *nothing* is 'fundamental' in the metaphysical realist's sense" (Pihlstrom, 95). Accordingly, the basic nature of reality must be understood as a constant flux of evolution, one which is somehow continuous with our world-constructing activities. As Pihlstrom (2011) argues, this creates what can be called *transcendental* pragmatism: "if we cannot expect metaphysics to deliver a view of the world in itself, we must carefully consider how exactly we humans contribute to 'constituting' the world, to 'structuring' it into what it is for us" (98). Pihlstrom further recognizes that this is consistent with a methodological appeal to phenomenology in the service of uncovering the (transcendental) conditions of respective experiences, habits, and inquiries. If we cannot take a view from nowhere, we must always include the limitations of our knowledge and the habits that inform it within our metaphysical analyses. As Putnam (2002) holds, there are no value-independent facts and no fact-independent values. This renders an inherent connection between metaphysics, epistemology, and ethics and implies that *all are latent in our habitual inquiry of the world.*

Though Harris did not write at length about the pragmatist works, their developments in philosophy of science and metaphysics are important here for three central reasons. First, because the past decade has witnessed a sudden outpouring of works dedicated to linking embodied cognitive science with pragmatism and I find that although the process metaphysical insights that lay therein have not yet been adequately developed, the results are already compatible with Harris's system. Second, although James's conception of *pure experience* is inadequate on its own, I argue that the metaphysical notion of "*neutral monism*", to which he appealed provides perhaps the most important adjustment to Harris's system. Finally (in Chap. 8), I argue that many of Harris's realist conclusions about consciousness and reality are best maintained if they are subdued by the above epistemic reservations of contemporary pragmatism. The resulting system, I maintain, remains consistent with the pragmatist themes of linking phenomenology, epistemology, ethics, and metaphysics into a non-reductive schema, but Harris's work provides a far more rigorous paradigm across the natural sciences than has hitherto been endorsed by the proponents of pragmatism.

1.1.5 Whitehead

Alfred North Whitehead's (1861–1947) metaphysics grew out of a career steeped in logic, mathematics, and theoretical physics, the results of which rendered one of the most complex and revered process philosophies of the modern age. First working with Whitehead's ideas during his early research at Oxford, Harris found Whitehead's "philosophy of organism" to provide some much-needed scientific and logical rigor for the earlier philosophical systems he most admired. Nevertheless, he ultimately maintained that to the extent that Whitehead had succeeded, he had merely revised Hegel's arguments concerning dialectical evolution, while his idiosyncratic terminology appeared burdensome and unnecessary.

Whitehead's greatest philosophical work *Process and Reality* (1929, 1978) was written for the 1927 Gifford Lectures in Edinburgh. Counterintuitively, there are no items in Whitehead's ontology called "processes". Rather, process refers to the way in which the basic things, "events" come into existence and cease to exist. Conscientious of Zeno's paradox, Whitehead argued that there must be a becoming of continuity, but no continuity of becoming. This implies that events have no proper parts (i.e. atomism) and temporal instances are mere abstractions. Though, events must come about all at once, Whitehead admits that we can perform "genetic" analyses concerning the antecedents of a given event, which can occur in and reveal stages of its "becoming". This genetic analysis became the central method of Whitehead's system and relied upon an appeal to *internal relations*, that is, that events are what they are because of their relations to one another.

For Whitehead, every new event is generated from its own universe (an "ingression"), which is comprised of two domains: (1) all "*eternal objects*", ideal kinds or universals outside of space-time; and (2) all concrete events that are accessible to or "*prehended*" by the new event, which are necessarily restricted by the speed of light and thus occur in its backward light cone. As Simons explains, the genesis of events is deliberately described in quasi-psychological terms as *striving* to come into existence (or *concrescence*) by *surveying* all eternal objects via *conceptual prehension* (2009, 186). Events select among the eternal objects and all other events in the universe which should be *ingressed* in virtue of their *feeling* and the decision is made based upon the sum of negative and positive prehensions regarding a particular manifestation.

Whitehead's (1929, 1978) concepts of the "society" serves as an anticipation of what Harris would later propose as the "unifying principle" (introduced in Chap. 2). For Whitehead, the "society" meant any (becoming) actual occasion (or interrelated *nexus* of entities) serving as a (layered) environment that influences the becoming of its constituent events. Importantly, the members or occasions of a "society" exist in virtue of the *laws* of their society and bring their laws into being (ff. 91). Whitehead posited a supreme category of *creativity* on the basis that no two events ingress the same universe, but that each comes to be in a unique and creative way. By implication, what Whitehead calls the basic cell *of becoming* may have different epochs, each with their own laws or spatiotemporal dimensions. Whitehead's genetic analysis is thus ecological in the sense of treating each event as an organism in interdependent relation with its environment. Because his cosmology is composed of a hierarchy of *perceptions* (beginning with its pre-conscious analogue of prehension), Whitehead attributes sentience to everything down to electrons, which means he is ultimately a panpsychist. Within this schema, consciousness is a late phase and God is the ultimate concrescent *"event" within* nature.

For Harris, the above struck him as easily reducible to a variant of Hegelian idealism. Although Whitehead does not provide an explicit account of the method of process, Harris claims that what he does reveal is evidently dialectic in form. In his effort to preserve realism, Whitehead avoids a logical account of process, but Harris claims, he posits the stages of concrescence as that of an organic development, which Hegel called *Geist* (1993, 253). Harris claims that Whitehead's conception of the process of concrescence occurs through the transition from one "epoch" to the next via prehension, while the principles involved are the same throughout:

> The phases are those of dative ingression, conformal physical feeling, conceptual feeling, and comparative feeling. The first two are virtually one, for conformal feelings are simply the veridical prehensions of the data. So we have a Hegelian triad: first immediacy, then distinction and differentiation, and finally articulated synthesis. (1993, 254)

In its final draft, Harris explains that the thesis he had developed at Oxford "had argued that the concept of evolution could build a bridge between traditional Idealism and Realism" (2015, 176). He goes on to say that towards this end, reading Spinoza through Whitehead's works provided a framework that ruled out reductionist and dualistic depictions of mind. While, in a later work he elaborates how reading Hegel through Whitehead revealed a way of positing an idealism in realist terms:

> In his mature thought, as we have seen, Nature is the process of concrescence throughout which principles of definiteness, the objects of conceptual prehensions, which are the germ of mind, are everywhere immanent [...] and this mentality develops until it emerges as consciousness, which attains

> its ultimate and complete realization in the consequent nature of God. So Whitehead's philosophy reconciles the Realism and the Idealism of his day. (1993, 256)

While Harris embraces Whitehead's methodological appeal to internal relations and appreciates his conception of creative evolution, he diverged from Whitehead's philosophy in numerous ways. Principally, Harris endorses both cosmological and biological teleology, while Whitehead rejected it. We find further disagreement when considering the topics of universals and the very conception of concrescent reality. For Whitehead, universals are "eternal objects" located in the mind of God, who is an actual entity. According, the divine actuality primordially orders the eternal objects into ideal patterns arranged in a hierarchy of complexity. For Harris however, it can be said that there exists only one universal and that it is enacted through the whole of physical evolution—what he believed could unite abstract and concrete domains. Contrary Whitehead, for Harris the universals identified by thought are understood as shadows (so to speak) of a unified process that constitutes ultimate reality. While the universe-as-a-whole is construed as *organic*, Harris rejects the notion that individual occasions may be deemed either sentient or atomic. For Harris, sentience (just as consciousness) is an accomplishment of a system that has achieved a sufficient level of organization.

1.1.6 Gestalt Psychology

At the end of the nineteenth century there were many complex discussions concerning the nature of consciousness (e.g. Brentano, Russell, and Husserl) and the necessary methods for studying it (e.g. Wundt, G.T. Fechner, and James). Though many of these researchers read and learned from one another initially, over time each diverged from one another and inspired their own approaches to philosophy and psychology. These included analytic philosophy, pragmatism, phenomenology, and eventually behaviourism. By the 1950s behaviourist psychology had reached its peak and managed to reject the disciplines of idealism and phenomenology as idle introspection. Proponents of behaviourism, such as John Watson and Wilhelm Wundt distrusted the methods of introspection under the assumption that it led to self-absorption and arbitrary observations. By the late twentieth century these schools were hardly communicating anymore, having been entrenched within their own tradition and preferring to argue amongst themselves.

Harris was certainly unusual among philosophers in the early twentieth century for not only dedicating himself to studying both analytic and phenomenological philosophies of mind, but for also making a concerted effort to follow the contemporary research in psychology. Though he appreciated the increasing rigor of scientific psychology, Harris found the philosophical assumptions of the early twentieth-century empiricists deeply inadequate for understanding the natures of consciousness and reality. Specifically, gestalt psychology provided an important logic crucial in bypassing the behaviourist programme and developing Harris's own philosophy of mind.

The main figures of gestalt psychology were Max Wertheimer (1880–1943), Wolfgang Kohler (1887–1967), and Kurt Kafka (1887–1941). Proponents rejected the hitherto accepted Cartesian framework that mind, and matter were *separate substances* (dualism), as well as the theory that experiences were essentially *bundles of sense data* (atomism). Gestalt theory began as an effort to establish organizing principles of experience, whereby meaning and understanding are established not in isolatable increments, but as immediate *forms* that appear within human perception. "Gestalts were not imposed on experience by the mind but were discovered in experience. Gestalts were objective, not subjective [...] physically real, natural self-organizations in nature, in the brain, and in experience, all of them isomorphic to one another" (Leahey 2000, 272). Originally guided by six principles of experiential organization, gestalt psychology introduced concepts that were radical at the time and developed an empirical research programme whose ripple effect can still be seen in phenomenology and cognitive science today.

Following gestalt psychology, Harris maintains there is an isomorphism between phenomenal and physiological wholes and so we ought to look for "laws of organization [...] not physical forces but dynamic principles governing the way in which phenomena group themselves into patterns and parts are drawn together into wholes" (1965, 393). Although it would be more accurate within Harris's system to exchange "isomorphism" with "homeomorphism", the connection between gestalt theory and Harris's metaphysics is so significant that one could summarize his work as an *effort to establish the underlying process, by which such wholes are generated within phenomenology and across all the natural sciences.* Nevertheless, Harris encountered disagreements within the gestalt theory as well.

While gestalt psychology was an advance over behaviourism, Harris's approach differed in his rejection of representationalism. In his *Formal, Transcendental and Dialectical Thinking* (1987), Harris followed Whitehead's, contention that while logic and science serve as excellent tools, "exactness is a fake" and so there is need for synthetic, transcendental, and dialectical thought. Here, Harris appealed to J.J. Gibson's (1950) *Perception and the Visual World* to both endorse many of the gestalt psychological conceptions and argue that mind does not represent an external world. Harris claimed that sensory stimuli are always perceptually ambiguous, which

> require elaborate organization and interpretation in reference to a context both spatial and temporal and to a body of already acquired objective knowledge in order to become perceptual objects. There is thus no scientific evidence of any process of copying by the percipient organism of objects in its environment nor any transmission of precise replica of external things. (66)

A further disagreement between Harris and gestalt psychology concerned their early endorsements of *methodological* reductivism. Though proponents of gestalt psychology maintained the "reality" of mental phenomena, they also argued that the psychological descriptions and predictions should (at least in principle) be reduced to physiology because the theoretical language of the former can be mapped, one-to-one, into theories of the latter (i.e. "isomorphism"). While the gestalt psychologists provided some important theoretical tools for understanding mental and neurological processes, Harris ultimately drew a great deal more from other sciences to achieve his aim of framing mind within nature. This meant that Harris was largely an outlier in his effort to synthesize idealist metaphysics, scientific data, and phenomenological methods in such a way that left room for teleology and emergence within his system.

1.1.7 Phenomenology

The discipline of phenomenology originated with the works of Franz Brentano (1838–1917) and his student Edmund Husserl (1983, 1989) as part of the broader category of continental philosophy. This tradition began in Europe and included a number of mutually complementary schools such as existentialism, *hermeneutics* (theory of interpretation), *postmodernism* and *post-structuralism*. Building on Husserl's work, his student

Martin Heidegger (1889–1976) first proposed a phenomenological ontology (2010) and later philosophers such as Sartre (1905–1980), and Merleau-Ponty (1908–1961) made key contributions to phenomenology in the following decades. Harris appreciated Husserl's methodological developments and Merleau-Ponty's emphasis on the role of the body in his theory of mind. Nevertheless, he ultimately found many of the earliest proponents of phenomenology inadequate for their failure to articulate a sufficient ontological foundation for their theories of consciousness.

What characteristically distinguished the phenomenological study of mind from the analytic approach was that the latter endeavoured to answer particular questions about the nature of mind by outlining a number of metaphysical positions (e.g. dualism, functionalism, identity theory) before any attention is paid to experience. Phenomenology brackets these metaphysical issues and tries to clarify the structures and conditions (the *how*) of experiences themselves. What is distinctive of Harris's approach is that while embracing this move to bracket metaphysical assumptions, he ends up re-establishing an ontological foundation in and through his examination of the conditions of experience. This foundation draws from the holism of gestalt psychology, the emergence of systems theory, and a fundamentality of process. Though phenomenologists have not historically been overt advocates of process ontology, because its methods presuppose many of the above (a–e) questions, numerous phenomenologists today, such as philosopher Evan Thompson (2007), have stated that their approach does indeed imply an ontology of process. From the above, we can see that numerous classificatory strategies have been employed in process philosophy, so if contemporary phenomenologists are indeed committed to this otology, they must argue among themselves about which one is to be preferred.

Contemporary process philosophers and phenomenologists all tend to agree that the world is an assembly of physical, organic, social, and cognitive processes that interact at and across levels of dynamic organization. This is diametrically opposed to the empiricist and materialist paradigm, which is committed to foundationalist ontologies of substances and particles. While the Western philosophical tradition of Parmenides (atomism) focuses on "what there is", process philosophers have developed taxonomies of dynamics and identifies basic entities based upon what each "does'" (i.e. its function or purpose). Phenomenologists and process philosophers tend to agree that "activity-like", non-developmental processes or "goings-on" should be distinguished from "accomplishment-like" processes. They also appreciate that processes take up space and take time, but may not be identified with, nor individuated by *the contents of space-time regions.* On this view, an entity should be individuated in terms of its functioning within a context. Within this broad framework, process philosophers and phenomenologists can debate how such a world of processes relates to the human mind, and how the dynamic nature of reality relates to our scientific theories more specifically. One of my central arguments in the chapters to follow is that Harris has anticipated many developments of contemporary phenomenology and proposed further ontological clarifications and scientific applications for a phenomenological methodology.

Husserl calls the phenomenological point of departure the epoché (ἐποχή epokhē), which he considers the moment where all judgments about the existence of the external world are suspended. As Harris explains, this method emphasizes how objects appear to our awareness, not how they are in themselves:

> The content must be regarded simply as it is presented primordially with its own direct and immediate evidence. [...] In the case of material objects, he often seems to suggest that this is direct sensuous perception although there are massive reasons, many of which Husserl himself provides, for denying that direct sensuous perception ever presents us with a material object as "itself given". (1987, 95)

All objects of awareness, Husserl contends, occur within the context of a horizon or lifeworld (*di Lebenswelt*) that endows the object with meaning via relations to contrasting objects. Quite agreeably, Harris finds that Husserl's requirement of such "systematic relations between different presentations, both of the object concerned and of others with which it is associated or connected" to be suggestive of "a coherence theory of truth" (96). On Harris's account, while the traditional phenomenological tools and arguments provided a great deal of support for his approach to ontology, they were ultimately found lacking:

> The constitutive activity which Husserl seeks to describe is apparently one of interpretation and judgment, informed and supported by a background of knowledge, a context (or "horizon") systematically constructed in the course of experience according to principles of order which are a priori—that is, inherent in the intentional performance of the transcendental subject. This is all very Kantian and, so far as it goes, is none the worse for being so. But the system, whether of categories, constitutive activity, or experience

> as a finished product emerges from Husserl's exposition with any degree of clarity only in some few cases [...] For the most part, the principles are not definitively revealed nor the logical structure of the intentional operations described in sufficient detail or with the requisite degree of precision. (1987, 102–103)

Specifically, Harris goes on to claim that despite the protests from his followers, "Husserl's transcendentalism is ineradicably subjective" (113). Harris argues that even if the subjectivism of traditional phenomenology is avoided, a much deeper problem awaits:

> Accordingly, my transcendental ego is not the same as my psychophysical self, although I say "I" in both cases, for the content of each differs from that of the other. The transcendental "I" is prior to all possible objects, whereas my psychophysical self is an object experienced in part introspectively and in part through outer sense. (Ibid.)

The problem for Husserl and contemporary phenomenologists, is that the intersubjective world that gives rise to objectivity is essentially constituted by one's own transcendental awareness. If we also wish to maintain that *some* objective world has given rise to our transcendental awareness, then each becomes necessarily prior to the other (the *priority problem*).

Faced with the challenge of relating mind to the process of nature, Harris drew inspiration from Hegel, which meant that he endorsed an innate capacity in consciousness to apprehend the process of its own genesis. Indeed, these "Hegelian" sympathies were responsible for inspiring Harris to take an interdisciplinary approach to consciousness nearly 40 years before the collaborative methodology gained mainstream popularity. Harris eventually maintained that physics, biology, and psychology had revealed a holistic metaphysics and a world that was "itself a manifestation of the concrete universal particularizing itself as a scale of forms as Collingwood and Hegel had foretold" (2015, 176).

The primary theoretical link between process philosophers and phenomenology is an agreement that processes give rise to and provides the conceptual support of *emergence*, which can deal with a host of phenomena that are presumed *irreducible*. On Harris's account, an appeal to process may naturalize mind and normativity, as well as provide common conceptual structures applicable to each of the natural sciences, thereby solving the above priority problem.[2] In the following section I take a closer look at how proponents of process ontology and phenomenology have appealed to various twentieth-century theoretical developments in the special sciences to support their positions.

1.2 Process-Phenomenology and Contemporary Science

On the substance metaphysics account there is a reductionist interpretation of processes in which, what ultimately exists are not people, galaxies, and societies, but only the relations among fundamental particles. This reductive account thereby implicitly appeals to a promissory note to the effect that eventually the human mind, like all complex phenomena, will be understood in terms of fundamental physics. Hence, without process we are left with mere presupposition of respective (identifiable) substances or structures that are assumed to be exhausted through static definitions. Alternatively, the strongest arguments for process philosophy and phenomenology alike, over that of its competitors in recent years, has been that their scope of explanations stretch across a wider range of phenomena, account for more relationships, and generate better philosophical descriptions.

As Harris encountered through much of his career, both process philosophers and phenomenologists have faced a common dismissal in mainstream discussions due to long-standing biases in favour empiricist and analytic assumptions. These assumptions initially led to an outright rejection of metaphysics and cosmology in the early twentieth century, but later accounts accepted only the strictest ontologies of materialism, reductionism, and atomism. In an effort to capture some of the philosophical motivation behind Harris's approach and what this means in terms of real-world scientific theories, it is essential to become familiar with a number of problems that have traditionally defined Western philosophy. To Harris's mind (and for all sympathetic approaches), taking phenomenology and process ontology together provides a means of advancing beyond five dichotomies typically inherent in analytic metaphysics:

1. *Abstract-concrete*—process metaphysics abandons the substance-metaphysical principle that concrete entities are determinate while general or indeterminate entities are abstract and thereby incapable of initiating or undergoing change. For a process-phenomenological

approach, both abstract and concrete elements must be individuated by attending to (the dynamics of) the wider system in which they occur (addressed in Chap. 2).

2. *Part-whole*—recent appeals to systems biology, relativity, and quantum theories have been considered by many process philosophers (including Harris), to signal the demise of classical atomism and empiricism. To many philosophers and scientists alike, complex biological systems, particle entanglement, and relativity suggests a kind of emergent holism in nature, in which analysis can only be undertaken upon processes of relation across multiple spatiotemporal domains (addressed in Chap. 5).
3. *Fact-norm*—the dichotomy of "fact" and "norm" can be bridged by maintaining that some natural processes realize a low degree of normativity. For example, in far-from-equilibrium systems each component process presupposes every other for its own occurrence. In such contexts, organizational dependence among individual processes is not a matter of linear causation but *constraints* (i.e. formal and final causes) that reveal how some processes *function for the sake of" their system-as-a-whole* (examined in Chap. 6).
4. *Mind-matter*—most important for the present text, following the contention that there exists an underlying process for both mind and matter, any strict division of these domains in terms of properties or substances is abandoned. This "gestalt switch" dissolves the traditional mind-body problem by permitting basic constituents of reality to be understood as dynamical relations capable of exhibiting both "mental" and "physical" aspects according to context (addressed in Chap. 7).
5. A priori-a posteriori—a long-standing challenge in Western philosophy (particularly empiricism) has been the effort to relate analytic reasoning conducted prior to empirical observation with synthetic reasoning, which brings together data that has been collected from the world. This brings to the fore some questions concerning both the justification of our reasoning and the degree to which humans have any innate knowledge. A process-phenomenological account may provide a means of bridging this divide by reconceiving the natures of thought and observation (considered in Chap. 8).

As noted in the previous section, the main challenge for those drawing from phenomenology was the criticism that it amounts to mere introspection. From the late 1960s, Hubert Dreyfus was nearly the only voice advocating for the utility of phenomenological methods from within the analytic tradition. Following Merleau-Ponty's (1962, 1963) lead, more phenomenologists began to argue that mind was deeply related to *dynamics of the living* being *as a whole*, which eventually opened a new line of research in the cognitive neurosciences in the 1970s. By the 1980s research had returned to a focus upon experience from both phenomenological and analytic camps. The nature of consciousness was further raised to the status of "scientific question" with further help of authors like Thomas Nagel (1974), Dan Dennett (1991), and David Chalmers (1996), who in one way or another motivated the problem of consciousness. Nevertheless, analytic and phenomenological methods differed significantly.

The most general way to capture the difference between an analytic and phenomenological approach to the study of mind is that while the former advocated naturalism, the latter tended to be non- or even anti-naturalistic. Here, methods of analytic philosophy and cognitive sciences became continuous as each emphasized computational models based on the analogy that mind is like a computer. An essential feature of the cognitivist approach has been the promise of reverse engineering mind in the form of *artificial intelligence* (AI). Importantly, this aim depends upon the assumption that there exists an identifiable logic (computation), which plays the same role in the functionality and instantiation of mind, irrespective of its material medium, for example, silicon or biology.

From early in his career, Harris was strongly opposed to the philosophical presuppositions of AI research, believing that reverse engineering based on language was a mistake. He further rejected behaviourism, empiricism, and any position that posited units of mind or compared consciousness to computation. This rendered Harris among the few who had a deep appreciation for science, held phenomenological and idealist sympathies, but rejected both dualism and panpsychism. Though Harris was vastly more critical of the analytic discussions of mind and nature, his method can be characterized by a development upon phenomenology. Hence, a deeper inspection of phenomenological concepts and methods will be key to understanding Harris's system and his later use of respective scientific theories.

1.2.1 Phenomenological Methods

Whereas an experimental psychologist would try to explain an experience by appealing to biological structures and (third-person) processes, the phenomenologist will attempt to clarify the subtlest of details that make the experience what it is from one's own (first-person) perspective. The phenomenologist notes that all conscious perception seems to have a certain structure, namely it is *intentional*. As a ubiquitous character of consciousness, intentionality implies that all perception, memory, imagination, judgement, and so on is *about* or *of something*. Hence, no consciousness occurs in isolation, but always arises in and through our engagement with a world and its constituents. Moreover, perception is not a simple reception of information, but always an interpretation that is context-dependent. "To say that perception involves interpretation doesn't mean that first we perceive some nebulous entity and then we add an interpretation—something over and above perception that bestows meaning on it. Rather, perception *is* interpretational" (Gallagher and Zahavi 2012, 8). So, for example, the meaning of a particular visual scene, like your house, will be different depending upon your associations, your embodied means of interaction (i.e. perceptual and motor limits), and pragmatic goals about the object.

Importantly, this led many phenomenologists to maintain that all perceptions involve tacit assumptions and anticipations about what remains unseen (perceptual incompleteness). This they hold, involves a "gestalt" feature to perception, that is, a salient object of attention that is ***distinguished from a background and its unseen facets inferred***. Additionally, a moment of conscious awareness involves tying various perceptions together into a unified whole, which means consciousness is fundamentally temporal. This is because making different perceptions through space involves extension through time. A temporal gestalt synthesis is achieved via *protention* (anticipation of the future) and *retention* (memory of the previous moment) across an array of perspectives. Further, an analysis of the *phenomenal state* of the perceiver, that is, "what it is like" to have a particular experience must be taken into account.

Phenomenology provides a methodology of investigation that has been central to the tradition in both historical and contemporary neurocognitive contexts and will provide important insights in many of the discussions to follow. The phenomenologist is concerned with phenomena and their conditions of experience. The phenomenon is the immediate givenness of an object of attention, but the reality of the object is not considered to be somehow behind its appearance, as if the appearance hides the real object. While the distinction between appearance and reality must be maintained (since some appearances are false), phenomenologists do not understand this as a distinction between two separate realms (i.e. between phenomenology and science), but a distinction in how objects can appear with more or less accuracy. The natural sciences operate under the tacit presupposition that a mind-, experience-, and theory-independent reality exists, just waiting to be discovered. This is rejected by phenomenologists, Harris included (1987, ff. 108), in place of which a phenomenological method has been proposed:

- Epoché—Husserl posited this first step in a phenomenological method to bracket our acceptance of the ***natural attitude***, that is, that there exists an objective physical world, along with any other metaphysical presuppositions. The purpose is to suspend any potentially dogmatic attitudes about experience and reality. Importantly, this is not an all-or-nothing task and once achieved, it must be continually enacted or sustained. Contemporary phenomenologist who follow Merleau-Ponty maintain "consciousness ultimately calls for a transcendental clarification that goes beyond common-sense postulates and brings us face to face with the problem concerning the constitution of the world" (Gallagher and Zahavi 2012, 26). By "transcendental" they invoke a post-Kantian framework, in which our cognitive apprehension of reality is more than a mere mirroring of a pre-existing world; it is a reflection on the necessary conditions of a reality that already includes consciousness. This approach is complementary with science because it provides a means of better clarifying the "constitution", or genesis of objectivity—if indeed such a thing is possible at all.
- *Phenomenological* reduction—the purpose of the reduction is to analyse the correlational interdependence between specific structures of subjectivity and modes of givenness. This move is meant to clarify *how* the objects appear to our consciousness when they are imagined, remembered, hallucinated, dreamed, and so on. Here, we do not simply focus on the objects precisely as they appear, because we also want to clarify our subjective accomplishments and the intentional structures that are at play. An implication of this method is the rejection of any attempt to fit consciousness into a pre-established natural

framework, since this treats mind as yet another object that can be analysed with the natural attitude:

> Too frequently the assumption has been that a better understanding of the physical world will allow us a better understanding of consciousness; rarely has it been thought that a better understanding of consciousness might allow for a better understanding of what it means for something to be real. (Gallagher and Zahavi 2012, 27)

- *Eidetic variation*—Husserl proposed this step as an imaginative exercise of drawing out the invariant characteristics of things in our experience. We might ask, "what can vary about X such that it would still remain *essentially* the same?" Although this can be applied to objects we experience in the world, it can also be directed towards cognitive acts themselves, such as imagining, dreaming, empathizing, and so on. However, Husserl was explicit that cognitive phenomena are characterized by an essential vagueness and he consequently argued that any attempt to classify and define them with the same kind of precision that one might find in say, geometry, would do an injustice to their true nature.
- Intersubjective *variation*—phenomenologists ought to seek corroboration on their analyses. This concerns the replication of some phenomenal structures over time and across subjects. The effort to establish the invariant structures and conditions of experience cannot be achieved on an independent basis since, in doing so, we would have no way of safeguarding against our taking idiosyncrasies to be universal. Additionally, insofar as we rely upon language to conduct our analyses, we are already endeavouring upon a collective exploration.

From a phenomenological perspective then, some ubiquitous aspects or structures of perception can be established: (1) intentionality, (2) *gestalt character*, (3) *perspectival incompleteness*, (4) *temporality*, and (5) *phenomenal quality*. Traditionally, phenomenology does not attempt to develop a naturalistic *explanation* of consciousness, nor does it seek to uncover its biological genesis, or neurological basis, but is concerned with attaining an *understanding* and proper *description* of the experiential structure of our embodied mental life. As with the traditional account, contemporary phenomenology provides a means of avoiding dogmatism, thereby shunning metaphysical and theoretical prejudices as much as possible. In this way, scientific, analytic and phenomenological approaches can be mutually informative.

On these points Harris (and I) would surely agree. However, Harris's phenomenological approach retains a distinctive Hegelian feature. On Harris's account, by emphasizing process ontology, the *genesis* and *maintenance* of consciousness are reintroduced within the context of *evolution* (broadly construed) as an additional (6) universal feature of experience. This move, I maintain, may be understood as an elaboration of the third phase of classical phenomenological investigation:

1. *Static phenomenological analysis*—determines the various types of experiences (parts of the invariant structure called "transcendental subjectivity") and its objects, for example, the common experiential structures that are involved in how we learn a song, musical categories, and so on.
2. Genetic phenomenology—studies how the cognitive structures and the respective types of content develop through time. This presupposes prior knowledge of intersubjectivity and some lawful system(s), for example, syntax, semantics, and imagination. Here, initial rule-following such as categorically articulated movements for a particular skill, habitualize until they establish a pre-reflective level of motor intentionality (called "secondary passivity" or automatization by Husserl).
3. *Generative* phenomenology—concerns the historical, social, and cultural becoming of human experience. As static phenomenology is restricted in scope with respect to genetic phenomenology, genetic phenomenology is restricted in scope with respect to generative phenomenology. Whereas genetic phenomenology focuses on individual development without explicit analysis of its generational and historical embeddedness, the subject matter of generative phenomenology is the historical and intersubjective becoming of human experience.

I propose that Harris's approach effectively broadens and enriches (3) by arguing that the fundamental evolutionary process that has resulted in consciousness may be recognized within the generative dynamics of consciousness itself. This means Harris includes within this historical analysis a phenomenological account of how one's sense-of-self and world-as-a-whole co-emerge in and through *evolution*. He contends that an explication of this process may provide a non-reductive ontological foundation for phenomenology and guiding principles within the metaphysics of science.

Bringing together decades of research through the history of metaphysics and the natural sciences, Harris summarizes his conclusion in *The Foundations of Metaphysics in Science* thus:

> It now transpires that a polyphasic unity is in essence a scale of forms progressively realizing, with continuous increase of adequacy the principle immanent in whole and part. It is a continuum of activity or process issuing in such a scale, in forms, or phases, of which are mutually related (i) as degrees of realization of the principle, (ii) as distinct specifications of its generic nature, (iii) as mutual opposites and (iv) as the reconciliation of oppositions lower in the scale. Each sublates the lower phase, all of which are carried up into and preserved in mutual interdependence in the higher phase that supersedes them. (1965, 483)

Harris found these "polyphasic unities" to be identifiable at every scale of nature, including the space-time system, the biotic sphere, and consciousness, each more complex, coherent and integrating through time. As will be discussed in Chaps. 2 and 5, Harris maintained that this "scale of forms" was essential to both phenomenology and scientific theory.

To his great frustration when this text was finally published in 1965, it made nearly no impact on the philosophical community. Nevertheless, for the rest of his career Harris believed that this was his most important work. Interestingly, in the *British Journal for the Philosophy of Science* (1966), W.H. Thorpe's review maintained that few philosophers have gained a deep understanding of the recent explosion of scientific discoveries across disciplines, but that if more were to do so we would witness "great conceptual advance". He goes on to write, "If this happens much of the credit will be due to the thinking of a small band of philosophers among which Errol Harris takes a high place" (263).

Continuing this line of research, Harris along with many contemporary phenomenologists and process philosophers, have appealed to self-organization in the fields of chaos and complexity. The common theme here has been an emphasis upon irreducible and emergent phenomena in the contexts of cosmology, evolution, and neuropsychology. Although older versions of speculative process ontology maintained a directed *creativity* in nature, modern versions only permit downward causation via dynamic organization. Here, authors deny the physicalist interpretations that rely upon substance but also refute any dualist qualities in their causal account.

1.2.2 General Systems Theory

In explaining the historical paradigm shift that permitted *general systems theory* (GST) to emerge in a number of scientific fields, Hooker (2011) summarizes with a point that is central to Harris's cosmology:

> In short, in place of a dis-unified collection of assorted boxes of elements, dynamical complex systems approaches offer a rich, unified conception of the world and its science based on interrelating the dynamical models being employed in each—albeit a complex unity of differing but mutually interacting processes and levels governed by condition-dependent laws. (20)

This idea of rejecting individual units for models of interdependence echoes Hegel, Whitehead, and Harris's appeal to *internal relations*, in which entities obtain their properties *in virtue of their dependent relations*.

Following Nicholas of Cusa's lead, Austrian biologist Ludwig von Bertalanffy held *ex omnibus partibus relucet totum*, that is, each part reflects the whole. Believing that such organic systems could not be reduced to the sum of their parts via an analysis within physics, he proposed GST. In his earlier (1950) work, Bertalanffy maintained that there is a hierarchy from the parts to the whole of organic systems and that these levels are only clearly distinguished via the relations of the organism as a whole. Bertalanffy's original conception of which was not really a theory per se, but an interdisciplinary method of understanding systems as *innumerable kinds as wholes*. Modern GST aims to synthesize the more classical analytic view with both descriptive and organismic (or normative) methods.

Following Bertalanffy's early work, proponents aspire to establish a unity of science through a common dialogue between the special sciences. Hofkirchner and Schafranek (2011) lay out three interrelated aims of GST that exemplify this transdisciplinary approach in terms of its implications for epistemic practices, ontology, and ethics:

1. *Epistemic aim*—bring together analytic and scientific methods for investigating the organismic realm by postulating common principles between different disciplines.
2. *Ontological aim*—relate real-world systems to one other by postulating *isomorphisms* between them.
3. *Ethical aim*—reconcile the world of facts with the world of values by considering GST a tool for intervening in systems in humane, non-mechanistic, and non-reductionistic ways (179).

Here we can already see some efforts to surmount a number of the analytic metaphysical dichotomies (1–5) mentioned above. Additionally, Bertalanffy's *perspectivism*, which attempted to take a middle ground between constructivism and reductionism, is continuous with the kind of phenomenology employed by both Harris's holism and some contemporary cognitive scientists. As Hofkirchner and Schafranek continue, "the idea of interdependence became the central criterion for comprehending wholeness" (183). They go on to say, a "system in this regard is essentially an identity preserving phenomenon, something that keeps constancy amid change; this idea underlies the organismic notion of a whole and reveals itself in an act of cognition" (188). Many have been hopeful that such an approach may thus dissolve the dichotomy of world and mind.

A famous benchmark in the history of GST, Scottish engineer *John Scott Russell* was first to observe and describe *solitons*. In 1834, Russell found that within canals, a wave could be generated that remained robust against perturbations and could conserve its energy for miles (Scott 2007, ff. 43). Solitons have since been identified in innumerable phenomena and will be crucial in later discussions of emergence. GST recognizes a kind of downward constraint (via order and control parameters) in certain systems that results from their capacity for self-organization. If GST could be contextualized within a proper (process) metaphysics, some have held, it may thus be possible to vindicate an account of formal and final causation sufficient to naturalize norms and mind alike. One of the most important works to begin the project of extending GST to a study of consciousness was the *Embodied Mind*, by Thompson, Rosch, and Varela (1991). Though not yet considered by current proponents of embodied cognition, Harris's works aimed to marry GST and phenomenology within a process metaphysics of mind and nature since 1965!

Process philosophers are generally divided into teleological (typically theological) views on the one side and naturalistic (non-directed) views on the other. This division can be traced back to the original divide concerning interpretations of biological evolution that either relied upon or rejected appeals to God, for example, Hegel. Both flavours of process philosophy have in common the view of evolution as an invaluable phenomenon for understanding the self-creative works of nature. Both views also recognize *emergent* qualities of natural phenomena across scales. By emphasizing the phenomenological vein in Harris's system, I maintain that this debate may be entirely side-stepped in the discussions of metaphysics and philosophy of science that follow. However, there remains a need to demarcate teleological explanation (e.g. galaxy formation, autocatalytic sets in chemistry, individual organisms, ecosystems, or cultural organization).

Throughout his works, Harris has taken pains to argue for a teleological conception of nature-as-a-whole. He does so by linking arguments from cosmological and biological evolution with those for the nature of consciousness. The result has been a unique interpretation of consciousness within cosmology. His development of these ideas came in two important parts. First, in October of 1986, Harris attended a symposium at Colorado State University that was intended to address the question "Does the new physics need a new metaphysics?" Here he received his first inspiration for a project that would endeavour to unite his philosophies of physics, biology, mind, and knowledge into a single cosmology. Harris recalls that the scientists who participated in the conference "unanimously" called for a "holistic metaphysic" of the kind he had been working on for most of his career (1991, vii).[3] Further inspiration came when presenting a paper at George Mason University on the *Anthropic Principle*, which was so well received that he was inspired to elaborate his arguments into a monograph: *Cosmos and Anthropos*.[4]

Although the conceptualizations of GST have been utilized to establish philosophical frameworks for mind and biology, the implications have not yet been extended into a metaphysical cosmology and this link is precisely what distinguishes Harris's system. According to Harris, following a process ontology and phenomenological methodology means endorsing the view that consciousness is necessary within the cosmos. It is my aim to evaluate his arguments to this end in light of recent research within the very fields and theories to which he appealed over the course of his more than 70 years of philosophical work.

1.3 Overview of the Arguments to Follow

The purpose of this book is to argue that Harris's metaphysics both anticipates and provides a means of bridging the theories of Bohmian quantum mechanics (QM), systems evolution, and 4E cognition. I maintain that taking this approach serves to unite the methods of phenomenology with the metaphysics of process and connects proponents of both to a range of current theoretical and empirical projects in the natural sciences. In the remainder of this chapter I will briefly note the arguments that will be advanced over the course of the following seven chapters to achieve this end.

In Chap. 2, the phenomenological and epistemic principles of Harris's holism will be presented and compared with contemporary theses from a range of differing fields. Here, I refrain from providing direct arguments in defence of Harris's system, but instead, aim to demonstrate that like Harris's thesis, Bohm's implicate order and enactivism each rely upon the same principles of process ontology, dialectical relations, and scale of forms. In addition, I argue the key insight that Harris provides is that these underlying metaphysical principles may inform our interpretation of phenomena ranging through physics, biology, and psychology. I conclude that what results from reading Harris's work through current enactivism and Bohmian theory is an original phenomenological and process-based metaphysics that holds great promise for dissolving the hard problem of consciousness.

The aim for Part II is to critically consider both the epistemic and the ontological coherence of Harris's holism in cosmology. In Chap. 3, I assess Harris's appeal to relativity, unity in the standard model of cosmology, and entanglement in quantum mechanics. Harris's central claim is that empirical evidence from these domains support a number of central pillars of his metaphysics, including dialectical relations and the *unifying principle*. Additionally, I introduce Harris's appeal to anthropic reasoning and develop some of his largely implicit motivations for linking the philosophical discussions of cosmos and mind.

In Chap. 4, I examine and elaborate Harris's proposal for an original version of the anthropic principle. Towards this end, I evaluate four previously developed versions of the anthropic principle in the context of four multiverse theories. In recent years, many cosmologists have argued that these multiverse theories provide sufficient reason to avoid the stronger conclusions of anthropic reasoning that posit a teleological necessity of consciousness. Building upon Harris's reasoning with insights from phenomenology and process ontology, I propose a means of avoiding the philosophical force of multiverse theories. These discussions culminate with the introduction of Harris's (paradigmatic) teleological anthropic principle. I conclude that Harris's updated system avoids many appeals to multiverse theories as a means of defeating such a teleological account, but his original ontological conclusions must nevertheless confront numerous unforeseen challenges to be addressed in later chapters.

The aim of Part III is to read Harris's metaphysics through contemporary complexity theory to outline a model of evolution that is applicable to both cosmological and biological domains. In Chap. 5, I examine in detail Harris's conception of cosmogenesis as characterized by teleology and self-organization. I propose that *if updated,* Harris's arguments for a scale of forms may dovetail with emergentist and process metaphysics that have been proposed in recent years. I contend that insofar as enactivism follows the deeper metaphysical commitments of complexity and GST (with Harris), their ontology is not only consistent with, but entails a teleological version of the anthropic cosmological principle.

In Chap. 6, I outline Harris's process philosophy of biology and assess its accuracy in the face of recent proposals from complex systems theory. Here, Harris's central claim is that previously outlined conceptions of evolution and self-organization are again warranted, thereby extending teleology into the biological domain. Towards this end, I exhibit Harris's theoretical anticipation of autopoiesis, neo-Lamarckism, the Gaia theory, and symbiogenesis. The goal of this chapter is to present and clarify Harris's stance on key topics in philosophy of biology so as to lay a theoretical bridge to metaphysics of mind, to be discussed in the remaining chapters.

In Part IV, I examine Harris's approach to the problem of consciousness. I confine my focus to only the central thread of his argument concerning efforts to naturalize subjectivity. Here, I highlight Harris's anticipation of empirical research and metaphysical arguments developed by contemporary enactivists for the nature of mind. In Chap. 7, I assess Harris and Damasio's respective appeals to Spinoza's conception of ideatum as a model of embodied cognition and contrast this approach with more recent arguments from embodiment cognitive science. This line of thought is extended to Spinoza's concept of conatus in order to clarify how Harris's understanding of self-awareness bears upon corresponding views from the embedded theory of cognition.

In Chap. 8, I outline Harris's appeal to dynamic field theory in neuroscience and his anticipation of theories depicting consciousness as "phase of matter". Here I consider possible implications this proposition has for enactivism. I reconsider the resulting *dialectical holist* view of teleology in face of the hard problem and develop some consequences of this theory for a naturalization of knowledge. To this end, I maintain that current pragmatist contentions are especially supportive of Harris's system and lead to some far more defensible conclusions than he originally maintained. Finally, I suggest that some recent developments of *Klein Bottle Logic* serve as a suitable analogue of Harris's phenomenology and extends enactivism from a philosophy of mind into cosmology. I conclude that insofar as the resulting metaphysics of *dialectical holism* remains consistent, it provides both methodological and theoretical principles towards revising neutral monism and naturalizing phenomenology.

1.4 Conclusion

This brief tour of process philosophy and phenomenology has revealed many of the major inspirational contributions that gave rise to Harris's metaphysics. While the resulting system is incompatible with all but a very specific thread of theories, the position sketched thus far should be consistent with any philosophy that appreciates the following three moves:

1. Traditional metaphysical assumptions of the *objective materialist paradigm* (i.e. a metaphysics based on static analysable substances, objects, states of affairs, or abstract entities) are dispensable theoretical presuppositions to be replaced by a pragmatist line of *inquiry*.
2. Phenomenology is an indispensable methodology that can provide a means of continually liberating us of our biases in the construction of scientific theories and may yet provide guidance in our effort to conceptualize the relation between subject and cosmos.
3. Solutions to a number of important philosophical problems (e.g. the relation between natural evolution, subjective experience, and norms) can *only* be adequately addressed by attending to their common underlying features via the framework of process metaphysics.

In this chapter I have outlined how Harris aimed to synthesize a range of metaphysical systems that extended through the course of Western philosophy. Appealing to a phenomenological methodology, Harris distilled numerous guiding principles, which led him to a host of compatible theories across the natural sciences. These empirical views then in turn filled in the contents of his metaphysical framework. This resulted in a teleological process-phenomenology, in which the epistemic process of reflection becomes increasingly identified with the self-evolving process of nature itself. In the following chapter, I more formally introduce the principles of Harris's phenomenological method, ontology, and epistemology, and subsequently identify these principles within contemporary philosophical theories.

Notes

1. This quote comes from Plutarch *On the E at Delphi* 392B.
2. Some have argued that process ontology can also provide a bridge across Eastern and Western philosophy (Capra 2010, ff. 330; Herfel and Rodrigues 2011).
3. When this project was completed Harris extended this portrayal of holism and the anthropic principle into social, moral, and religious domains, resulting in *Cosmos and Theos* (1992). Alas, analysis of Harris's philosophy of religion will have to wait for a future book.
4. Two years prior, Grier noted that Barrow and Tipler's (1989) discussion of the relation between the "cosmological anthropic principle" and the "tradition of teleological explanation" has been anticipated by Harris's earlier work: "In Harris's work, stretching back over several decades, one can find a very sophisticated philosophical defence of just such a development in scientific theory" (1989, xiii).

Works Cited

Aristotle, Philip H. Wicksteed H. Wicksteed, and Francis M. Cornford M. Cornford. *The Physics*. Cambridge, London: Harvard University Press. W. Heinemann, 1957.

Bertalanffy, Ludwig V. "An Outline of General System Theory." In *British Journal for the Philosophy of Science*, 134–165. 1(2), 1950.

Capra, Fritjof. *The Tao of Physics: An Exploration of the Parallels between Modern Physics and Eastern Mysticism*. Boston, MA: Shambala Pub, 2010.

Chalmers, David J. *The Conscious Mind: In Search of a Fundamental Theory*. NY New York, Oxford University Press, 1996.

Coope, Ursula. "Aristotle: Time and Change." In Le Poidevin, Robin., Peter Simons., Andrew McGonigal., and Ross P. Cameron. *The Routledge Companion to Metaphysics*. 39–46. New York, NY: Routledge. 2009.

Dennett, Daniel C. *Consciousness Explained.* Boston MA, Little Brown, 1991.
Dewey, J. *The Philosophy of John Dewey*, ed. John J. McDermott. University of Chicago Press, 1981.
Gallagher, Shawn, and Dan Zahavi. *The Phenomenological Mind.* 2nd Ed. New York, NY: Routledge Press, 2012.
Gibson, J. J. *Perception and the Visual World.* Boston: Houghton Mifflin, 1950.
Grier, P. T. (Ed). *Dialectic and Contemporary Science: Essays in Honor of Errol E. Harris*: University Press of America, 1989.
Harris, Errol, E. *The foundations of Metaphysics in Science.* London: Allen and Unwin Pub, 1965.
———. *Fundamentals of Philosophy: A Study of Classical Texts.* Holt, Rinehart and Winston, 1969.
———. *Salvation from Despair: A Reappraisal of Spinoza's Philosophy.* Martinus Nijhoff, The Hague, 1973.
———. *Formal, Transcendental and Dialectical Thinking: Logic and Reality.* State University of New York Press, 1987.
———. *An Interpretation of the Logic of Hegel.* Washington DC: The University Press of America, 1983.
———. *Cosmos and Anthropos: A philosophical Interpretation of the Anthropic Cosmological Principle.* Atlantic Highlands, NJ: Humanities Press International Inc, 1991.
———. *The Spirit of Hegel.* Atlantic Highlands, NJ: Humanities Press International, Inc, 1993.
———. *The Substance of Spinoza.* Amherst New York: Humanity Books, an imprint of Prometheus Books, 1995.
———. *Bound in Shallows: Autobiographical Reminiscences.* Marquette University Press, 2015.
Hegel, G. W. F. (1977). *Phenomenology of Spirit.* Trans. A. V. Miller. Oxford: Oxford University Press.
Heidegger, Martin, Joan Stambaugh, and Dennis J. Schmidt. *Being and Time.* Albany, N.Y. Bristol: State University of New York Press Excelsior University Presses Marketing distributor, 2010.
Herfel, William, Gao, Y. and Rodrigues, Dianah. (2011). *Chinese Medicine and Complex Systems Dynamics.* In Handbook of the Philosophy of Complex Systems. 675–719. https://doi.org/10.1016/B978-0-444-52076-0.50023-7.
Hofkirchner, Wolfgang, and Schafranek, M. *General System Theory.* In Cliff Hooker. ed. *Philosophy of Complex Systems.* 177–194. The Netherlands, Amsterdam: Elsevier Pub, 2011.
Hooker, Cliff, ed. *Philosophy of Complex Systems.* The Netherlands, Amsterdam: Elsevier Pub, 2011.

Husserl, Edmond. *Ideas Pertaining to a Pure Phenomenology and to a Phenomenological Philosophy. First Book.* Trans: F. Kersten. The Hague: Martinus Nijhoff, 1983.

———. *Ideas Pertaining to a Pure Phenomenology and to a Phenomenological Philosophy. Second Book*, 1989.

James, William. "A World of Pure Experience." In *James (1975–88)*. Vol. i, 21–44, 1976 [1904b]

Kant, Immanuel, Paul Guyer, and Allen W. Wood. *Critique of Pure Reason.* New York: Cambridge University Press, 1998.

Kirk, Geoffrey S. III. "Natural Change in Heraclitus." *Mind.* 35–42. Volume LX, Issue 237, January, https://doi.org/10.1093/mind/LX.237.35, 1951.

Leahey, Thomas H. *A History of Psychology: Main Currents in Psychological Thought.* (5th Ed). Saddle River NJ: Prentice-Hall, Inc, 2000.

Merleau-Ponty, Maurice. *Phenomenology of Perception.* Trans: Colin Smith. London: Routledge Press, 1962.

———. *The Structure of Behaviour.* Trans. A. Fisher. Pittsburgh, PA: Dusquene University Press, 1963.

Nagel, Thomas. "What is it Like to be a Bat?" In *The Philosophical Review.* 435–50. LXXXIII, 4 (October), 1974.

Peirce, Charles S. *Collected Papers of Charles Sanders Peirce*, eight volumes. Eds. C. Hartshorne, P. Weiss and A. W. Burks. Cambridge: Harvard University Press, 1931–1958.

Pihlstrom, Sami. "Metaphysics." In Sami Pihlstrom, ed. *The Continuum Companion to Pragmatism.* 92–100. New York, NY: Continuum International Publishing Group, 2011.

Putnam, Hilary. *The Collapse of the Fact/Value Dichotomy and Other Essays.* Cambridge, MA: Harvard University Press, 2002.

Scott, Alwyn C. *The Nonlinear Universe: Chaos, Emergence, Life* (The Frontiers Collection). Heidelberg Germany: Springer Pub, 2007.

Simons, Peter, (2009). "Whitehead: Process and Cosmology." In Le Poidevin, Robin., Peter Simons., Andrew McGonigal., and Ross P. Cameron. *The Routledge Companion to Metaphysics.* 181–190. New York, NY: Routledge.

Spinoza, Baruch. *Spinoza: The Complete Works.* Morgan, M. L. ed. Translations by Samuel Shirley. Hackett Publishing Company, Inc, 2002.

Thompson, Evan. *Mind in Life: Biology, Phenomenology, and the Sciences of Mind.* Cambridge, MA: Belknap Press of Harvard University Press, 2007.

Tsanoff, R. A. (2015). The Great Philosophers. Second Edition. New York: NY. Harper & Row. ISBN-10:1296829669

Varela, Francisco J., Evan Thompson, and Elenore Rosch. *The Embodied Mind: Cognitive Science and Human Experience.* Cambridge MA: MIT Press, 1991.

Whitehead, Alfred North. *Process and Reality.* New York: Free Press. Cambridge: Cambridge University Press, 1929, 1978.

CHAPTER 2

Harris's Holism and Contemporary Analogues

2.1 Harris's Holist Metaphysics

While Harris's metaphysical publications provided him with a successful career and appreciation from idealists, because his position was counter to the empiricist worldview of his time, his ideas were largely ignored by mainstream philosophers. Despite his relatively original style and terminology, I argue that in the face of contemporary science, there are tremendous theoretical riches to be gained by studying his metaphysics.[1] Common across most of his works, Harris begins by laying out what may be called a *phenomenological* foundation, an account of what he takes to be *synthetic a priori* first principles.

Harris proposes that certain propositions about wholeness are fundamental for both scientific observation and philosophical investigation alike. Responding to Kant's question "what makes *a priori* synthetic judgments possible", Harris proposes "*a priori* synthetic judgments would be established by the demonstration of the existence of wholes with internal relations between their parts, for once the principle of organization is known, universal and necessary judgments about the structure and its parts would be possible *a priori*" (1987, 75). For Harris, such a "synthetic whole" provides one's sense-of-self, laws of thought, and rules for individuation. He maintains that without such presuppositions we would have no figure-ground distinction or object-focused attention. Metaphysics, he

J. Schofield, *A Phenomenological Revision of E. E. Harris's Dialectical Holism*,
https://doi.org/10.1007/978-3-030-65029-2_2

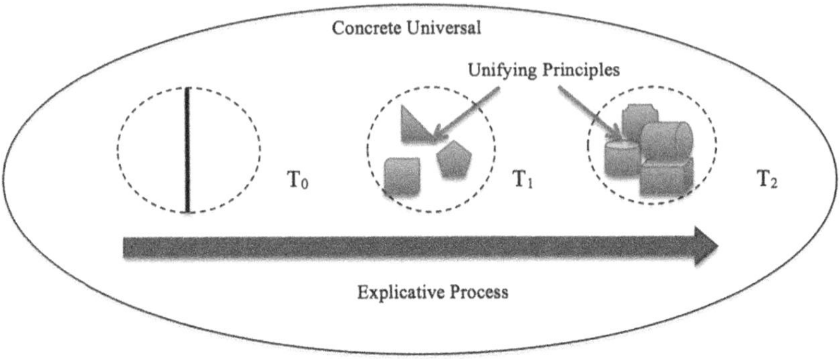

Fig. 2.1 Explicative process, unifying principle, and Concrete Universal. At t_0 Nature or Being is not homogenous but differentiated in a primitive way, here symbolized by a simple *asymmetry*. At t_1 differentiation establishes unifying principles characterized by structures and process of corresponding domains. At t_2 increasingly complex unifying principles emerge that build upon those of a previous stage. Harris refers to the transitions in complexity as steps in a scale of forms, each partially reflective of the whole

says, "is the comprehension of the whole and the exposition of the principle of structure by which it is pervaded" (1988, 11). The task of metaphysics then is to organize our experiences, both scientific and subjective, *towards* a single unity, "not simply to reveal the presuppositions of science but also, and more significantly, to trace the process, and presumably, to detect the reasons, for their changes" (1988, 15).

A synthetic whole of the kind Harris has in mind will be instantiated iff four axioms are satisfied. I first summarize these criteria before addressing each in the following sub-sections:

(i) The components are interdependent with one another within their system, that is, the thesis of *internal* or *dialectical relations.*
(ii) The whole is dynamical (process ontology).
(iii) The *differentiated* components interact in accordance with and give rise to what Harris calls a *unifying principle* (global constraint or dynamical feedback).
(iv) As a result of (i)–(iii), the whole undergoes self-differentiation, what Harris calls *an explicative process* through a scale of increasingly complex *forms.*

2.1.1 *Internal Relations and Process Ontology*

For Harris, "dialectical logic" is the logic of *system*, the proper logic of science and metaphysics, one that both includes and surpasses transcendental and formal logics (1987, 131). Formal logic, according to Harris, can only posit external relations—relations that are external to their terms. This means the same terms can remain unchanged "in different external relations and the same relations can obtain between different terms" (1987, 132). Harris argues however that if relations are external, "they fall between their terms, and then they fail to relate them or to bring them together [...] a new relation must be sought between each of the terms [...] so relatedness degenerates into an infinite regress" (1987, 133). Accordingly, the first criterion runs as follows:

> (i) *Internal relations*—in any synthetic whole, the physical entities therein will be related in such a way that the nature of the terms depends on their mutual relations and vice versa.

For Harris, an internal relation is one that determines the natures of the relata and arises because of those entities being so related (1991, 24). As Blanchard writes about Harris's thesis considering the relations between two entities A and B: "[i]f the absence of that relation would leave A as it was, the relation is external. If, when the relation was absent, A would be different, or would cease to be, the relation is internal" (1989, 3). Hence, any changes in entities or internal relations are concomitant and thereby alter the system to which they belong.

Towards a clarification of this axiom, it is important to understand what it does not claim. First, it is crucial to notice that Harris does not consider internal relations between individual entities but only within or among a system of entities. Second, this position is not claiming that *all* relations modify or in some way influence their terms, that is, *that all relations are internal*.[2] This would entail "the idea that whenever there are some things, there exists a whole that consists exactly of those things—i.e., that there is always a *mereological sum* (or "fusion") of two or more parts" (Achille 2019). Here we can have a gerrymandered set of entities, in which case alteration of the set does not change the entities therein and vice versa. Hence, Harris would agree that any given relation such as *X* is taller than *Y* is not going to somehow impact the nature of the respective terms (e.g.

Y changes when X's height is increased). These relations are indeed external, conventional, and extrinsic.

In the context of a synthetic whole Harris writes, "the membership of each part will be conditioned by its relations to the other parts, and it will be the member it is because of these relations and because the other parts are as they are" (2000, 112). Internal relations are here understood to be ***intrinsic to the nature of the entities*** and according to Harris, all proper entities must have some such relations. The argument is that any ***identifiable*** entity must be part of a system and that its nature depends upon certain relations within that system: relation *R* is internal iff it obtains between entity *X* and system *Y*, such that alteration of *R* results in an alteration of *X* and *Y*. One particularly important implication of this contention is that internal relations become necessary for identifying or ***individuating*** an entity. In other words, it follows that what are typically considered intrinsic properties or essences of an entity are, phenomenologically speaking, ***abstracted from a system of internal relations.***[3]

Importantly, Harris claims internal relations are essential to both ***entities in the world and the entities of experience***. Concerning the world itself, in his *Foundations*, Harris anticipated the ***holomovement*** (discussed below) maintaining that "[t]he-rest-of-the-universe is a constituent of every quantitative assessment—in short, the whole is immanent in every part and determines the dimensions of the part" (1965, 107). He goes on to claim that quantum mechanics has shown us "[w]hole and part are mutually determining and no detail could be other than it is without making some difference, however slight, to all the rest" (ibid.). Harris later uses this claim to capture his philosophy of science: "If the results of contemporary science are taken seriously we must acknowledge holism to be pervasive throughout the world, and all relations actually obtaining between real existents and their constituent parts to be internal" (1987, 133). On the other hand, Harris maintains that in the case of observation, isolated sensations are not even perceivable: "Only structures, the simplest of which is a figure-and-ground Gestalt, are ever cognized [...] Accordingly, the cognition of any object whatsoever requires a synthesis of a manifold—a grasping of a plurality in unity" (1987, 77). Of course, these are bold claims and will be reconsidered in due course.

As discussed above, following Whitehead (1978), Harris endorsed the axiom that everything we consider to be concrete existence is more accurately understood as a process of becoming (Seibt 2018):

(ii) *Process ontology*—the nature of any entity is continually brought forth in virtue of interactions an entity has with its *other* (i.e. the whole to which it belongs), interactions which are necessary for its being.

Initially this can be framed in phenomenological and epistemic terms: if bodies, elements, or entities do not interact and relate in particular ways, we the observers cannot justifiably attribute any *properties* to them (nor can we be sure they exist). Towards an ontological conception, Harris claims the boundary of a whole is the locus of conflict that arises between a given whole and its *other*, that is, its background or whatever is not the system. This conflict brings about what Harris describes as a vacillation "between denying its other and affirming its inevitable dependence upon it" (1991, 18).

In any whole the elements into which the system differentiates are considered to be finite in that each is limited and defined by what it excludes and what negates it:

> The result of this conflict is relative chaos and contingency in the unfolding of relations among the elements within a complex, which is overcome, and unity re-established, only when identity in and through differences is acknowledged, so that the effort of the finite to maintain itself and to persist in its own being succeeds only when it reaches out (so to speak) beyond its own limits to embrace its other and unite with it in mutual complementation in a larger and more comprehensive whole. (1991, 19)

While this "dependence" upon the other is described as "self-preservation", unlike Whitehead, Harris rejects the claim that mind (or prehension) occurs in abiotic systems. Such "persistence" and "preservation" are rather features of natural evolution.

The upshot of positing internal relations as dynamic is that entities are dependent upon but also in conflict with the others within a common system. Reconciliation of oppositional and conflicting natures between wholes takes place through the development of new wholes, which leads to further differentiation. For Harris, internal relations provide a paradigmatic framework by which structures of phenomenology and processes of nature may be more clearly understood.

2.1.2 *The "Concrete Universal" and "Explicative Process"*

In Harris's system the Concrete Universal and the explicative process are the most important terms to understand if one is to follow his arguments regarding evolution and the nature of consciousness. "Traditionally", Harris claims, abstract universals are understood as "a class under which its particulars are subsumed", while particulars are understood to be concrete, demonstrative qualities that compose individuals (1991, 23). Specifically, according to Newtonian physics and the empiricist philosophy, particulars were concrete and universals were abstract. If we are to maintain internal relations however, Harris claims that we must also posit a third criterion:

> (iii) *Concrete Universal*—for the widest sense of Being, an overarching principle must specify the physical relations that are possible and that govern the dynamical trajectory of components therein. (1954, ff. 233; 1965, ff. 467)

"A principle of this kind is universal because it prevails throughout the system and is universal to its parts. It integrates them into a single concrete whole" (1991, 24). Within a synthetic whole, Harris holds that components must exist within a "continuum" organized into a space-time interval. For without which, the respective elements could not interact, there would be no coherence, and the whole would be homogenous. Harris claims that a "homogenous whole" is contradictory because if its components are completely uniform and indistinguishable, there can be no continuity of relations and thus no identifiable whole. Within say, a 2D series, there cannot be total irregularity: "For absolute irregularity means a total lack of continuity between the terms of the series, so that once again the continuum would be dissolved" (ibid.). Positing such a system as *totally homogenous* or *totally random* amounts to what Harris calls the "fallacy of spurious homogeneity" (1965, 462). If the parts do not differ, at least in position, "space collapses to a single point; and a single point apart from and unrelated to other points is *nowhere*, and so no point in space" (1987, 138). As a result, Harris holds randomness is always relative to order and so "order is prior to disorder; and the primary form of order is continuous seriality in a heterogeneous but graded scale of overlapping terms" (139–40). This Harris claims, will be true for every system.

What Harris calls a *unifying principle* (AKA a "system", "gestalt", "scale", or "field") is considered a partial instantiation (or phase) of the

Concrete Universal. He reasons that the determination of the parts of such principles is the same as the determination of these principles by the overarching Concrete Universal. "If every element in the whole is a particular specification of its universal principle of order, they must all have something in common, yet equally they must all differ" (1987, 135). Harris takes an abstract mathematical space, or ordered manifold as the simplest example of the unifying principle.[4] Originally, he maintained that in mathematics "a function at once signifies the uniqueness and cohesion of the totality and regulates the distinguishing relations of the elements of which it is composed and in and through which the organizing principle is expressed" (1965, 83). A space of this kind is considered the simplest model of a synthetic whole, as it "implies an interlock between parts that are systematically interrelated (as in a jigsaw), so that their mutual relations are governed by a principle of order or organization that pervades the entire structure" (1991, 17–18).

Importantly, Harris denies that such forms can really exist as abstractions. He writes, "this structural principle is nothing, or at best a mere abstraction, unless it is embodied and expresses itself in the multiplicity of the elements comprising the system" (1987, 144). For example, Harris appeals to the widespread proliferation of fractal geometry identified in the development of natural systems where, he argues, overarching forms are iterated across spatiotemporal scales (1991, ff. 35). Here, the materially instantiated *form* and *function* of a system is identified as a ***unifying principle*** in that it imposes an organization that determines the unity of its parts, but also depends upon the specific material manifestation of its parts. Only as materially instantiated systems can the unifying principle reflect the Concrete Universal to some degree of adequacy (2000, ff. 106). In this way, Harris's holist metaphysics is compatible with a kind of *ontic structural realism* (OSR) that takes dynamical relations and their structures as fundamental and interdependent.

> (iv) *Explicative process*—when applied to the whole of Nature, internal relations, process ontology, and the Concrete Universal result in the generation of a *scale of forms*.

The "self-differentiation of system", Harris writes, involves the "explication of a totality", an "interplay of unity and diversity" that must be understood as "a perpetual dynamic activity", but one "prior to all temporality and process because it is already involved in any succession or movement" (1987, 144–45). In other words, temporal change is believed to be a facet, not an exhaustive depiction of this process. In this process, each whole that arises supersedes, includes, and transforms those beneath it, while implicating those that may yet develop: "progressing by successive steps, from a primitive element up the scale of degrees of more adequate manifestations of the universal principle, the totality that is immanent in every element and every phase of the process develops" (1991, 20). Metaphorically, one can imagine this progression as a complete picture guiding the formation of a jigsaw puzzle, as convex and concave of its pieces are brought together in a unique way. Hence, the process is creative and self-referential.

The inaccuracy of this metaphor, however, is that in nature this process is ongoing; the immanence of the whole in its parts drives the components from contradiction to supplementation, augmentation, and completion—*the picture emerges with increasing clarity as more complex relationships come into being* (2000, 108–9). "So the process of successive unifications of opposites is propelled toward the generation and eventual achievement of greater wholeness" (113). Harris holds that a "scale of this kind is dialectical because it proceeds through opposition and distinction which is at the same time complementary, interdependence and mutual identity" (1991, 20). The Concrete Universal (or the *form of all forms*) thus exerts a conatus that "differentiates itself" into each of its respective phases or unifying principles (24).

Henceforth, the scale of forms is to be denoted the *explicative process* (𝕰): the evolutionary process of self-differentiation attributed to the Concrete Universal whereby new *wholes* (or phases), permitting new modes of internal relations are generated. This, Harris argues, is a common principle of cosmological and biological evolution, as well as phenomenological development. He claims that this conception of material self-differentiation can be vindicated in contemporary science:

> [E]nergy is matrix and its laws are the principles of motion in the physical world. It manifests itself as curvature in the space-time continuum [...] It manifests itself likewise as gravitation governing the movement of matter in space, the formation and the motion of the heavenly bodies, and the structure of the universe generally [...] in elementary particles, in atoms, in molecules, and in chemical bonds and affinities between diverse substances, right up to the complex macromolecules which make possible the self-reproductive chemical cycles typical of living activity. (1987, 151–52)

Harris's prime example of a unifying principle is the self-organization of living being. In his *Foundations*, Harris proposes *auturgy* as a term distinguishing living from non-living systems by their adaptive capacity to do self-work (1965, 180). Here for the first time in the scale of forms, "metabolic transformations" permit a system "to create and sustain individual structure, spontaneously adjusting itself to external circumstances and adapting its operation amid changing conditions so as to counteract influences that would otherwise bring the sustaining processes to an end" (1991, 64). This he calls "self-determination" and, in agreement with Hans Jonas, "the first form of freedom" that "foreshadows consciousness" (ibid.).

Crucial for later discussions, Harris holds that a direct implication of auturgy is *embodied* cognition, a conclusion that he derived from Aristotle and Spinoza. Accordingly, the mind and body are understood to be two aspects of one *living* system; that is, the mind is the *form* of the body and (from the later philosopher) the body is the primary *object of intention* for the mind (e.g. 2006, 98–99). Harris maintains that neither description of phenomenal experience, nor the actual neural discharge deserves primacy:

> It is not merely a matter of describing one set of facts in two different ways, but of one reality manifesting itself in two different forms [...] Neither is reducible to the other. The entity which is both has two forms of being, just as (analogously) a triangle exists as three straight lines on a plane forming three angles, and as a distinctive three-cornered shape. (1973, 84)

Harris claims that the "most widely canvassed theories of perception today seem to agree in rejecting the notion that the percept is an immediately given atomic and simple datum that can be isolated from all others and from other elements in experience" (2000, 85). Rather, Harris holds that perception is more appropriately understood as an "achievement", the end result of an active and "discursive process" that presupposes a given body of knowledge. This achievement, he argues, is not to be seen as intellectual, involving the manipulation of atomic symbols. Sympathetic to Merleau-Ponty, Harris maintains, "perception is a functioning of the body as a whole, and not just the sense organs" (2000, 87).

Harris claims that there is no zero or pure beginning in the scale of forms and neither does it continue on unfulfilled forever. Rather, the Concrete Universal is implicit at each phase and becomes complete only when fully actualized in self-awareness. Thus, he maintains that consciousness of the Concrete Universal is itself an essential phase of the explicative process:

> [W]holeness, by its very nature, involves dynamic and dialectical self-specification, by way of self-enfoldment (with consequent overlap of specific forms). It tends towards intensification of centreity, increasing self-sufficiency and widening comprehension, and culminates as an all-embracing awareness of an all-encompassing world. (1991, 26)

Roughly, this is how Harris's metaphysics encapsulates his theories of life and mind. While he refers to his position as *monistic*, as I argue below, by embedding consciousness in the scale of forms, Harris has effectively provided a means of re-visioning *neutral monism* (NM).[5]

2.1.3 Epistemic Implications

The central contention that made many of Harris's mainstream contemporaries wary was his persistent criticism of the epistemology and methodology of empiricism. Harris claims the world-view of contemporary science contradicts the long-standing atomistic, mechanical, and dualistic theories about knowledge and nature. For Harris, scientific practice reveals an unavoidable commitment to a holistic metaphysics that contextualizes empirical observation. Consequently, the implicit assumption that formal and mathematical logic are the gold standards of reason is brought into question. He calls for a "scientific philosophy" to be "carried out conjointly by scientists and philosophers in collaboration" (1965, 32). The goal of this project is to critically assess the methods and fit disparate branches of science "into a coherent world-concept" (ibid.). Following this line of thought, much of Harris's career was devoted to proposing such a scientific philosophy, in which the appropriate worldview is dialectic, holistic, organismic, teleological, and hierarchical. This, he claimed, requires a "logic of system" (1987, ff. 131).

Harris claims the verification principle upon which empiricism has been built is "untenable" because sense perception does not and cannot be of immediate self-evidence, but relies upon inference (1970, 183). In addition, he holds that appealing to abstraction so as to isolate terms of philosophical analysis cannot avoid this challenge. Harris's reason for this is that perception depends on the identification of a context for its interpretive strength, one which is a result of thought's *discursive activity*. Harris claims, "[e]xperimentation is, in fact, the question that the scientist puts to nature, and no question arises unless some theory is already being entertained." Moreover, even the simplest observation is also "interpretation in the light of prior knowledge" (2000, 13). Here Harris takes issue with the verificationist picture of scientific practice and maintains that classical logic is incapable on its own, to facilitate or reflect the development of scientific philosophy. He argues this is because no objects of experience are ever based strictly on *atomic observables*, but phenomenologically depend upon presuppositions and paradigm.[6]

> We must notice that this priority of structure to individuation is not just an empirical discovery but is essential to any and every principle of individuation. What distinguishes and defines individuals is their relations to one another; and every relation of whatever sort presupposes a background system appropriate to it. (1965, 456)

For example, Harris takes a given "mark" on a given "body" (e.g. a thermometer) as having no meaning unless it is correlated with other bodies in a meaningful way, which presupposes an observer/experimenter with a finite complex of assumptions regarding relationships between elements of the *system as a whole*. Harris stresses that no direct observation is possible without a "connection between matters of fact and universal rules of correlation" (2000, 29). Echoing Peirce's conception of *abduction*, he claims theories about the world are "*constructive* of system rather than either inductive or formally deductive, and it demands a logic akin to both yet identical with neither" (20).

Laws (or principles of order) are considered *synthetic*: "General laws are not revealed by direct observation, and if derived from inductive inference, they cannot be confirmed by any one particular observation, nor are any number sufficient" (2000, 26). In this vein, by denying the possibility of *a priori synthetic* knowledge as empiricism has done, Harris finds we lose not only metaphysics but also any sort of organizing principle necessary for rendering a coherent epistemology. As a result, all propositions become tautological on their own terms and unrelated to one another. Without such a synthesis, philosophers and scientists alike are no longer dealing with a "coherent reality", are unable to appeal to past experiences, and lack a basis for their claims about the world.

Again, to make his point clear, Harris maintained that the *unifying principle* is a structure of both knowledge (e.g. scientific theory) and the world (e.g. self-organizing systems):

> A scientific theory is itself a principle or organization systematizing the observed facts within a specific field of experience. It formulates laws that determine and interrelate the facts, and it interrelates both facts and laws according to logical principles, so as to form a coherent whole. The logical principles are precisely those we are seeking: namely the laws of systematic and coherent thinking. (1987, 142–43)

A theoretical system then is an example "*par excellence*" of systematic order and its parameters are the logical principles upon which the theory relies. Of utmost importance for the process of scientific development, facts are always, "organized by a principle or law, or a set of such principles themselves interrelated according to some higher principle" (1970, 350). Principles thus designate the domain in which the facts are expected to obtain and "provides the rule of inference" (ibid.).

Again, echoing the pragmatists, Harris claims the aim in scientific progress is to remove contradiction and continuously reveal the principle of order in a given domain, *which emerges as a new theory*:

> Theoretical knowledge of the ideal as existent is declared to be unattainable, but the idea of it exercises an indispensable regulative influence upon the quest for knowledge and an imperative demand for its realization in practice. This demand, as it were, spills over from practice into theory, so that the postulation of an absolute system as the standard of all truth is the product of the practical conatus. In consequence, the ultimate standard of value coincides with the ultimate standard of truth and the separation of fact from value, of theory from practice, is overcome. (1987, 119)

He admits certainty would require insight unto "a complete systemic whole of reality" (2000, 32), because in "any given situation the observable facts are infinite in number" (88). By implication, Harris claims a theory, "is always partial, truncated, and incomplete. It is true only in a slight degree, for what it omits and fails to specify, when brought into the limelight, modifies, if only by explication, its intrinsic meaning" (1987, 205). Though this level of certainty is never reached, Harris maintains that human endeavours get ever closer to it as we achieve more holistic methods of observation and synthesis—that is, by accounting for a greater range of relations.

Scientific practice is hereby believed to reveal the metaphysical presuppositions scaffold all practices and perception, which takes the very form of the dialectical whole. Over the course of continued application to observable phenomena, the form of this system is overturned by increasingly coherent alternatives. An alternative is *preferable* insofar as it exhaustively encapsulates the former systems, thereby accounting for the limits of their organization. Hence there is a continuous effort to reveal the foundations of rival camps and vindicate them to some degree, putting each in their rightful place within a hierarchy.

The essential points of Harris's system established heretofore may be summarized thus:

1. Every proper system is organized by a unifying principle that arises from and instantiates some set of internal relations upon its constituents.
2. Properties of natural systems (unifying principles) depend upon and are derived from processes of internal relation that they bear to the wider system to which they belong (ultimately the Concrete Universal).
3. Process ontology combined with internal relations results in the directed genesis of unifying principles (a *synthetic* scale of forms 𝕮).
4. The resulting "dialectical whole" is inherent in science, philosophy, and phenomenology; that is, Harris claims this structure-process is fundamental to our knowledge (paradigms) and to Nature (ontology).

The remainder of this thesis is devoted to fleshing out the arguments and implications of Harris's system that respond to the anthropic principle and the hard problem of consciousness. Towards this end, I next take a preliminary exploration of what are today some of the most sympathetic positions with respect to Harris's holism: *enactivism* and Bohm's implicate *order*.

2.2 Analogies with Autopoietic Enactivism

Though enactivism is relatively new, first appearing in Maturana and Varela's *Autopoiesis and Cognition* (1972) and later canonized in Varela, Thompson, and Rosch's *The Embodied Mind* (1991), its primary tenets can be traced to Bertalanffy's (1950) *outline of general systems theory* and a host of phenomenological theories developed by Husserl (1983, 1989),

Merleau-Ponty (1963), Gibson (1979) and their followers. Starting with Capra's (1996) synthesis of autopoiesis and complexity theory, numerous works have either elaborated or contributed to a common 4E foundation (Newen and Gallagher 2018), including: *neurophenomenology* (Varela 1996; Thompson et al. 2005); *extended mind theories* (Clark and Chalmers 1998); *participatory spirituality* (Ferrer 2002); *collective mind theories* (Jankovic and Ludwig 2018); *extended neutral monism* (Silberstein 2009); *radical enactivism* (Hutto and Mylin 2013); and *enactivist education theories* (Reid 2014; Li and Winchester 2014).

While all these branches of the *enactivist paradigm* (Stewart et al. 2010) bear significant commonality with Harris's system, this text is concerned to evaluate the extent to which Harris anticipates and broadens the metaphysical foundation for *autopoietic enactivism* (AE). Today, AE can be characterized as endorsing a version of the "4E thesis":

(1) *Enactive thesis*—cognition is not a state or substance, but a self-organizing process. This thesis takes issue with *representationalism*, which claims that mind is composed of a symbolic manipulation system somehow related to the brain.
(2) *Embodiment thesis*—cognition obtains as a feature of an organism as a whole. This is a rejection of classical *modularity*, which posits cognitive systems (among others) as distinct and limited to the brain.
(3) *Embedded thesis*—cognition depends upon *ongoing* social and environmental interaction. This thesis rejects *nativism*, which maintains an individualistic and independent view of cognitive development.
(4) *Extended thesis*—mind is (at least partially) constituted by environmental features, rather than being limited to processes or structures of the brain as is the case for classical cognitive science.

While many proponents of the enactivist paradigm listed above endorse some version of these four theses, proponents of AE also add two of their own:

(5) *Autopoiesis*—maintaining a double biconditional between *cognition* (c), intentionality (i), and *life* (l): $\forall$ c, i, l (i $\leftrightarrow$ l) $\wedge$ (i $\leftrightarrow$ c) $\therefore$ (l $\leftrightarrow$ c). If one obtains, the other two do as well.
(6) *Neurophenomenology*—a non-reductive methodological programme, which replaces the effort to reduce experience to some

neural correlates of consciousness, with the ongoing clarification of homological relations across each domain.

In the following sections I show that AE is built upon many of the same metaphysical principles as Harris's holism, including analogues of *internal relations*, *process ontology*, and *unifying principles.*

2.2.1 The Metaphysics of Autopoiesis

The theory of *autopoiesis* (self-creation) was developed by neuroscientist Francisco Varela and his mentor, biologist Humberto Maturana (1972), with the aim describing and explaining the self-emerging, self-maintaining operation of individual cells. Since its early use however, the theory has been applied to the organization of virtual processes, the living body, and social systems. The hotly debated implications of this theory are far-reaching but the most important for the present purposes concern what proponents consider *a deep continuity of life and mind*: "life and mind share a set of basic organizational properties, and the organizational properties distinctive of mind are an enriched version of those fundamental to life. Mind is life-like and life mind-like" (Thompson 2007, 128). I argue that this theory is the most significant modern manifestation of Harris's unifying principle in the philosophy of biology.

According to Reid's (2014) synthesis of Maturana and Varela's works, the necessary and sufficient conditions of an autopoietic system are now: (i) *self-production*—synthesizing its own components; (ii) *embodiment*—physically instantiated; (iii) *self-organizing*—recurrently maintaining itself in a steady state; (iv) *composite unity*—being distinguishable from a background medium; (v) *interactionally open*—exchanging matter and energy with its environment; and (vi) *dynamic*—recursive and emergent (ff. 153). In concrete terms, this implies that the system maintains a semipermeable boundary, a reaction network within the boundary, and that *the two are interdependent* (thereby excluding candles, whirlpools, etc.) (Thompson 2007, 103).

Additionally, AE supporters maintain that when the reaction networks within such a system are interdependent with one another and coupled to an environment, the resulting dynamics are sufficient for *adaptivity*, which is believed to demonstrate minimal cognition *and* intentionality. This is because the system is considered to be *about* its own preservation. "Cognition" is therefore considered to be "behaviour or conduct in relation to meaning and norms that the system itself enacts or brings forth on the basis of its autonomy" (Thompson 2007, 126). To support this point, AE appeals to complex systems theory to explain the shift from chemical to intentional as a *dynamic co-emergence*:

> Dynamic co-emergence means that a whole not only arises from its parts but that the part also arises from the whole. Part and whole co-emerge and mutually specify each other: A whole cannot be reduced to its parts, for the parts cannot be characterized independently of the whole; conversely, the parts cannot be reduced to the whole, for the whole cannot be characterized independently of the parts. (Thompson 2007, 38)

Proponents argue that dynamic co-emergence is applicable to the relation between the organism and its components as well as the relation between the organism and its environment. In *The Embodied Mind* (1991) Varela et al. claimed "organism and environment are mutually enfolded in multiple ways, and so what constitutes the world of a given organism is enacted by that organism's history of structural coupling" (202). They claim that such interactive "coupling" is constrained by the organism's phylogenetic and ontogenetic *histories*, which permits changes to the organism's physiological structure, and results in the sensation of a meaningful world. Sensory and motor structures are claimed to be fundamentally interconnected and interdependent, such that each influences the form and function of the other. By implication, qualities of the environment do not exist independently of the organism, just as those of the organism cannot exist independently of their environment: *both organism and environment define each other through time*.

The result is what can be called an ongoing *sedimentation effect*, in which past modes of coupling to an environment provide the means for, but also constrain future coupling. Hence, the enactivist slogan that cognition is like "laying down a path in walking" (Thompson 2007, 180). Thompson argues that organizational-operational closure generates a circular relation between an autonomous system and its milieu. Just as the organism cannot be defined independently of its environment, proponents of AE argue that the organism cannot be decomposed into *mechanical relations*, in which each part can be independently analysed; in each case proponents appeal to *dialectical relations*:

> [A] dialectical relation is one in which (i) A determines B, and B determines A (bi-directional dependence or reciprocal determination); and (ii) neither A nor B is analysable into discrete, causally efficacious elements that stand in a one-one correspondence (nondecomposability). Further, dialectical relations are dynamic, not static. Hence (iii) A alters B, and B alters A. (68–69)

Here we can plainly see that the AE appeal to dynamic co-emergence and dialectical relations in their theories of mind and life follows the same reasoning as Harris's *internal relations* and *unifying principle* (e.g. auturgy), which are crucial pillars in his own metaphysics.

2.2.2 *Embodiment Theory and Neurophenomenology*

While there are many forms of embodied cognition today, the basic idea is that the physical characteristics of an agent's body play significant causal and constitutive roles in fulfilling an agent's cognitive processes (Wilson and Foglia 2017). Accordingly, the body can serve as a *constraint*, *distributor*, and *regulator*, of cognitive functions. This means a given bodily structure provides both a *means* and *restrictions* for corresponding functions and perceptions. AE proponents typically endorse a version called *sophisticated radical embodiment*, which (i) denies that an individual's mind reduces to the individual's brain; (ii) posits that an individual's non-neural body is a constituent of her mind; and (iii) accepts the implication that there can be no formal boundary for cognitive processes (Chemero 2009). The remainder of this section introduces this kind of embodiment and its associated research programmes.

A key inspiration in the development of embodied cognition has been Gibson's *ecological psychology* (EP) concerning visual perception. "For Gibson, perception is strongly coupled with action, because most of the invariants within the optic array do not appear except against a background of change. Perceivers actively scan, move within, or otherwise manipulate the environment" (Shapiro 2012, 121). Cognition is here described as a *meaning making* process resulting from living bodies *actively fitting* into their environment. This approach is in direct contrast with the Fodorian trichotomy consisting of (i) perception (modular input); (ii) cognition (information processing based on syntactical rules); and (iii) agency (modular output). Embodiment here rejects the claim that cognition consists in the manipulation of symbolic (brain-instantiated) representations of an external world (Vörös et al. 2016, 191).

Accordingly, the world is not merely soaked up "as is" by observation, rather observation involves active participation, and how the world appears depends upon how agents conceive, move, and feel their body through time. As Harvey et al. (2016) explain, embodied perception

> depends not only on current sensation but also on the way sensation, and current movement, is systematically related to upcoming and potential sensory events, on the basis of the structure of the environment and the organism's neuromuscular expertise [...] This expertise comprises anticipatory sensitivity to the contingencies of sensory events in relation to potential or ongoing bodily movement, and is developed by means of individually and socially regulated experiences during ontogeny. (236)

Sensorimotor feedback thus involves changes in biological structures and perceptual sensitivity, which creates layers of reflexes in anticipation of various social and environmental relations. These dynamics constitute a wide array of cognitive capacities, which can themselves become the objects of attention, reflection, and emotional association in a biography founded upon bodily awareness.

Examples of such embodied memory use the term *conceptualization* to address how the body can be a part of one's cognitive processing, while *replacement* addresses how the body produces cognitive abilities outside of traditional schema of cognitive science (Shapiro 2012). For instance, *deictic pointers* have been proposed in support of *replacement*. Ballard et al. (1997) have found that subject's eyes would jump to the position or *deictic points* corresponding to where original stimuli were presented within their visual field in memory tasks, for example, the position of coloured blocks after a momentary presentation. Hence, an embodied theory of learning finds that memory tasks are more efficiently carried out with the aid of bodily cues. While this doesn't rule out representations, it does give credit to placing representational content in a secondary status relative to the physiological response, which is an adaptive function developed for the purpose of lessoning the burden of working memory.[7]

This thesis conflicts with the classical positions that understand cognition as consisting "in computational operations on informationally meagre symbolic representations in order to produce new representations that serve cognitive functions such as perception, language comprehension, problem solving and so on" (Shapiro 2012, 122). By depicting cognition as characterized by bodily feedback loops, enactivists have attempted to walk a "razor's edge" between solipsism (or idealism) on the one hand and either direct or indirect (representational) realism on the other (Maturana 1992, 241).

With computational and representational approaches deemed inadequate, embodiment theorists turn to *dynamic systems theory* (DST), which avoids reducing cognition to mere input/output processes:

> A dynamical system cannot be explained (or its behaviour predicted) based on the behaviour of the separate components that make up the system or in terms of synchronic, static, or purely mechanical interactions of parts within an isolated slice of time [...] In a dynamical system parts interact in a non-linear fashion, reciprocally determining each other's behaviour through a process of self-organization in which the parts remain dynamically coordinated to each other over a period of time. (Gallagher and Zahavi 2012, 89)

This theory implies that dynamics across brain, body, and environment become so coupled that there is no clear way to distinguish each from the others in terms of cognitive involvement. Accordingly, "[i]nputs are described as perturbations to the system's intrinsic dynamics, rather than as instructions to be followed, and internal states are described as self-organized compensations triggered by perturbations rather than representations of external states of affairs" (Thompson 2007, 11). *Meaning* is no longer seen as one-to-one structural correlations and causations posited by the representationalist, but is repeatedly redefined through living.

Although many AEs have appealed to Gibson's theory of *affordances* to support their theory of perception, according to Fultot et al. (2016), careful consideration reveals an incompatibility between the two camps. They maintain situation *X affords* activity *Y* for organism *Z* on occasion *O* iff *X* and *Z* are *mutually* compatible on dimensions relevant to *Y*.

As Chemero (2009) elaborates, this "mutuality" involves a *symmetry of information* between organism and environment:

> [T]he principle of symmetry is that (1) the environment specifies the information available for perception and the information available for perception specifies what is perceived, and (2) what is perceived specifies the information available for perception and the information available for perception specifies the environment. (120)

This means there is a "1:1 correspondence" between the environmental information and the information perceived. In virtue of its self-organized thermodynamic constraints, the *XYZO* system holds the primacy of agency—not the (autopoietic) organisms. Ecological psychology (and Radical embodied cognition (REC)) thereby maintain a ***direct realism*** about affordances and an observer-independent reality for ***meaning***. Phenomenology thus becomes superfluous on EP's account of consciousness, which strongly conflicts with AE proponents, for whom this symmetry does not hold. Providing some defence for AE, Chemero explains that whereas the "environment-to-perception fit is at least partly causal, the perception-to-environment fit is primarily normative" (121). This asymmetry between environment and agent on the basis of value-construction implies that (rejecting 1) ***agency*** must be more narrowly defined and if so, phenomenological methods are still needed to specify the agent-environment relation.

Proposing a notion of organism-environment ***interdependence***, Thompson (2007) has argued for a middle way between behaviourism and cognitivism, maintaining that "as a skilful activity of the whole animal or person, perceptual experience emerges from the continuous and reciprocal (non-linear) interactions of sensory, motor, and cognitive processes, and is thereby constituted by motor behaviour, sensory stimulation, and practical knowledge" (256). The environment, then according to AE, is not ***directly perceived*** by the organism—as in Gibsonian ***affordances***, REC, and ***sensorimotor enactivism***—but neither is it represented. Rather, organism and environment are enacted via sensorimotor and chemical ***feedback loops***. Phenomenology cannot be reduced to external information since there remains a normative valence for the subject and thus an ***asymmetric*** relation between agent and world.

Since its inception (Varela et al. 1991), AE has been unconcerned with the traditional ontological arguments concerning the mind-body relation. Varela (1996) considered Chalmers's (1995) "hard problem" as a mere artefact of the prevailing paradigms in philosophy and neuroscience, maintaining that the solution was not a theoretical fix or extra ingredient, but rather an overthrow of the paradigm that gave rise to it. He proposed ***neurophenomenology*** as a methodology towards this end. For proponents, this method is not considered a solution, but a way to "live away the problem" by way of ***pragmatic*** heuristics (Varela et al. 1991; Varela 1996; Thompson et al. 2005). Their working hypothesis is that phenomenological accounts of the structure (i.e. the ***how***) of experience and accounts of neuro-cognitive processes informed by Dynamic field theory (DFT) can be ***mutually informative*** and equally constraining:

> Isomorphism suggests an epistemological parallelism in which the biological and phenomenological accounts run on parallel tracks, with no mutual interaction or influence. Neurophenomenology, following Merleau-Ponty's lead, proposes a more daring idea—that biology and phenomenology can stand in a mutually enlightening, explanatory relation. (Thompson 2007, 358)[8]

Following Varela, biology and phenomenology are not considered identical to one another, but as correlated dynamics, each contributing to an ongoing means of clarifying the previously *unrecognized variables* of the other. While the intention of this *pragmatic* methodology remains invaluable, I maintain that on its own, this method is insufficient to avoid the deeper metaphysical issues and that it has further consequences for philosophy of science than has hitherto been recognized.

Given the results of Sects. 2.2 and 2.3.1, it appears clear that Harris's rejection of representationalism and direct realism combined with his conception of a pragmatic approach to mind-body holism renders his system deeply compatible with radical embodiment as illustrated above. Going forward, I maintain that the neurophenomenological methodology and Harris's scale of forms can each provide invaluable insight into the other and may yet provide a sufficient ontological response to the "*explanatory gap*".

2.2.3 *Embedded and Extended Theories*

Going one step further than the embodied thesis—to what some proponents of enactivism contend are its logical entailments—the *embedded* (or situated) thesis maintains "cognition deeply depends on the natural and social environment" (Wilson and Foglia 2017). The embedded mind proposes that cognition relies upon elements of a subject's environment with which they are in constant interaction. Beginning with Clark and Chalmers (1998), proponents of the *extended mind hypothesis* have argued cognition is *constituted* by spatiotemporally extended worldly engagements if those engagements fulfil the same functional roles as they would within the brain. Though Harris did not discuss these theses, I maintain that his metaphysics directly imply the following versions.

Here, physical alterations of an environment, including both language use and physical artefacts, circle back to reshape agents' cognitive structures. Relying upon GST, proponents of the embedded thesis maintain cognition can be adequately understood only by connecting biological, psychological, and social-ecological accounts of perception and behaviour into an interconnected and dynamical whole (Clancey 2009, 12). As Noë (2009) explains, the embedded thesis builds upon the organism-environment feedback loop discussed above by adding a dimension that is both ecological and cultural:

> The point is that just as a rake extends our reach and has the potential to extend our body schema, so language extends our capacities for thought and therefore can be treated as extending our mind schema. Insofar as language is itself socially manufactured and shared by linguistic communities, then to that extent our cognitive powers require for their very exercise the existence of a sociolinguistic environment. Our minds cross out of the skull and get supported in a shared sociolinguistic scaffolding. (88)

The extended mind hypothesis begins with many of the same premises as the embedded thesis, bringing attention to processes of either offloading information into an environment through intentional alteration (e.g. writing) or using environmental features as tools for cognition (e.g. a stick for balance) that reduces strain on the agent's internal cognitive faculties. According to Rowlands (2010), for the enactivist the extended mind hypothesis is not an argument for identifying environmental structures with cognitive states, but rather including what we *do with* environmental structures (e.g. manipulations, exploitations, or transformations) in cognitive processes: "It is the manipulation of environmental structures that forms part of the cognitive process; and so these structures form part of the process only insofar as they are part of the process of their being manipulated" (67).

Following Merleau-Ponty's example of how a walking stick can exemplify a plastic nature of bodily boundaries, Favela and Chemero (2016) provide a scientific argument for how our being-in-the-world can be reshaped by the incorporation of artefacts. Appealing to DFT, they argue that "the locations of the quantities tracked by the variables in dynamical models are not constrained to be on just one side of the skin-environment boundary" and that "certain kinds of dynamical systems are non-decomposable, in that the equations that model them cannot be solved separately" (p. 66). An implication of this coupling is that cognitive systems *exhibit interaction-dominant dynamics*: "A system exhibits interaction-dominant dynamics when the interactions among the components dominate or override the dynamics that the components would exhibit separately. Systems with interaction-dominant dynamics [...] are not modular" (p. 67). They argue that when an artefact is used to facilitate some mode of sensation (e.g. BrainPort), there exists interaction dominance and thus cognition is functionally extended into these artefacts.

As will be discussed in Chap. 6, the ontological implications of this kind of reasoning has brought both AE proponents (Thompson, W. 1988; Thompson, E. 2007) and Harris (1987, 254) to invoke Lovelock and Margulis's (1974) "*Gaia hypothesis*", maintaining that the biosphere as a whole may demonstrate life and cognition. Methodologically speaking, some have argued (again, in unwitting agreement with Harris) that extension theories entail significant implications for what disciplines hold authority over consciousness research:

> [I]f extended cognitive science is an important component of scientific psychology, then neuroscience cannot subsume or replace scientific psychology. Because extended cognitive science consists in laws, generalizations, and regularities concerning brain-body-environment systems, neuroscience is not appropriate as a base theory. (Chemero and Heyser 2009, 71)

As will be discussed further in later sections, from the above introductions I conclude that both Harris and AE find common ground in a kind of *formal governance* that plays a central role in their theories of biology and cognition. For Harris, mind and life are depicted as instances of unifying principles, whereas for AE they are emergent phenomena instantiated by non-linear feedback loops. By grounding his holist theory in a metaphysics of Nature however, Harris also extends his theses into the philosophy of physics.

2.3 Analogies with Bohm's Holism

In this section I take a preliminary look at the *implicate order*, a philosophy of physics reliant upon metaphysical theses analogous with Harris and AE. The implicate order was developed by American educated theoretical physicist David Bohm (1917–1992) in the early 1970s. Bohm's intent was to offer a holistic alternative to reductive and mechanistic frameworks of mind and matter.[9] In his *Wholeness and The Implicate Order* (1980a) Bohm synthesized these ideas into a coherent metaphysics and for the following ten years, applied his system to a host of topics beyond physics. In this section, I first review the central tenets of the implicate order in quantum mechanics, before turning to the theory's metaphysical implications concerning the nature of consciousness.

Bohm maintained that contrary to its intended aim of rendering philosophical accounts based upon a contemporary scientific framework, *physicalism* had become dependent upon an out-dated understanding of matter. As a result, any explanation of mind on this basis is inevitably and fundamentally flawed. In his synthesis and extended application of the implicate order, Pylkkänen's (2007) work aptly explains the starting point for Bohm's position as a dissatisfaction with the disconnected picture of nature revealed by respective domains of natural science. For example, whereas relativity theory entails continuity, strict causality/determinism, and locality, quantum mechanics requires non-continuity, non-causality and non-locality.

Consequently, the basic concepts of the two most fundamental physical theories directly contradict each other. If we are to reconcile their differences and establish a new and more fundamental theory, Bohm maintains the "best place to begin is with what they have basically in common. This is undivided wholeness. Though each comes to such wholeness in a different way, it is clear that it is this to which they are both fundamentally pointing" (1980b, 84). The goals of his *implicate order* were thus to reinterpret the results of quantum experiments and bring the contemporary sciences in line with one another in order to provide an account of consciousness.

2.3.1 *Irreducible Holism and Internal Relations*

To obtain such a holistic theory of nature, Bohm set a course balanced between Einstein and Bohr—the two contrary philosophical interpretations of physics for much of the early twentieth century (Bohr, ff. 212). In its inception, the implicate order was intended to share Einstein's conviction that an objective reality exists independently of us, but also retain sympathies for Bohr, who maintained that the meaning of an experimental result and the form of experimental conditions are an inseparable whole that requires our *participation*.[10] Pylkkänen explains that what Einstein and Bohm had in common was a particular dissatisfaction

> with the extremely empiricist, "positivistic" feature of the usual quantum theory, which did not allow one to discuss reality beyond the observations.

> Observations, in turn, were fairly limited (e.g. a spot appearing in a photographic plate), so it seemed that quantum theory was providing a truncated, fragmented view of reality. (14)

In line with Bohr (1959), Bohm recognizes that our understanding of the world is confounded by how we look at it. More precisely, "our scientific instruments can affect our assumptions about perceivability; and these in turn can affect our assumptions about what is conceivable" (Pylkkänen, 57). So, what is discovered is the ***product*** of the experimental system as a whole:

> [D]escription of the experimental conditions does not drop out as a mere intermediary link of inference, but remains inseparable from the description of what is called the observed object. The "quantum" context thus calls for a new kind of description that does not imply the separability of the "observed object" and "observing instrument." Instead, the form of the experimental conditions and the meaning of the experimental results have now to be one whole, in which analysis into autonomously existent elements is not relevant. (1980a, 169)

In his earlier work, Bohm considered whether probability in quantum theory is a result of our ignorance of what variables to use in describing a system, as is the case in classical physics. He maintains that examining these classical situations may provide theoretical guidance concerning the ontology of quantum theory:

> For example, in thermodynamics we measure the pressure, temperature, and volume of a given system. In very small regions of space, especially near the critical point, we find that these quantities no longer obey an equation of state exactly, but instead exhibit large random fluctuations about a mean value that is predicted by the equation of state. Hence, the deterministic laws of thermodynamics break down and are replaced by laws of probability. This is because the thermodynamic variables are no longer appropriate for the problem and must be replaced by the position and velocity of each molecule, which are, from the viewpoint of thermodynamics, hidden variables. (1951, 29)

For Bohm then, both quantum and classic quantifications are averages of "hidden variables", whose causal laws cannot be observed by only lower-level methods but require a description in terms of some ***non-local*** relations. Interestingly, this also anticipates theories of emergence via *phase transition* (discussed in Sect. 5.3). Bohm first applied this idea to the behaviour of individual particles and in later years used the resulting framework as means of explaining the nature of consciousness.

The *de Broglie-Bohm theory* (AKA *hidden variables interpretation of quantum mechanics*) was first proposed by de Broglie (1924) and later adopted by Bohm (1951).[11] Contrary to Copenhagen or orthodox quantum mechanics, this interpretation renders the randomness of quantum phenomena a result of our ignorance, whereas reality is considered determinate in virtue of *hidden variables* of a *pilot wave*. As philosopher of physics, Tim Maudlin explains, a deterministic theory can produce "definite non-probabilistic predictions only given the precise initial state of a system. If one is unsure about the precise initial state, then that uncertainty will infect predictions about the evolution of the system" (2011, 108). Since Bohm's theory presupposes an uncertainty about the exact initial state of particles, he claims that the probability function is used for the initial configuration, which produces the exact same predictions as orthodox quantum theory.

Maudlin (2011) explains that Bohm's ontological interpretation has three corresponding principles, which are directly opposed to those of an orthodox interpretation:

1. *The wave-function is incomplete*—because particles always have definite positions, fields always have definite values, which means systems sometimes have definite observable properties even when not in eigenstates of the associated operators; for example, a particle can have a position even if its wave-function is not an eigenstate of the position operator. Additionally, two systems can have identical wave-functions but be in different physical states because the values of the extra physical parameters that are not reflected in the wave-function may differ.
2. *There is no wave collapse*—the wave function always evolves according to Schrödinger's equation and never collapses; it does not represent the states of systems, but the state of a quantum field (on a higher-dimensional configuration space) that influences the physical system.
3. *The dynamics are deterministic*—since the wave-function is always governed by Schrödinger's equation, the system evolves deterministically. Unlike the orthodox view, particles are also *physically guided* by this wave. "The potentials determine how the wave-function evolves and the wave-function determines how the particles move" (111).

Bohm's ontology depicts particles similar to classical mechanics in that they always have definite positions and trajectories. However, he also posits a physical object that is described by the wave-function, which is not made up of particles, always evolves according to Schrodinger's equation, and determines how the particles will move. Two equations follow:

(i) The Schrodinger equation determines how the wave-function changes:

$$ih\frac{\partial\psi}{\partial t} = -\frac{h^2}{2}\Delta\psi + V\psi$$

(ii) Bohm's equation specifies how the particle position will change given the wave-function:

$$\frac{dq}{dt} = hIm\frac{grad\psi}{\psi}(q)$$

Bohm's equation reveals that the wave-function determines the velocities of the particles and as the Hamiltonian operator determines how the wave-function will evolve, when combined with the position of the other particles, this determines the evolution of the entire system.

What Bohm contributed to the mathematical interpretation was an ontology in which reality is composed of ***explicate*** and ***implicate orders***. He argued that relations of the implicate order can be invoked to explain (observed) quantum phenomena in the same way that an "ensemble" of glycerine within cylinders can be invoked to explain the distortion and reformation of ink droplets through time (1980a, ff. 186). Particles for example, are not continuous entities that move "through" time and space, but rather emerging explicate structures, cyclically unfolding from and enfolding back into a vast sea of background (vacuum) energy.[12] Whereas the explicate order consists of relatively separate and externally related entities typical of classical physics, the implicate order provides an unseen and unified ground from which the ***individuals*** arise (Pylkkänen, 24).

To free us from the dominant mechanistic worldview, Bohm suggests that rather than conceiving the world as analysed into individual parts via a lens, a "hologram" provides a holistic paradigm of nature (1980a, 224–25). He maintains that like a hologram, each separate and extended form in the explicate order is enfolded in the whole and in turn, the whole is enfolded in its forms, but "only in a limited and not completely defined way" (Bohm 1987, 193). Moreover, the way in which an explicate form enfolds the whole is considered essential to it and to how it behaves: "So the whole is, in a deep sense, internally related to the parts. And, since the whole enfolds all the parts, these latter are also internally related, though in a weaker way than they are related to the whole" (ibid.). For example, Pylkkänen holds, "there is an internal relationship between an electron and its environment, or between two electrons in a non-local relationship" (53). So, the incident of an electron represents the explicate, while the form of an electromagnetic field is that of implicate order. Likewise, the general theory of relativity depicts the universe "in terms of a unified field" (specified by non-linear equations); thus, space-time cannot be analysed in terms of separable points (Bohm 1980a, 158–59).

In sum, just as Harris and proponents of AE, the implicate order relies upon common conceptions of holism, internal/dialectical relations, and a balance between realism and constructivism. Accordingly, "there is a kind of objective wholeness, reminiscent of the organic wholeness of a living being in which the very nature of each part depends on the whole" (Bohm & Hiley, 1993, 177). Importantly Bohm's appeals to holography and enfolded information are really metaphors for guiding his readers towards his preferred theory of nature, what he called the *holomovement*.

2.3.2 𝕰 and the "Holomovement"

Harris was first introduced to Bohm's *Wholeness and the Implicate Order* when he attended a meeting of the *Southern Philosophical Society* in the mid-1980s. He recalls, "I had not read Bohm's book, but when I did, I was gratified to discover that a distinguished physicist had independently confirmed (apparently without knowing what I had written) the sustained thesis of my *Foundations of Metaphysics in Science*" (2015, 243). Like Harris's 𝕰, Bohm's holomovement is intended as a framework for both Nature and consciousness, each of which I briefly outline in this section.

Bohm's basic proposal "is that what *is* is the holomovement, and that everything is to be explained in terms of forms derived from this holomovement" (1980b, 86). His ontology is essentially considered a *movement* or cyclic process of development from enfolded potential to unfolded explication, which then returns to implicate form:

> This enfoldment and unfoldment takes place not only in the movement of the electromagnetic field but also in that of other fields, such as the electronic, protonic, sound waves, etc. [...] Moreover, the movement is only approximated by the classical concept of fields (which is generally used for the explanation of how the hologram works). More accurately, these fields obey quantum-mechanical laws, implying the properties of discontinuity and non-locality. (1980b, 85)[13]

Consequently, the "particle", for example, is understood as an abstraction "from the more fundamental movement of unfoldment and enfoldment" (Pylkkänen, 70). The way in which this process unfolds is always constrained or guided by the form of a more inclusive whole or physical system in what many have called an "*infinite tower of pilot waves*".

Bohm's response to the EPR experiment characterizes his conception of holomovement. In the EPR experiment, each of two electrons that were previously combined in a *spin singlet state* are observed after they have separated and are far enough apart that they can be measured before light could pass between them and therefore (due to relativity) exclude causal relations. Upon spin measurement, particles nevertheless demonstrate non-local correlations. For Bohm, the motion of a particle is guided by its pilot wave (quantum potential), which supplies "*active information*" of an environment as a whole, including its entanglements, thus permitting the correlated measurements (1990, 280). He explains the EPR experiment can be metaphorically understood by considering *two cameras* recording the *correlated* (2D) movements of a single (3D) fish within a tank. In the case of either particle or camera recording, each constitutes "a system as a projection of a 'higher-dimensional' reality" rather than *individual* phenomena (1980b, 96).[14]

Providing an analogy (and possible elaboration) of Harris's 𝔈 and Concrete Universal, Bohm proposes that there must likewise be a "super-quantum potential" (super-implicate order) that provides *form* (active information) to the unfoldment of the first-order quantum potential (implicate orders). If so, the first-order quantum potential would act only as an approximation when the action of the super-quantum potential could be neglected. "One could go on to suppose a series of orders of superquantum potentials, with each order constituting information that gives form to the activity of the next lower order (which is less subtle)" (1990, 283). In Bohm's model, our (observable) explicate order is considered one such sub-whole obtained from a higher-level implicate order (1987, 196–97).

In line with pragmatic fallibilism, awarding primacy to an immeasurable holomovement implies that there is "no meaning to talk of a fundamental theory, on which all of physics could find a permanent basis, or to which all the phenomena of physics could ultimately be reduced" (1980b, 131). Rather, we must always "be ready to discover the limits of independence of any relatively autonomous structure of law, and from this to go on to look for new laws that may refer to yet larger relatively autonomous domains of this kind" (1980b, 86). This *holonomic principle* does not however preclude the relevance of analysis, indeed it grants the "possibility of describing the 'loosening' of aspects from each other, so that they will be relatively autonomous in limited contexts (as well as the possibility of describing the interactions of these aspects in a system of heteronomy)" (1980a, 198). Nevertheless, the laws of respective sub-totalities are *abstractions* that may be surpassed by identifying the *origin* of their corresponding domain, which leads to yet *deeper laws*, none being inherent or of an ultimate level (Pylkkänen, 91–2).

Extending the implicate order to biology, Bohm maintains that just as the elementary particle demonstrates an internal relationship with its environment, living systems too, are constantly exchanging matter and energy with their environment such that we are unable to make a sharp distinction between the two. The idea is that we find a common process of enfoldment and unfoldment that gives rise to different structures, depending on how an environment *informs* the entity (Pylkkänen, 86). For Bohm, as in AE, life is properly understood as a totality that includes relations of organism and environment. Specifically, Bohm maintains, the holomovement is "life implicit", in that "it is the ground both of 'life explicit' and of 'inanimate' matter, and this ground is what is primary, self-existent and universal. Thus, we do not fragment life and inanimate matter, nor do we try to reduce the former completely to nothing but an outcome of the latter" (1980b, 102).

Aiming to avoid Cartesian dualism, just as epiphenomenalism, Bohm is again in agreement with AE, but he goes even further to contend "the implicate order applies both to matter (living and non-living) and to consciousness" (1980b, 103). He goes on to write, "it can therefore make possible an understanding of the general relationship of these two, from which we may be able to come to some notion of a common ground of both (rather as was also suggested in the previous section in our discussion of the relationship of inanimate matter and life)" (1980b, 103–4).[15] For Bohm, the notion of "active information" provides "a rudimentary mind-like behaviour of matter, for an essential quality of mind is just the activity of form, rather than of substance" (1990, 281). The common principle of both mind and matter then, is considered the formative or governing force of "active information", which Bohm maintains can serve "as a kind of link or 'bridge' between these two sides of reality as a whole" (1990, 282). With this move, Bohm's theory of mind is perfectly compatible with the neurophenomenological methodology in that cognition and matter can be conceptually unified by attending to their *homologous processes of organization*.

This depiction of mind and matter being constrained by a boundless series of implicate orders seems *prima facie* to imply that "freedom of will" is out of the question, but this conclusion misses an essential point. The implicate order is at each point incomplete and perpetually unfolding, not merely actualizing potentialities that were somehow previously established, but defining itself through time. This Pylkkänen claims, makes "genuine creativity" possible on a fundamental level (126). What purportedly makes this possible are the roles of "recurrence and stability" that permit the existence of "relatively independent sub-totalities". Human beings are considered such "sub-totalities", whose relative independence implies that the individual has some freedom. "Recurrence and stability are required if the sub-totalities are to have any independence and autonomy in relation to the holomovement. Otherwise, the unending flux of the holomovement would destroy any independence, including the one required for freedom" (ibid.). So, for Bohm, there is *relative* freedom for every sub-totality.

One formulation of such *relative freedom* articulated in a joint unpublished work of Bohm and Pylkkänen (1990–1991) appears to reiterate Harris's conception of onto-epistemic coherence as well as AE's conceptions of enaction. Here, they proposed a theory of "overall coherence" applicable to both mind and nature. Pylkkänen explains that according to this view, truth implies "logical consistency", but also "coherence of the state of being of the individual who is concerned with truth, as well as the environment" (2007, 154). Here, the domain of truth is *neither* internal *nor* external to mind, but is a fundamental movement constantly overcoming incoherence that involves both domains: "As with consciousness, each moment has a certain explicate order, and in addition it enfolds all the others, though in its own way. So the relationship of each moment in the whole to all the others is implied by its total content: the way in which it 'holds' all the others enfolded within it" (Bohm 1980a, 263). ***Truth is thus the degree to which a*** subject ***coherently enfolds their world through time.*** As Pylkkänen concludes, positing the holomovement as fundamental for mind and nature implies that although complete coherence may be unreachable, "movement toward coherence is possible, and is indeed the essence of cognition" (155).[16]

From the above there appears clear agreement between the implicate order and Harris concerning dialectical relations, process ontology, a pragmatic coherence theory of truth, and the scale of forms. For AE and Bohm, the result of maintaining these tenets has been the formation of two partially overlapping holist theses of nature and mind. What Harris provides, in broader strokes, is a means of systematizing these and other complementary arguments across a range of domains. I now turn to deeper commonalities, both epistemic and ontological, that will serve as paradigmatic guides in the following chapters.

2.4 Towards a *Converging* Paradigm

At this point the theoretical similarities between Harris's holism, AE, and the implicate order should be in reasonably sharp relief. On the topics of individuation and neutral monism we find still greater commonalities.

Writing on the issue of body-based perception and learning, Gallagher (2014) articulates what may serve as an ***AE theory of individuation***:

> [I]t is through our interactions with others that we learn what objects are significant or valuable. We learn to understand the world along these lines of significance and value, and often objects that fall outside of such lines don't even register. In the same way that expert training hones the perceptual system so that experts are able to perceive things that non-experts fail to perceive, in some sense, we all become experts in everyday life through our interactions with others. (239)

This clearly implies individuation depends upon the social system within which we are embedded. So, our brain "attunes to and responds to its environment in a way that enacts a meaning relative to the particularities of its embodiment" (243). Consequently, we find what enactivists have called a ***reflexivity*** underlying all our epistemic practices, which implies the impossibility of obtaining what Nagel (1986) calls "a view from nowhere".

AEs thus align themselves with the aims of *second-order science*, in positing a fundamental circulation between "lived experience" and "scientific understanding", which attempts to ***account for the observer and their writing of a theory, within the theory itself*** (Vörös et al. 2016, 191). The purpose of which, is to account for and (partially) ***go beyond*** the biases of our present theoretical constructs. Individuation of an entity is thus embodied and socially enacted, requiring that the phenomenological process of this enactment be repeatedly examined ***ad infinitum***.

Indeed, a similar notion is identifiable in Bohm's works. Although the implicate order is considered fundamental to consciousness, we are less aware of it, Bohm maintains, because we have become habituated to functioning in such a way that emphasizes the explicate order. He claims our emphasis on nouns, for example, has conditioned us to perceive a fragmented world of static objects. His remedy the *rheomode*—a shift in semantic primacy from subjects and objects to verbs. His aim, in part, was to draw our attention to what he calls the *act of levation*—the way in which we raise an entity to ***relevance*** from its background:

> So when relevance or irrelevance is communicated, one has to understand that this is not a hard and fast division between opposing categories but, rather, an expression of an ever changing perception, in which it is possible, for the moment, to see a fit or a non-fit between the content lifted into attention and the context to which it refers. (1980a, 43)

This implies that what appears to be *irrelevant* in a given context can be seen not only as useful, but as eventually necessary for an *ongoing process* of description, distinction, and definition of any phenomenon because the implicate order of which, is an ***undivided whole***.[17]

In physicist Karen Barad's (2007) *agential realism*, agency is addressed from a very different angle (postmodern feminist philosophy), but she has—without recognizing the connections to Bohm or enactivism—articulated the very point that binds the two systems and highlights the core of Harris's onto-epistemology. Following Bohr's philosophy-physics, Barad maintains that agency is created through participatory practices of "***diffraction***" that cut an entity from its background. The observed or diffracted entity is thereby inseparable from the diffracting observer and their means of observation (118–19). "Reality is composed not of things-in-themselves, or of things-behind-phenomena, but of things-in-phenomena. Because phenomena constitute a non-dualistic whole, it makes no sense to talk about independently existing things as somehow behind or as the causes of phenomena" (205).

In agreement with AE and Harris, according to Barad, objectivity can be achieved by reproducing marks on bodies left by an "intra-active cut". This cut is determined by the diffractive system as a whole, which is, by relations among agents and tools (both marker and marked bodies) within which the object (or agent) of our investigation comes into being (320):

> [L]aboratory manipulations, observations, interventions, concepts, and other human practices have a role to play, they do so as part of the material configuration of the world in its intra-active becoming [...] Hence [...] what is at issue is not knowledge of the world from above or outside, but knowing as part of being. (341)

She goes on to contend that by implication, all bodies, "including but not limited to human bodies, come to matter through the world's iterative intra-activity—its performativity. Boundaries, properties and meanings are differentially enacted through the intra-activity of mattering" (392).[18] Barad's use of *diffraction* is thus the *experimental* counterpart to *social enaction* and *analytic levation*.

Whereas autopoiesis attributes agency only to a self-organized system that we might say, continually enacts, levates, or diffracts *itself* over time, for Barad agency can be attributed to non-organic, non-living and even totally inanimate objects as well. This is possible just in case *their identity, functionality and continuity is preserved (i.e. enacted) by their external entanglements/relations over time.* "Agency [...] is the enactment of iterative changes to particular practices—iterative reconfigurings of topological manifolds of spacetime-matter relations—through the dynamics of intra-activity" (235). In line with Harris, Barad maintains the interdependent *relations* of enacted phenomena hold both scientific and philosophic primacy (104; 141; 205–6). Taking all camps together, *diffraction* is really a broadening of enactivism as anticipated by Harris's earlier (1965 and 1970) works. Again, Barad's thesis relies upon analogues of dialectical relations, the unifying principle, and process ontology.

If the above may serve as the common epistemic points of convergence, the (preliminary) ontological implications may be outlined as follows. As Thompson (2015) recognizes, in an AE ontology, "understanding how consciousness is a natural phenomenon is going to require a radical revision of our scientific concepts of nature or physical being" (103–4). He goes on to advocate for what can best be described as neutral monism:

> [S]uch an understanding would replace our present dualistic concepts of consciousness and physical being, which exclude each other from the start, with a non-dualistic framework in which physical being and experiential being imply each other or derive from something that is neutral between them. (105)

Though they make no reference to Thompson's above contention, concerning the metaphysical implications of enactivism, Silberstein and Chemero (2015) have gone some way to elaborate upon what sort of neutral monism is entailed. They argue that because *individuals* in DST can have parameters on each side of their boundary, the non-linear behaviour of the agent is *coupled to and partially constituted by the relations it bears to its environment* (e.g. via motion parallax). They recognize that enactivists must hereby dispense with qualia, essences, properties, and substances so as to reject the categorical divisions of inner/outer, self/world, subjective/objective, and so on. Silberstein (2009) proposes that proponents of an such *extended neutral monism* (ENM) should instead appeal to *ontological structural realism* (OSR).

Treating the brain-body-environment as a nondecomposable dynamical system, they argue, means the "phenomenological world of experience is neither in the 'head' nor in the 'external world'—it is fundamentally relational" (2015, 190).[19] They claim that the "world and the self co-emerge, in so far as the activities of the living system determine what aspects of the physical environment it engages" (191). So, they conclude that by rejecting *neural correlates of* consciousness as sufficient for consciousness and rephrasing the subject/object cut as a self-consistency relation, ENM provides a means of naturalizing phenomenology: "there is only one reality (the field of pure 'experience' as James might say)" (193).

Though Harris (and I) would agree with most of Silberstein and Chemero's contentions in articulating a metaphysical basis for enactivism, they leave several issues unaddressed. For instance, they do not reconcile the significant conflict between emergence (upon which enactivism relies) and ENM. More generally, their adoption of James's (1904) argument for a 'pure experience' is *admittedly* problematic and appears to only succeed in pushing the problem of consciousness up one level. In this regard, they make limited effort in spelling out what the *neutral base* is by positing a relational ontology, but much more is needed for this to provide a metaphysically satisfactory response to the hard problem.

As Pylkkänen recognizes (2007, 37; 39), Bohm's (1980a) philosophy of physics posits a ground for "neutral monism" in the holomovement.[20] Bohm maintained, "the more comprehensive, deeper, and more inward actuality is neither mind nor body but rather a yet higher-dimensional actuality, which is their common ground and which is of a nature beyond both" (265). Just like his interpretation of the EPR experiment, explicate mind and matter are again considered *correlated phenomena* that are abstractions from a common *implicate identity* in the holomovement. As Harris has sympathetically recognized, treating the holomovement as a "common ground" is deeply reminiscent of Spinoza's system, in that mind and matter are *two aspects* of a common underlying *Substance* (here a process) (1995, 34–5).

In agreement with both Harris and enactivists, Bohm recognizes that proponents of this view cannot regard human beings as individuals in the conventional sense. That is, we ought to treat the human in the same way that we treated the individual particle—*as arising from an implicate system*:

> [I]f we are to give an adequate account of what actually happens and this must eventually include other people, going on to society and to mankind as a whole. So it will be ultimately misleading and indeed wrong to suppose, for example, that each human being is an independent actuality who interacts with other human beings and with nature. Rather, all these are projections of a single totality. (1980b, 116–17)

Hence, all parties appear willing to agree with Favela and Chemero's (2016) contention that the "proper object of study" concerning consciousness, is the "extended agent-environment system" (67).

When I asked Evan Thompson of the possible metaphysical and cosmological implications of AE with regard for Bohm and Harris's systems, he replied as follows:

> (1) The enactive approach, as I see it, has its main home in cognitive science and is concerned to answer questions about the nature of cognition and how cognition requires autonomous agency. This viewpoint doesn't seem to me to presuppose any particular cosmology. That being said, the overall background conception of coupled systems arising through dependent origination could be taken as a cosmological view, or at least as a kind of principle that would need to inform an "enactive cosmology." (2) The enactive

> approach strikes me as based on or presupposing a metaphysics of process rather than a metaphysics of substance. And a metaphysics of process can naturally use as methodological tools the techniques of dynamic systems theory in physics. (Personal communication: September 20, 2013)

Here we clearly have a thesis of internal/dialectical relations and co-emergence based upon a metaphysics of process, which I contend are perfectly analogous to the principles of Harris's holism. As mentioned above, Harris maintained that the synthesis of scientific data necessarily presupposes metaphysical tenets, the question is the extent to which the theorist is aware of these tenets, which ultimately culminate in some conception of Nature *as a whole*. I maintain that a closer inspection of Harris's system can indeed provide an "enactivist cosmology" and simultaneously clarify the "neutral monism" to which AEs and Bohmians adhere.

2.5 Conclusion

This preliminary investigation has shown that Harris's metaphysics (i.e. the marriage of internal/dialectical relations, embodiment, process ontology, emergence, coherence, and neutral monism) bears a common framework with AE and Bohm's implicate order. In the following chapters I provide further evidence to the effect that while AE and the implicate order address the philosophies of cognitive science on the one hand and physics on the other, their inevitable metaphysical implications are the same and have already been significantly detailed in Harris's work. I propose the title "*dialectical holism*" not as reference to Harris's holism, but as an umbrella term for these three systems taken as mutually constraining pillars of a single metaphysical paradigm.[21]

The primary goals for the remainder of this text will be to present the arguments for Harris's holism, evaluate their applications across a range of domains in philosophies of science and mind, and elucidate their implications. The result of this work will be to either vindicate or reveal possible weaknesses of a number of similar contemporary positions and establish whether such a *dialectical holist paradigm* can survive Harris's original formulation. With Thompson's personal communication in mind, I claim that what follows works towards an "enactivist cosmology". To address these tasks, I next consider topics in the philosophy of physics, with a focus on the *cosmological anthropic principle* in particular.

Notes

1. This chapter draws primarily from works published in the later stages of Harris career because these works provide the clearest introduction to his metaphysics. Arguments to the effect that Harris has anticipated respective philosophical and scientific theories will be provided in later chapters.
2. This was the very position that Moore (1919, 3) (perhaps unnecessarily) argued against.
3. It may also be noted that mass, a measure commonly referred to as intrinsic, is itself relational: the relative concentration of matter in a given inertial frame and speed of which is again relative to light and other mass-bodies.
4. For example, in Escher and Vermeulen's (1989) *plane division* there is an internal relationship between angles, faces, internality, and externality of an object. A more modern example is *cellular automata*, in which the nature or *state* of each cell depends upon its relations to an immediate neighbourhood and that neighbourhood further depends upon its relationship to the total system. The unifying principle can also be exemplified as the *update*, in that each cell interacts with those in its neighbourhood and the way in which they do so is determined by (and thus demonstrates) its overarching law(s). For further details see Wolfram's Mathworld: http://mathworld.wolfram.com/TotalisticCellularAutomaton.html.
5. For an introduction to Neutral Monism see Stubenberg (2018). For a recent historical analysis and revision of this position see Banks (2014).
6. Harris recognizes that the claim that "nothing exists except matter is not empirically verifiable" (2000, 2). He claims *verification* consists in systematically assembling diverse and corroborative evidence, the interconnections of which make denial of the resultant theory "impossible" without resulting in a "breakdown of the entire conceptual scheme" (93).
7. James's ideo-motor theory was instrumental in numerous embodiment conceptualizations. Specifically, his case of *confronting a bear*, illustrated that we are afraid *because* we run, with the initial physiological response being a conditioned reflex that causes a specific emotional (and thus cognitive) response (James 1884).
8. For example, this type of methodology has revealed that phantom limb syndrome arises due to a failure of sensorimotor feedback, in that the brain sends signals to which the body cannot respond (Ramachandran and Hirstein 1998). It has been shown that by utilizing this feedback with the aid of mirrors to simulate the missing limb, it is possible to significantly alter one's *phenomenal self-model* and correlated neural pathways, thus altering or even eradicating the phantom limb (Metzinger 2003, ff. 470).
9. Bohm's early work found many empirical applications, for example, the theory of plasmas and "Bohm diffusion". Nevertheless, before he was able

to complete his PhD thesis, it was taken from him by the US government because it provided insights in the Manhattan Project—he was awarded his degree anyway. Since his philosophy emphasized wholeness and *hidden variables* his position was counter the mainstream account of reality (i.e. the Copenhagen interpretation), so his system was marginalized. Due to Bohm's communist affiliation in the early 1950s, he was also unable to find work within the US for many years. For further details see Junior (2015, ff. 17).

10. The Copenhagen interpretation of quantum mechanics consisting of Bohr's philosophy physics as well as von Neumann's later elaboration is considered the *orthodox* version today. Stapp (2011) claims that this view is fundamentally *subjectivist,* "in the sense that it is forthrightly about relationships among conscious human experiences, and it expressly recommends to scientists that they resist the temptation to try to understand the reality responsible for the correlations between our experiences that the theory correctly describes" (13).
11. See Bell (1966, 1987) for further details.
12. White et al.'s (2016) recent work at the NASA Johnson Space Centre involving *measurement of impulsive thrust from a closed radio-frequency cavity in vacuum* has demonstrated evidence in favour of EM drive technology that favours Bohm's interpretation: "In short, the supporting physics model used to derive a force based on operating conditions in the test article can be categorized as a nonlocal hidden-variable theory, or pilot-wave theory for short" (10).
13. The central idea of Bohm's causal interpretation of quantum mechanics is that, in addition to the classical potential (V), there is a new quantum potential Q. The difference between classical and quantum behaviour is thus considered the operation of Q and the classical limit is the point at which the effects of Q become negligible. For Bohm, particles are affected by the quantum potential. Because ψ is contained both in the numerator and in the denominator for Q, it is independent of the strength or intensity of the quantum field but depends only on its form, which is a key difference from Newtonian physics. The particle is thus analogous to a ship being automatically guided by its sonar: the ship's power is independent of the power of the sonar but is deterministically guided by the sonar's form (1987, 185). This view is distinct from the Copenhagen account, in which there is no corresponding way to discuss such phenomena as it occurs *in the world.*
14. More specifically, as Collins (2006) notes, accounting for Bohm's EPR holism requires that "the quantum potential must be non-local, meaning that it cannot be thought of as a field spread throughout space—that is, with spatiotemporal parts. Instead, it is written as a field inhabiting the

3N-dimensional configuration space of particles in the system. (For example, assuming there are at least 10^{80} particles in the universe, it would inhabit at least a 3×10^{80}-dimensional space!)" (334–35).

15. To support this point, in a few instances Bohm followed Pribram's now largely neglected holographic theory of consciousness (1971), maintaining memories are recorded all over the brain rather than in a localized region. Hence, a response from memory ought to "fuse with the nervous excitation coming from the senses to give rise to an overall experience in which memory, logic, and sensory activity combine into a single unanalysable whole" (1980b, 105–6). While this kind of holism has been retained in enactivist arguments against modularity, the "holographic theory of consciousness" goes far beyond my present scope.
16. In Sect. 6.3.1, I suggest that combining Harris and AE philosophies of biology results in an argument for seeing minimal mind and evolution as processes of increasing coherence and freedom in a way that follows exactly from Bohm's contentions. In Sect. 7.4.1, I examine Harris's appeal to Spinoza's notion of embodiment and again find that the primary function of mind turns out to be coherence towards increasing freedom. Unfortunately, space prohibits a direct discussion of the philosophy of free will.
17. For the last decade of his life, Bohm extended his system by developing dialogic tools to facilitate ways of communication that emphasized wholeness and movement over atomistic mechanism, that is, "sociotherapy". This was considered a means of discovering ever-greater holistic and coherent truth among participants, for example, *Thought as a System* (1992) and Bohm (1996), Although these topics are underappreciated by most contemporary writers on his philosophy, this method falls out of Bohm's metaphysics and clearly parallels both Harris's emphasis on the dialectic and AE's interest in interdependent coupling and empathy. Unfortunately, space prohibits my further addressing these topics.
18. I have neglected to mention any of the mathematical formalism or their implications regarding Bohmian mechanics as contrasted with agential realism. There are undoubtedly many discrepancies between these systems as Barad draws from Bohr, whose ontological account is notoriously at odds with Bohm, whom attempted to maintain an objective reality with Einstein. A reconciliation of direct realism versus construction will reappear in many of the following sections.
19. For an introduction to ontic structural realism see Ladyman (2014).
20. Bohm holds that the unbroken whole of reality involves mental and physical poles with *active information* evident at every level. It must be remembered, however, that such an analogy of the magnet in which there are "two poles" is a result of abstract thought: "the deeper reality is something

beyond either mind or matter, both of which are only aspects that serve as terms for analysis [...] To pursue this approach further might perhaps enable us to extend our knowledge of both poles into new domains" (1990, 285).

21. To my knowledge, Harris used the label only once in all his works: "we should not beg the question if we tested the metaphysic of empiricism by its coherence, nor if we appealed to experience for evidence of the truth of dialectical holism" (2000, 238).

Works Cited

Achille, Varzi. "Mereology", *The Stanford Encyclopaedia of Philosophy.* Edward N. Zalta ed., 2019.

Ballard, Dana H., Mary M. Hayhoe, Polly K. Pook, and Rao, R.P.N. "Deictic Codes for The Embodiment of Cognition." 723–767, *Behavioural and Brain Science.* 20, 1997.

Banks, E. C. (2014). *The Realistic Empiricism of Mach, James, and Russell: Neutral Monism Reconceived.* UK: Cambridge University Press.

Bell, John S. "On the Problem of Hidden Variables in Quantum Theory." Rev. Mod. Phys., 38: 447–452; reprinted in Bell 1987 and in Wheeler and Zurek 1983, 1966.

Bell, John S. *Speakable and Unspeakable in Quantum Mechanics.* Cambridge: Cambridge University Press, 1987.

Barad, Karen. *Meeting the Universe Halfway: Quantum Physics and the Entanglement of Matter and Meaning.* Duke University Press. 2007.

Bertalanffy, Ludwig V. "An Outline of General System Theory." In *British Journal for the Philosophy of Science*, 134–165. 1(2), 1950.

Blanchard, Brand. "Harris on Internal Relations." In Grier, Philip ed, *Dialectic and Contemporary Science: Essays in Honour of Errol E. Harris.* University Press of America, 1989.

Bohm, David. *Quantum Theory.* Englewood Cliffs, NJ: Prentice-Hall Inc., 1951.

Bohm, David. *Wholeness and the Implicate Order.* New York NY: Routledge, 1980a.

Bohm, David. "The Enfolding-Unfolding Universe and Consciousness." In Lee, Nichol, ed. *The Essential David Bohm.* 2003. 78–120. New York, NY: Routledge, 1980b.

Bohm, David. "The Causal-Ontological Interpretation and Implicate Order." In Lee, Nichol, ed. *The Essential David Bohm.* 2003. 183–197. New York, NY: Routledge, 1987.

Bohm, David. "A New Theory of the Relationship of Mind and Matter." In: Philosophical Psychology, 271–286. 3: 2, 1990.

Bohm, David, and Basil J. Hiley. *The Undivided Universe: An Ontological Interpretation of Quantum Theory.* New York NY: Routledge, 1993.

Bohm, David. *Thought as a System*. New York: Routledge Pub, 1994.

Bohm, David. *On Dialogue*. New York, NY: Routledge, 1996.

Bohr, Niels. "Discussion with Einstein on Epistemological Problems in Atomic Physics." In Paul, A. Schilpp, ed. *Albert Einstein: Philosopher-Scientist*. 199–241. NY: Tudor Publishing Co and the Library of Living Philosophers, Inc., 1959.

Capra, Fritjof. *The Web of Life: A New Scientific Understanding of Living Systems*. New York, NY: Anchor Book Pub, 1996.

Chalmers, D. J. (1995). *The Conscious Mind: In Search of a Fundamental Theory*. NY New York, Oxford University Press.

Chemero, Anthony. *Radical Embodied Cognitive Science*. MA: MIT Press, 2009.

Chemero, Anthony., and Charles J. Heyser. "Methodology and Reduction in the Behavioural Neurosciences: Object Exploration as a Case Study." In Bickle, John., ed. *The Oxford Handbook of Philosophy and Neuroscience*. 68–90. New York NY: Oxford University Press, 2009.

Clancey, William J. "Scientific Antecedents of Situated Cognition." In Philip Robbins., and Murat Aydede, eds. *The Cambridge Handbook of Situated Cognition*. 11–34. New York, NY: Cambridge University Press, 2009.

Clark, Andy, and David Chalmers. "The Extended Mind." In *Analysis*. 7–19. Vol. 58, No. 1, 1998.

Collins, Robin. "Contributions from the Philosophy of Science." In: Clayton, ed. *The Oxford Handbook of Religion and Science*. 328–344 New York NY: Oxford University Press, 2006.

de Broglie, Louis. *Phil. Mag*. 47: 446, 1924.

de Broglie, L. (1927) Solvay Conference.

Favela, Luis H., and Anthony Chemero. "The Animal-Environment System." In Coello, Yann, and Martin H. Fischer. *Perceptual and emotional embodiment*. 59–74. London: Routledge, 2016.

Escher, M. C., and J. W. Vermeulen. *Escher on Escher: Exploring The Infinite*. New York: Harry N. Abrams, 1989.

Ferrer, Jorge. N. *Revisioning Transpersonal Theory: A Participatory Vision of Human Spirituality*. Albany: State University of New York Press, 2002.

Fultot, M., Nie, L., and Carello, C. "Perception-Action Mutuality Obviates Mental Construction." In Tom Froese., Sebastjan Vörös, and Alexander Riegler, eds. *Exploring the Diversity within Enactivism and Neurophenomenology*. 298–307. *Constructivist Foundations*: Vol 11, 2, 2016.

Gallagher, Shawn, and Dan Zahavi. *The Phenomenological Mind*. 2nd Ed. New York, NY: Routledge Press, 2012.

Gallagher, Shawn. "Making Enactivism Even More Embodied." In Komendziński, Tomasz., Przemyslaw Nowakowski, and Witold Wachowski, eds. *Enactivism: Arguments and Applications*. 232–247. Avant Vol. V, No. 2. ISSN: 2082-6710, 2014.

Gibson, J. J. *The Ecological Approach to Visual Perception*. Boston: Houghton Mifflin, 1979.

Harris, Errol. E. (1954). *Nature, Mind, and Modern Science*. London: Allen & Unwin Pub.

Harris, Errol E. *The foundations of Metaphysics in Science*. London: Allen and Unwin Pub, 1965.

Harris, Errol E. *Hypothesis and Perception: The Roots of Scientific Method*. London: Allen & Unwin Pub, 1970.

Harris, Errol E. *Salvation from Despair: A Reappraisal of Spinoza's Philosophy*. Martinus Nijhoff, The Hague, 1973.

Harris, Errol E. *Formal, Transcendental and Dialectical Thinking: Logic and Reality*. State University of New York Press, 1987.

Harris, Errol E. *The Reality of Time*. Albany NY: State University of New York Press, 1988.

Harris, Errol E. *Cosmos and Anthropos: A philosophical Interpretation of the Anthropic Cosmological Principle*. Atlantic Highlands, NJ: Humanities Press International Inc, 1991.

Harris, Errol E. *The Substance of Spinoza*. Amherst New York: Humanity Books, an imprint of Prometheus Books, 1995.

Harris, Errol E. *The Restitution of Metaphysics*. Amherst New York: Humanity Books, an imprint of Prometheus Books, 2000.

Harris, Errol E. *Bound in Shallows: Autobiographical Reminiscences*. Marquette University Press, 2015.

Harvey, M. I., Gahrn-Andersen, R., and Steffensen, S. V. "Interactivity and Enaction in Human Cognition." In Tom Froese., Sebastjan Vörös, and Alexander Riegler, eds. *Exploring the Diversity within Enactivism and Neurophenomenology*. 234–245 *Constructivist Foundations*: Vol 11, 2, 2016.

Husserl, Edmond. *Ideas Pertaining to a Pure Phenomenology and to a Phenomenological Philosophy. First Book*. Trans: F. Kersten. The Hague: Martinus Nijhoff, 1983.

Husserl, Edmond. *Ideas Pertaining to a Pure Phenomenology and to a Phenomenological Philosophy. Second Book*. 1989.

Hutto, Daniel D., and Erik Mylin. *Radicalizing Enactivism: Basic Minds without Content*. Cambridge MA, MIT Press, 2013.

James, William. "What is an Emotion?" First published in Mind, 9, Retrieved on Sept 17, 2014 from: Classics in the History of Psychology. 188–205. http://psychclassics.yorku.ca/James/emotion.htm. Christopher D. Green. York University, Toronto, Ontario, 1884.

James, William. "A World of Pure Experience." In James (1975–88). Vol. i, 21–44, 1976 [1904].

Jankovic, Marija., and Kirk Ludwig. *The Routledge Handbook of Collective Intentionality. New York, NY: Routledge Press*, 2018.

Junior, Olival. *The Quantum Dissidents: Rebuilding The Foundations of Quantum Mechanics (1950–1990.* Heidelberg: Springer, 2015.
Ladyman, James. *Structural Realism. Stanford Encyclopaedia of Philosophy.* Edward N. Zalta ed., 2014.
Li, Q., & Winchester, I. (2014). *Enactivism and Freedom Education.* (pp. 113–136). In Komendziński, T., Nowakowski, P., & Wachowski, W. (Ed). *Enactivism: Arguments and Applications.* Avant Vol. V, No. 2/2014 ISSN: 2082-6710.
Lovelock, James E. and Lynn Margulis. *Atmospheric Homeostasis by and for the Biosphere: The Gaia Hypothesis.* Blackwell Munksgaard, 1974.
Maturana, Humberto. R., and Varela, Francisco. J. *Autopoiesis and Cognition the Realization of the Living.* Dordrecht: Holland/Boston. D. Reidel Pub, 1972.
Maturana, Humberto. R., and Varela, Francisco. J. *The tree of knowledge: The Biological Roots of Human Understanding* (Rev. ed.). Boston; New York: Shambhala; Distributed in the U.S. by Random House, 1992.
Maudlin, Tim. *Quantum Non-Locality and Relativity: Metaphysical Intimations of Modern Physics.* Third Ed. Malden, MA: Wiley-Blackwell, 2011.
Merleau-Ponty, M. *The Structure of Behaviour.* Trans. A. Fisher. Pittsburgh, PA: Dusquene University Press, 1963.
Metzinger, T. (2003). *Being No One: The Self-Model Theory of Subjectivity.* Cambridge, Mass: MIT Press.
Moore, G. E. (1919). *External and Internal Relations. Proceedings of the Aristotelian Society.* 20 (1919–20) 40–62. Reprinted in Philosophical Studies and in *G. E. Moore: Selected Writings 79–105.
Nagel, Thomas. *The View from Nowhere.* New York: NY. Oxford University Press, 1986.
Newen, Albert B., and Shaun Gallagher. *The Oxford Handbook of 4E Cognition.* Oxford: Oxford University Press, 2018.
Noë, Alva. *Out of Our Heads: Why You are Not Your Brain and Other Lessons from the Biology of Consciousness.* NY: Hill and Wang, 2009.
Pribram, Karl. "Languages of the Brain." In G. Globus, et al. eds. (1976). *Consciousness and the Brain,* New York Plenum Press, 1971.
Pylkkänen, Paavo T. *Mind, Matter and the Implicate Order* (The Frontiers Collection). Heidelberg, Germany: Springer Pub, 2007.
Ramachandran, V. S., and Hirstein, W. (1998). *The Perception of Phantom Limbs.* (pp. 1603–1630) In: *Brain*: 121.
Reid, David A. "The Coherence of Enactivism and Mathematics Education Research: A Case Study." 137–172. In Komendziński, Tomasz., Przemyslaw Nowakowski., and Witold Wachowski, eds. *Enactivism: Arguments and Applications.* Avant Vol. V, No. 2/2014 ISSN: 2082-6710, 2014.
Rowlands, Mark. *The New Science of the Mind: From Extended Mind to Embodied Phenomenology.* MA: MIT Press, 2010.

Seibt, J. (2018). *Process Philosophy. The Stanford Encyclopedia of Philosophy* (Fall 2013 Edition), Edward N. Zalta (ed.), http://plato.stanford.edu/archives/fall2013/entries/processphilosophy/.

Shapiro, Lawrence A. *Embodied Cognition*. In Margolis, Eric, Richard Samuels, and Stephen P. Stich, eds. *The Oxford Handbook of Philosophy of Cognitive Science*. New York, New York, Oxford University Press, 2012.

Silberstein, Michael. "Why Neutral Monism is Superior to Panpsychism." In *Mind & Matter*. 239–248. Vol 7, (2), 2009.

Silberstein, Michael, and Anthony Chemero. "Extending Neutral Monism to the Hard Problem." In *Journal of Consciousness Studies*. 181–94. Vol 22, No. 3–4, 2015.

Stapp, Henry. *Mindful Universe: Quantum Mechanics and the Participating Observer* (The Frontiers Collection). Heidelberg, Germany: Springer Pub, 2011.

Stewart, John, Olivier Gapenne, and Ezequiel A. Paolo. *Enaction: Toward a New Paradigm for Cognitive Science*. Cambridge, Massachusetts: The MIT Press, 2010.

Stubenberg, L. (2018). *Neutral Monism*. In: *The Stanford Encyclopedia of Philosophy*. Edward N. Zalta (Ed.). http://plato.stanford.edu/archives/fall2014/entries/neutral-monism/.

Thompson, Evan, Antoine Lutz, and Diego Cosmelli. "Neurophenomenology: An Introduction for Neurophilosophers." In Brook, A., and Akins, K, eds. *Cognition and the Brain: The Philosophy and Neuroscience Movement*. 40–97. New York: Cambridge University Press, 2005.

Thompson, Evan. *Mind in life: Biology, Phenomenology, and the Sciences of Mind*. Cambridge, MA: Belknap Press of Harvard University Press, 2007.

Thompson, Evan. *Waking, Dreaming, Being: New Light on the Self and Consciousness from Neuroscience, Meditation, and Philosophy*. New York: Colombia University Press, 2015.

Thompson, William Irwin. *Gaia: A Way of Knowing – Political Implications of the New Biology*. Inner Traditions Bear and Company, 1988.

Varela, Francisco J., Evan Thompson, and Elenore Rosch. *The Embodied Mind: Cognitive Science and Human Experience*. Cambridge MA: MIT Press, 1991.

Varela, Francisco J. "Neurophenomenology: A Methodological Remedy for the Hard Problem." In *Journal of Consciousness Studies*. 330–50. Vol: 3, 1996.

Vörös, Sebastjan, Tom Froese, and Alexander Riegler. "Epistemological Odyssey: Introduction to Special Issue on the Diversity of Enactivism and Neurophenomenology." In Tom Froese., Vörös, Sebastjan., and Alexander Riegler, eds. *Exploring the Diversity within Enactivism and Neurophenomenology*. 189–203. *Constructivist Foundations*: Vol 11, 2, 2016.

White, Harold, Paul March, James Lawrence, Jerry Vera, Andre Sylvester, David Brady, and Paul Bailey. "Measurement of Impulsive Thrust from a Closed Radio-Frequency Cavity in Vacuum." In: *Journal of Propulsion and Power*, 2016.

Whitehead, A. N. (1978). *Process and Reality*. New York: Free Press. Cambridge: Cambridge University Press, 1929.
Wilson, Robert A. and Foglia, Lucia. "Embodied Cognition." *The Stanford Encyclopaedia of Philosophy* (Fall 2011 Edition), Edward N. Zalta (ed.), URL = http://plato.stanford.edu/archives/fall2011/entries/embodied-cognition/, 2017.

PART II

Dialectical Holism in Contemporary Cosmology

CHAPTER 3

Dialectical Holism in Space-Time and Particle Physics

3.1 Harris's Space-Time Holism

Harris claims that twentieth-century paradigm shifts in physics called for a holistic metaphysics. Following scientists like Einstein, Bohr, and Maxwell, Harris maintains physics "is no longer 'materialistic'". Matter, he claims, has been replaced by energy, which has been further transformed into space-time curvature. "The space-time (or material) field, moreover, is a structure of relations dependent on an act of thought …" or, an "arbitrarily chosen reference frame" (2000, 35). In this way, for Harris, the general and special theories of relativity reveal the necessity to posit a synthetic whole, or unifying principle reflective of the Concrete Universal.[1]

Harris recognizes that according to special relativity, as mass increases with velocity and velocity approaches the speed of light, mass approaches infinity. If the mass of a moving body depends upon velocity and velocity is relative to respective reference frames and the speed of light, there is no identifiable state of absolute rest. As an epistemic consequence "lengths, durations, and positions […] are mutually determinant. No statement about any one of them is free from entailment of statements referring to some or all of the others" (1965, 51). This means that "physical" parameters are abstractions obtaining only as interdependent relations within a synthetic whole. Measurements of either space, time, or velocity are here considered inseparable and interdependent, meaning "intrinsic physical

J. Schofield, *A Phenomenological Revision of E. E. Harris's Dialectical Holism*,
https://doi.org/10.1007/978-3-030-65029-2_3

properties of bodies are mutually dependent, and all physical relations are internal to their terms" (2000, 98). Hence, the space-time field arises from the interrelation of bodies, but as a unifying principle, this field also constrains the bodies within it.

The *principle of relativity* reveals, "laws of physics are invariant, with respect to the Lorentz transformations, for transition from one inertial system to any other" (1970, 120). Harris argues that the formalization of Lorenz transformations between reference frames in the special theory of relativity and tensor calculus in the general theory of relativity provides sufficient evidence to posit a unifying principle of space-time (1993, 144). "In short, space-time involves and imposes on its contents a principle of organization. One may say that space-time *is* such a principle, or a system of such principles. Spatio-temporal relations are *organizing* relations, and without them there is no physical space-time at all" (1965, 79). However, Harris maintains that "the space-time of physics is not the amorphous continuum of pure mathematics. It is the actual field of physical events and its geometry is revealed in the behaviour of the physicists measuring devices" (1965, 80). In this way, the dynamical object of space-time is not an independent substance, but neither is it a mathematical construction. Harris concludes that curved space-time demonstrates a "polyphasic unity" in virtue of the contrasting values of the field and is "the primitive type of a scale of forms—a scale (or procession) of changes of direction [...] governed by a principle of order" (1965, 470).

Concerning the structure of the universe, Harris contends that Faraday's discovery of the electromagnetic field and Maxwell's subsequent mathematical depiction of it provided a paradigmatic case of a unifying principle:

> The electromagnetic field in principle fills the whole of space and in accordance with the general theory of relativity, space-time itself is regarded as the metrical field from which the measured quantities are inseparable, and in which the elements that have them are inextricably embedded. Energy and mass thus become degrees of curvature in space-time. (2000, 98–99)

As a consequence, Harris believed that the curvature of space-time, due to the presence of matter, would bend the Universe upon itself to form a four-dimensional "hypersphere", one with finite volume but no spatio-temporal boundary. Considering the space-time field as an example of one particular unifying principle, Harris concludes that the physical world is to be understood as "a single whole [...] in which the primary principle of order determines all physical laws and governs all occurrences and relationships within it" (2000, 99).

Anticipating what has more recently been called the *temporal exclusion principle* (Pike 2009, 51–52), Harris takes a transition in time to be primitive: "time and change are identical, and as diversity manifests itself physically only either spatially or temporally, and spatial diversity is the infusion of time into space, time may be seen as essentially the principle of differentiation in the real" (1965, 471). In this way, Harris maintains that all unifying principles are *a form of time* (1965, 77). His conclusion is that if all unifying principles are internally related, partial reflection of the Concrete Universal, then "space-time as a single manifold, may prove to be the same as that which, on another plane and in a different way, characterizes human mentality" (1965, 84).

3.1.1 Prospects for Interpreting Space-Time as a Unifying Principle

One of the central goals for the philosophy of physics is investigation of what fundamental kinds of things exist in the world. Establishing or dissolving ontological theories of space-time understandably takes high priority. Today, the debate has come between "relationalist" and "substantivalist" theories of space-time, one which Harris neglected to address. As Pooley (2013) explains, the difference between the relationalist and substantivalist can be seen by their respective conceptions of the space-time coordinate system. The substantivalist contends the coordinates of space-time refer to an actual entity that exists among the fundamental objects in the universe. The relationalist argues the coordinate system assigns sets of real numbers to material events rather than to points of space-time, which means some coordinates are not assigned to anything at all (4). Upon orienting Harris's position with respect to this debate, it may be further possible to clarify *what kind of whole* he claims unifying principles to be.

Interestingly, both the substantivalist and the relationalist hold two theses in common. On the first theoretical agreement, numerous physicists and philosophers including Lange (2011), Morrison (1995, 2013), Yudell (2012), and Bangu (2013) have recognized *higher-order laws of nature* in the form of symmetry principles that provide key ingredients for the *Standard Model of Cosmology* (SM). For example, Yudell states, "Symmetry principles are laws about laws, and they seem to constrain or govern the laws just as laws constrain or govern particulars, and hence seem to deserve the title meta-laws" (2012, 1). Indeed, Lange (2011) argues that "Einstein's special theory of relativity incorporates a meta-law" that is considered a "symmetry principle" because first-order laws must be "unchanged (covariant) under a given [Lorentz] transformation" (1).

On the second commonality, both camps appear to share a commitment to internal relations. In response to Harris's argument, Blanchard (1989) has agreed that in GR, "every body is related to every other body in a four-dimensional space-time whole. No part of it [...] would be what it is without its spatial and temporal relations to the other parts" (6). As a relationalist, Rovelli (2007a) elaborates that general relativity "is the discovery that spacetime and the gravitational field are the same entity. What we call 'spacetime' is itself a physical object in many respects similar to the electromagnetic field" (1307). He claims that because the gravitational field is dynamical, "physical entities—particles and fields—are not all immersed in space and moving in time. They do not live on spacetime. They live, so to speak, on each other" (ibid.). Hence, general relativity does not deal with values at space-time points, but with values of dynamical quantities at space-time locations determined or localized only with respect by other dynamical quantities. According to Heil (2009) however, if objects are modes of space-time, their location and identity depends upon its substance, which means all space-time locations supervene upon modes of the field as a whole, i.e. events are internally related within the whole. For this reason, he thinks relativity has provided support for Spinoza's monism (318).

If sound, substantivist accounts would provide support for Harris's internal relations and a realist interpretation of (at least) one such unifying principle. However, Margaret Morrison's sustained thesis has been that history is ripe with examples of elegance and order being injected into nature via formalism without sufficient justification (e.g. the theories of Plato, Kepler, and Pythagoras certainly exemplify this blunder). For example, "symmetries from which we are able to derive a particular conservation law are themselves exact symmetries of the equations of motion, the Lagrangian, rather than exact symmetries of nature itself" (1995, 176). This kind of scepticism as to whether such overarching order exists in the world or is merely a product of our formalism is paradigmatic of relationalists. Pooley notes the ***kinematic argument*** in support of this view:

> Imagine a possible world W′ just like the actual world except that, at every moment, the absolute velocity of each material object in W′ differs from its actual value by a fixed amount (say, by two meters per second in a direction due North). [...] Two kinematically shifted worlds are observationally indistinguishable because, by construction, the histories of relative distances between material objects in each world are exactly the same. The worlds differ only over how the material universe as a whole is moving with respect to space. Since substantival space is not directly detectible, this is not an observable difference. (2013, 5)

Hence, if we are to take a conservative route, contemporary space-time theories do not directly provide support for a unifying principle as ontologically conceived. On the other hand, conceiving of space-time as a unifying principle may be considered a valuable *a priori* posit insofar as it is utilized in viable space-time theories. This would only grant the unifying principle (or meta-law) as an epistemic constraint, rather than an ontic level of nature.

After a thorough review of the major arguments both for and against substantivalism and relationalism, Pooley (2013) concludes that the former is "recommended by a rather straightforward realist interpretation of our best physics. This physics presupposes geometric structure that it is natural to interpret as primitive and as physically instantiated in an entity ontologically independent of matter" (10). As implied above, he goes on to write, "the only strong consideration in favour of relationalism is Ockham's razor" (ibid.). So, current theories can support a substantivist conception of space-time as instantiating a meta-law or unifying principle. For Harris however, conceiving space-time as "independent of matter" goes too far.

After an extensive historical analysis of space-time theories, Gunn (2011) maintains an interpretation of relativity that I propose is most adequately reflective of Harris's position. On the one hand, he contends space-time is passive to matter since matter is the source of gravity, which dictates the dynamical metric field of space-time. By this account, space-time can be considered an "ideal" construction derived from matter. On the other hand, Gunn argues,

> [T]he energy-momentum of non-gravitational matter cannot even be conceived independently of space and time [...] In addition gravitational waves require moving charges for the generation and the existence of their motion

[which] requires space-time to already exist. In these respects, then, matter would seem to be posterior rather than prior to space-time. (138)

Thus, he concludes that there exists a "co-dependency of space-time and matter" (ibid.).[2] Gunn reasons that the "simultaneous reality and ideality of each with respect to the other, indicates that [...] both exist together in something else which unites them both, and relative to which they are absolutely ideal, something like thinking and extension are in Spinoza's system" (138–39). For Gunn, matter, space, and time are abstractions from a more fundamental "self-determining" space-time field that is likewise an abstraction from something still more fundamental, which he calls *pneumetical force*. He concludes that after relativity theory, it is no longer reasonable to treat field theory as *mechanical*, nor to treat its dynamical principle as *analytical*, "in so far as 'analytical' refers to an explanation of the whole in terms of its parts" (145).

Indeed, Gunn's identification of "ideal" and "analytical" (abstract) concepts mistaken for reality is precisely in tune with Harris's appeal to what Spinoza called the *confused idea* (to be revisited in Chap. 7). Gunn's contention that space-time and matter are abstractions that each arise from a common self-determining process is isomorphic with the reasoning of both Bohm's *holomovement* and the enactivist theory of *co-emergence*. Such self-determination might be further evidenced by Belinski and Verdaguer's (2001) work, which uses the *soliton* of GST to provide a solution to Einstein's field equations, and Tomizawa and Mishima (2014), who applied soliton solutions to Faraday rotation and black holes. I maintain this line of reasoning can be subsumed within Harris's system, which posits that relativistic systems instantiate a unifying principle reflective of the Concrete Universal. An important consequence of which, is that the *substantivist* versus *relationalist* debate may be avoided. Despite their similarities however, Gunn later contends that Hegel's dialectic cannot be found in contemporary physics. However, Gunn is here referring to (what is perhaps a miss-reading of) the *thesis, antithesis, synthesis* sense of the word. As Harris's revision of the dialectic in terms of contemporary cosmological and biological evolution will be examined in Chaps. 5 and 6, a conclusion on this matter will have to wait.

Taking space-time as our paradigmatic example, it should now be possible to elucidate what kind of contemporary holism supports Harris's unifying principles. Initially, three options are apparent:

1. *Ontological holism*—maintains that some objects are not completely composed of basic physical parts.
2. *Property holism*—maintains that some objects bear properties that are not determined by the physical properties of their underlying constituents.
3. *Nomological holism*—maintains that certain phenomena instantiate laws that are not determined by fundamental physical laws governing the structure and behaviour of their underlying constituents (Healey 2016).

Consistent with Harris's rejection of substance (or pluralistic) ontology, because the "higher" scale cannot be substantively different from its composition and neither can there be basic physical parts, (1) must be rejected. From Harris's anticipation of meta-laws and paradigmatic appeal to internally related fields, the dialectical holist can agree with (2) just in case "properties" be replaced by *relational processes*. Option (3) fits with Harris's unifying principle because while the overarching meta-law is a result of its composition, as a whole, this composition instantiates a kind of order that is not found in nor deduced from an understanding of its parts.

Though viable when first proposed, recent research has shed a different light on Harris's theory that the topology of space-time is that of a hypersphere. Current models are limited because they exclude details prior to *decoupling* (cosmic background radiation era) and from the future states of our expanding universe. As a result, theories concerning cosmic topology could be verified if we happened to be living in a "small universe" whose size were the same as, or barely larger than the observable cosmos. Unfortunately, all "we can truly conclude is that the Universe is much larger than the volume we can directly observe" (NASA.gov 2014).[3] Our light cone thus creates an *epistemic limit*, by restricting what can be observed, beyond which any discussion becomes speculative. Cosmologists side-step this obstacle by extrapolating general theory of relativity to all of space-time (Smeenk 2013, 2). After a thorough examination of space-time topology, Manchak (2013) concludes "general relativity allows for a wide variety of global spacetime properties", including a range of possible topologies, and deciding among them remains an open issue (12). According to the most recent WMAP data however, the universe is flat with only a 0.4% margin of error. That is, observations from our *past light cone* have revealed a *Friedman-Lemaitre, Robertson-Walker* topology that is isotropic and homogenous. Hence, Harris's posit of the hypersphere is no longer in fashion and given recent data, is likely incorrect.

This section has demonstrated that general relativity can be interpreted to offer support for the thesis of dialectical relations and provides an example of the unifying principle. Importantly, the meta-laws that govern the space-time system are beyond empirical verification and serve as *synthetic* constraints upon our understanding of observable phenomena. As a result, Harris's claim that relativity theories provide evidence for an organizing principle can be granted only on epistemic (i.e. *phenomenological*) grounds. This is to say that in *discussing* physical phenomena of relativistic distances and speeds, such a principle is *pragmatically* invoked in order to account for the available observations of self-organization.

3.2 Harris's Appeal to Unity and Irreducibility in Quantum Theory

Towards a defence of the Concrete Universal, Harris appeals to the aim of contemporary physics to reduce all known fundamental forces to a unified *Theory of Everything* (ToE). Indeed, experimental particle collisions performed by the *large hadron collider* (LHC) have provided increasing support for the SM, encouraging many to believe that we may yet achieve the goal of unification. However, recent works have also raised substantial challenges to such unity and fierce debates continue in the philosophy of physics today. Harris also claims that quantum theory requires a "holistic logic" (as opposed to reductive analysis) that provides further evidence for the unifying principle on a microscopic scale. I maintain that although these two contentions of reducibility and non-reducibility are at odds with one another, these appeals can be shown to be consistent within Harris's system.

As we have seen, Harris claims that contemporary physics provides evidence in favour of dialectical relations in a multitude of ways. Paramount among them is the guiding theme of the SM that posits a family of particles, each relating to one another according to symmetry principles. For Harris, this shows that the "properties of each particle are a consequence of its relation to all the rest of the universe" (1965, 97). Hence, contemporary physicists are now "seeking a single primordial force, which subsequently differentiated itself into those now known as primary (strong, and weak, gravitation, and electromagnetism) according to a single fundamental law" (1988, 56). He maintains that since the electro-weak and strong forces have been unified via "gauge theory", all that remains is for the SM or a rival (e.g. string theory), to include gravity for a "final unification" (1991, ff. 39).

However, Harris has consistently claimed that explanation of Nature requires not a reduction, but an *explication* of its respective unifying principles:

> [S]elf enfoldment and new dimensions of complexification give rise to new types of whole and correspondingly new laws of action, so that different levels of integrated complexity produce different orders of reality, which may be mutually continuous without being mutually reducible. This is because the principle of organization in every totality determines its character throughout and in detail, and that principle is fully realized only in the completed whole. The ultimate key to explanation, therefore, is that which explains the most complex and highly developed, not that which applies only to the lowest level of complexity, even though the former is and must be, in a sense implicit or potential from the start. (1970, 369)

This passage has two important implications: (1) the epistemic implication is that we will remain unable to explain the whole of nature by one level of fundamental analysis and that the later developments of complexity provide the "ultimate key to explanation"; (2) the ontological implication is that nature is characterized by its "organized totality" of unifying principles.

In his defence, Harris appeals to Max Planck's works to contend that there exists an inter-connection between parts of a whole that emphasizes the importance of local context. Particles, he claims, "exist as distinguishables *only qua* features of the system and not as separate individual entities. They are not individuals and can be termed particular distincta only in terms of their participation in the articulated structure of the system to which they belong" (1965, 136). Harris maintained that the individual quanta are now more accurately understood as waves that cannot be considered in isolation but only as interdependent systems:

> Even free-moving particles attached to no particular group, such as neutrons expelled from atoms under bombardment, are seen as wave packets, presupposing some wave field in which the waves have become superposed. As soon as one speaks of waves, one is committed to the concept of periodicity and ordered structure. (2000, 157–58)

Such *periodicity and ordered structure* is for Harris demonstrative of a unifying principle. Likewise, in the case of a quantum system, Harris maintains, the whole assumes "priority over determination of the exact position, or the precise momentum, of particles within it" (1991, 31).

As discussed in Sect. 2.4, Harris endorses Bohm's theory of *hidden variables* (1965, 123), and later cites Bohm's (1980) *implicate order* in support of his holist interpretation of quantum mechanics. Accordingly, we must "regard what is measured and the measuring instrument as a single indivisible complex, within which what is measured comes to be" (1991, 32). For example, Harris invokes Bohm and Aharanov's modification of the EPR experiment as evidence that "events separated by a space-like interval are necessarily (even though not causally) bound together by faster than light influences" (1991, 33).[4] Hence, Harris believed that Bohm's theory supported the contention that the *intrinsic properties* of events and fundamental particles depend upon organizing principles of a system-as-a-whole. Nevertheless, he remained open to any interpretation of quantum mechanics that takes reality to be "a whole or structured totality in which individuality is constituted by form, and particulate existence is only provisional, approximate, temporary and adjectival to the dynamic pattern in which it occurs" (1965, 139).

Harris concludes that "space-time and energy are seen to be inseparable aspects of a single reality, as are energy and matter, wave and particle" (1991, 38). On this account, he finds that classical logic is insufficient to express the discoveries of twentieth-century science:

> There can be no atomic facts and no atomic propositions. The logic that assumes their existence may still be of use—if the subject matter to which it relates is to be regarded (for theoretical convenience) as a mere aggregate, but to the system of the real world, as science now conceives it, such logic must be wholly inappropriate. (2000, 101–2)

3.2.1 Reductive Versus Synthetic Unity

In this section I show that despite various supporting points, some of Harris's claims concerning the dialectical whole as overarching nature bear many more differences to the standard model than he was originally able or willing to recognize. On the one hand, a number of theorists, including Bangu (2013) and Carr (2006), are optimistic that experimental results, in combination with theoretical symmetry principles are getting us closer to a *grand unified theory*.[5] For example, Carr agrees with Davies's (1995) assertion of an organizing principle implicated by the coherence of the laws of our universe. He holds that there is an inherent beauty to the universe involving "mathematical elegance, simplicity, and inevitability". He goes on to say,

> [A]ll the laws of nature seem to be a consequence of a simple set of symmetry principles. For example, symmetrizing electricity and magnetism gives Maxwell's equations; symmetrizing space and time gives special relativity; and invoking gauge symmetries leads to the unification of the forces of nature. (Carr 2006, 147)

On the other hand, invoking symmetry appears to now require a bit more philosophical legwork (empirical observation being unavailable), to arrive at the kind of ontological conclusions Harris so confidently maintains. Morrison (2013) supports the interpretation that a grand unification in modern terms would mean the discovery that all known particles can be understood as "one and the same force" (382). She cautions however that unification in the SM is tenuous because of the use of *free parameters* "and the fact that the theory is an amalgam of three different symmetry groups SU(3) × SU(2) × U(1) rather than a single group" (383). Furthermore, gravity itself may not be a fundamental force as "there are no particles that couple to the gravitational field and act as force carriers; the effects of gravitation are ascribed to space-time curvature instead of a force per se" (ibid.). Consequently, it would appear that Harris was mistaken to believe that adding gravity to this system is a final step, because the weak, strong, and electromagnetic forces have yet to be reductively unified.

In assessing Harris's appeal to the SM, it is instructive to realize that there are different kinds of unification. For example, Barrow (2004) considers a range of principles by which scientific theories unify and advance. *Reduction* is "the discovery that the value of one constant of nature is determined by the numerical values of others" (407). In order to carry out a reduction of a given physical force, parameter, or phenomenon, we need already to know some likely set of constants, Barrow says, then we "develop a broader explanation that links their domains of application" (408). Barrow points out that a common example of a theory, which results from a reduction in the number of constants, is the unification of electricity and magnetism to form *electromagnetism*.

Morrison agrees that Maxwell's *electrodynamics* was the first unified field theory of physics and a rare example of reduction. The laws of physics were understood to take "the same form in all inertial frames" and "the relationship between electricity and magnetism could be represented by the transformation properties of the electromagnetic field tensor [...] Hence, the magnetic field appears as an effect of the transformation from one frame of reference to another" (393–94). The upshot is that in certain cases a single phenomenon can be mistakenly analysed as two or more different kinds until a reductive unification provides a framework for understanding their *underlying unity*. Though Harris did not distinguish between kinds of "unification" in the philosophy of physics, he nevertheless appealed to electrodynamics as a case of unification in support of his metaphysics—thereby implicitly endorsing reduction.

Barrow next highlights *elucidation*, which is said to occur "when a theory predicts that some observed quantity—a temperature or a mass for example—is given by a new combination of constants. The combination tells us something about the interrelatedness of different parts of science" (2004, 408). Barrow's example here is Hawking's prediction of the evaporation of black holes, as this requires combining our understanding of thermodynamics, light, gravity, and Planck's constant. This is also an instance of what Morrison calls *synthetic unity*. The most common form of reduction, Morrison claims that in such cases the respective phenomena remain independent but are described within one overarching theory. Here, she recognizes, there is no unification of the sort that would permit deriving the formalism of one theory from a more fundamental theory.

Morrison's example of synthetic unity is the electroweak theory, which results from combining the SU(2) group governing isospin/weak interactions (10^{-15} cm range) and the U(1) group of electromagnetism (∞ range). The SM results from the combination of quantum chromodynamics (QCD, SU(3) symmetry group) of the strong interactions with the electroweak theory. Morrison claims that such unity is not possible as a final form of the SM because in contrast to the SM, the electroweak unification

> was largely structural rather than substantial and as a result does not fit with the idea of reducing elements of the weak and electromagnetic force to the same basic entity [...] The SU(2) × U(1) gauge theory [...] specifies the form of the interactions between the weak and electromagnetic forces but provides no causal account as to how the fields must be unified [...] the unity that is supposedly achieved results from the unique way in which these forces interact. (399–400)

The two fields are thus integrated under an overarching framework that results from the interaction of their independent symmetry groups. This unity resulted in a reconceptualization of the electromagnetic potential and positing of additional dynamics. Such a synthesis, she maintains, preserves independence for each domain while providing a broader theoretical framework within which their integrations obtain (402). Consequently, it is widely accepted that electroweak unification results from formal operations rather than the discovery of empirical unity.

According to Barrow, *enumeration* is considered the holy grail of physics: "the calculation of the value of a constant of Nature from first principles, showing that its value is explained" (2004, 407). Though it has yet to unify the known constants, Bangu (2013) remains encouraged by methods appealing to mathematical symmetries.[6] He maintains that symmetry is an indispensable tool for particle classification and prediction, and notes two main benefits of this taxonomic function:

> The first is that order can be imposed over a huge variety of elementary particles. The second comes from the use of these schemes of classification as guides toward the fundamental ontology: while the known particles neatly fitted the schemes, physicists also perceived "gaps" in the schemes, that is, positions for which no corresponding particle was detected. The existence of these gaps suggests that the physical particles that would fill them might themselves exist. (290)

Hence, the ongoing revelation of symmetry in nature *implies* an elegant wholeness, and interrelatedness that fits the SM predictions. Nevertheless, many such predictions have not been and may never be observed. Moreover, the *underdetermination of theory by evidence* means that there will always be rival (possibly non-reductive) theories that can account for the available observations equally well.

Unfortunately, the SM is also wrought with "unnatural fine-tuning" of its parameters. For example, the mass of the Higgs was established because of fine-tuning, which was required to make it small enough for a number of other theories to remain consistent. This brings Morrison to conclude that despite the significance of the Higgs discovery, we still lack a truly cohesive and naturally unified model of nature. In this vein, new discoveries involving the Higgs at the LHC may overturn this model in favour of another:

> An examination of evidence from both experiment and theorizing suggests the following characteristics of unity: it is something that can be achieved in certain local contexts, it is characterizable in different ways, but cannot be extended to a "unity of nature" that is systematically defined. (407)

She maintains that grand unification is possible, but this will likely depend upon fundamental theories, phenomenological models, and *effective field theory*, which deal with the particles and scales that are important for a given investigation.

This section has shown that Harris was mistaken to appeal to the SM as a means of supporting his holism. Aside from the interrelation of particle types—which provides support for dialectical relations—the very notion of scales in nature that Harris advocated is not supported, but rather eschewed by this model. Whereas Morrison voices doubt that Nature can itself be unified in accordance with reductive theories, Harris went to great pains to show the necessity of utilizing synthetic (i.e. transdisciplinary) methods in providing a holistic theory of Nature. His mistake then was adopting too many theories that appear "holistic", without distinguishing among them. Consequently, there is a danger that Harris's conception of "unity" is threatened by contradiction to the extent that he both appeals to the standard model and maintains the *emergence* of unifying principles (e.g. space-time), which resist such reducibility. In Chap. 5, I conduct a more thorough analysis of emergence and reducibility to reconcile this issue, but the following section provides a first pass in the context of quantum mechanics.

3.2.2 *Entanglement as a Purported Example of a Unifying Principle*

In the first instance, to justify the central thread of Harris's appeal to quantum theory, only the most mainstream accounts need be invoked. According to Serway et al. (2013), in contemporary quantum theory the wave function is not merely associated with the particle, but "is determined by the particle ***and*** its interaction with its environment, so, like energy, it more rightfully belongs to a system" (1087). Similarly, when dealing with a pair of particles (or composite system) there are rules for understanding its state-space that adds further support to Harris's holism. These rules tell physicists how the state-spaces (H_A and H_B) for two respective particles obtains the state-space (or tensor product $H_A \otimes H_B$) of the pair: (1) if H_A and H_B are Hilbert spaces, $H_A \otimes H_B$ will be as well; but (2) the state of a composite system $H_A \otimes H_B$ is not uniquely defined by those of its components. Thus, it can be argued that there is "non-trivial" knowledge about complex composite quantum systems that does not supervene upon the knowledge about their components (Jenann 2015). Hence, the thesis of *separability*—which says the universe is determined by the intrinsic nature of each part—ultimately fails in the face of quantum entanglement.

According to Maudlin (2003), even an orthodox interpretation of the double slit experiment appears to depict a contradiction that implies a form of holism, for which classical logic is insufficient. Maudlin explains that in quantum theory "a disjunction [C (A ∨ B) ∧ C] can be true even through neither disjunct [(A ∧ C) ∨ (B ∧ C)] is true, so classical logic fails" (479). This is to say that in a double slit experiment, a particle must go through one of two slits, but that our observations reveal such determinations to be incomplete. Maudlin concludes that the deepest metaphysical innovation of quantum theory, whether Bohmian or orthodox, "lies in the holistic nature of the wavefunction, and the fact that the quantum state of an entangled system cannot be recovered from the quantum states of its parts" (485–86).[7]

Likewise, Barad (2007) notes that the apparent paradox of superluminal action-at-a-distance only arises due to a mistaken belief in "metaphysical individualism":

> [T]he mistaken assumption that there are individually determinate entities from the outset; this assumption, which is the basis for classical physics, is precisely what is being called into question here [...] in the EPR case the entities are not separately determinate individuals but rather inseparable parts of a single phenomenon. (315–16)

Barad contends that we are not dealing with pre-existing individually determinate entities with determinate spatial positions *communicating instantaneously at a distance* from one another, but "a phenomenon that determinately resolves the boundaries and properties of the entangled components in a way that gives meaning to the notion of individual" (ibid.). Hence, most accounts agree with Harris that unions are not reducible to disjunctions of corresponding facts about their parts.

More specifically, Bohm held that the trajectory of a particle through configuration space is locally determined in that only the value of the wave-function around that point plays a role. In cases like the two-slit experiment however, the particle goes through one slit, but the parts of the wave-function associated with both slits overlap in configuration space, which produces the resulting macro-scale interference fringes. According to Maudlin (2011), Bell's arguments concerning the measurement of spins of correlated electrons further reveals how superluminal causation appears explicitly in Bohm's theory (ff. 122). In an EPR experiment, the setting of each spin analyser affects the Hamiltonian and hence the evolution of the wave-function. In Bohm's theory, the wave-function determines the trajectories of the particles as-a-whole and so determines (global) superluminal influences from one wing of the experiment to another. Maudlin concludes that Bell's inequality proved that the thesis of *locality*—which says that events are non-influential at space-like distances—must be false for any quantum theory (193).

From the above, Bohm and Harris's contentions each entail some kind of *relational physics*. Endorsed by Rovelli and Smerlak (2007b) and Smolin (1997), relational quantum theory says the respective values of Ψ (p, p′, p″, …) are made determinate through the interaction of the quantum system with some external environmental system (*O*). This means "there is no meaning in saying that a certain quantum event has happened or that a variable of the system *S* has taken the value *q*: rather, there is meaning in saying that the event *q* has happened or the variable has taken the value *q for O, with respect to O*" (Laudisa and Rovelli 2013). Thus, concepts of absolute states, events, or values are denied, the isolated system becomes, meaningless and the physical world is described as a net of relations across disparate systems.

In this way, Harris's claims of irreducibility in quantum theory can be further supported by Teller's (1986) *relational holism*, in which quantum systems exhibit relations irreducible to and non-supervenient upon their underlying components. Relational holism has also been adopted by AE as a means of establishing a general theory of *emergence*. As Thompson (2007) writes:

> [C]ertain wholes possess emergent features that are not determined by the intrinsic properties of their most basic parts. Such emergent features are irreducibly relational. They are constituted by relations that are not exhaustively determined by or reducible to the intrinsic properties of the elements so related. (427–28)

For example, according to the property-assignment rules of orthodox quantum mechanics, in the *spin singlet state* of the EPR experiment particles are entangled in an anti-correlational state that cannot be decomposed into a product of separate states (i.e. of L- and the R-particles with individual spins). Thus, Berkovitz (2016) concludes, (1) *state separability* fails; (2) Teller's *relational holism* is vindicated because properties of the pair fail to supervene upon the intrinsic qualitative properties of the particles and their spatiotemporal relations; and (3) *process non-separability* is vindicated because the process that leads to each of the measurement outcomes is likewise non-separable.

Taking quantum theory and general relativity together, Maudlin concludes in agreement with Teller but counter Einstein's empiricist worldview that it's "not just that there are some relations which don't supervene on the intrinsic state of the relata but that the relata *don't have any intrinsic state at all.* It is not just that the whole is more than the sum of the parts but that the parts can't even be defined apart from the whole" (2011, 194). Contrary Harris however, Maudlin argues that while the EPR experiment's violations of Bell's inequality do not entail superluminal matter, energy, or signalling, *they do require superluminal causal connections and* information *transport* (221).

Importantly, Unger and Smolin (2015) point out that theories capable of accounting for the nature and effects of "'hidden variables' could only be relational views: they would show how the particles, fields, and forces studied by quantum mechanics relate, through reciprocated action, to features of the universe that have remained outside its scope of inquiry" (195).[8] So, Bohm's account is relational but not all relational accounts are Bohmian. Hence, while Harris shares with AE and Bohmians an appreciation of relational quantum theory, AEs and Harris are not (yet) necessarily tied to one another or to Bohm's realist ontological conclusions (I return to this issue in Chap. 4).

In an effort to further establish common ground among these camps, it is instructive to consider Fritjof Capra's holist philosophy of physics, one which Harris partially anticipated (1965) and later recognized to be supportive of his own system (1987). In the most recent edition of Capra's *Tao of Physics* (2010), the newly added summary of his paradigm captures what I take to be the central points of agreement between AEs, Bohm, and Harris concerning quantum holism:

1. *Part-whole relationships*—the classical scientific paradigm maintained that an understanding the relationships among the components exhausts the whole, but in the new paradigm the dynamics of the whole provide information not found in its components. "The whole is primary, and once you understand the dynamics of the whole, you can then derive, at least in principle, the properties and patterns of interactions of the parts" (329). Moreover, he claims that in quantum theory, *fundamental* parts reveal different properties, depending on their experimental context.
2. *Primacy of process*—the long-held view that some fundamental structures (e.g. forces and particles) are primary is here rejected: "the image of the universe as a machine is replaced by that of an interconnected, dynamic whole whose parts are essentially interdependent and have to be understood as patterns of a cosmic process" (331). He holds that the way we "cut" through this cosmic process determines what sort of "entities" we reveal.
3. *Epistemic science*—the above ontological shift entails the rejection of an objective analysis of nature and the necessity that "we can never speak about nature without, at the same time, speaking about ourselves" (331). This is to say that our epistemic process must be included within any depiction of the processes of nature. "In quantum physics the observer and the observed can no longer be separated, but they can still be distinguished" (ibid.).
4. *Anti*-foundationalism—the historical paradigm shifts that continue to undermine what was hitherto believed to be a *foundation* of science finally come to an end with the realization that our search for foundations is ill-conceived. Future, sciences may not need to "build our knowledge on firm foundations, and we may replace the metaphor of the building by the metaphor of the network" (332).
5. *Approximate descriptions*—although this implies that to understand any part, we need to understand the whole, we are nevertheless able to establish "approximate" knowledge. "Scientists do not deal with truth (in the sense of a precise correspondence between the description and the described phenomena); they deal with limited and approximate descriptions of reality" (334).

Essentially responding to many of the central questions of Sect. 1.1 (a–e), Capra's approach mirrors Harris's emphasis on holism and a singular process extending through both mind and nature, but here the emphasis is on the implications for knowledge and scientific practice. More specifically, Harris would agree with all but (3), claiming that phenomenology, of a certain kind, permits us to know the *whole* (I return to this in Chap. 8). Though proponents of AE have maintained a significant distance from engaging with Capra's work—likely due to his broad claims concerning "Eastern metaphysics"—it should be clear from the above that the paradigm is perfectly supportive of their own claims regarding nature, knowledge, parts and wholes.

This line of reasoning has demonstrated support for quantum phenomena being irreducible in at least epistemic terms and revealed a short list of mutually supportive theses.[9] Hence, Harris was right to claim that a prevailing holism of quantum mechanics requires a *logic* contrary to the analysis of separable entities and it is evident that he shares with AE, Capra, and Bohm a common thread of relational holism. Nevertheless, much more analysis is needed to clarify the ontological commitments among the above proponents. In order to draw out further details regarding Harris's philosophy of physics however, it is necessary to first consider how he used the above contentions to argue for the interdependence of cosmology and consciousness studies.

3.3 What Does Cosmology Have to Do with Consciousness?

Admittedly, the results of this chapter have not been particularly surprising thus far and neither would Harris have expected them to be. What these results have done however is prepare the way for a consideration of what is a guiding theme in his system. As discussed in Sect. 2.1, for Harris scientific practices are understood as the application of a whole system, one that depends upon the coherent arrangement of all its parts. This means that if any one theory is inconsistent with the worldview, the system itself must be reconsidered, rather than the part alone. He reasons that the scientific method is a "systematic explanatory scheme", a "whole" whose elements are "interdependent":

> Such a whole must be complete, and no scientific theory is complete until it embraces the entire universe. At this stage it becomes a cosmology [...] Every such worldview is a metaphysic; so the edict of empiricism against metaphysics is without a shred of justification, for intrinsic to all science and all experimental method a metaphysic is presupposed. (2000, 21)

Unfortunately, Harris did not spell out in sufficient detail why cosmology is a metaphysic. Nor did he explain the importance of considering theories of mind and cosmos together, but merely demonstrated it in his methods of investigation and attributed the approach to Hegel's dialectic phenomenology. Hence, the present and following section serve to fill in this gap and provide a clearer means of evaluating the merit of his later arguments.

Derived from its Latin origin Universe is composed of *uni* and *versus*, resulting in *universum*, meaning everything rolled into one. Hence, from its earliest Western conception the cosmos is a whole, as anything else becomes utterly *incoherent*. Following convention, I henceforth take the capital "Universe" to denote all that either has or will ever obtain in physical reality and "universe" or "world" to refer to everything within our cosmic horizon, or some region of the Universe distinguished by law-like parameters. Philosopher of cosmology George Ellis (2007) explains that cosmology involves the study of all that physically exists, not merely the study of all that we can observe; the aim of which being to delineate the overall structure and evolution of the Universe. "Thus cosmology is of substantial interest to the whole of the scientific endeavour, for it sets the framework for the rest of science, and indeed for the very existence of observers and scientists" (1184). On this view, cosmology provides an ontological framework for all scientific practices and a theoretical foundation for naturalistic phenomena, including life and conscious observers.

Smeenk (2013) further explains that "Cosmology confronts a number of questions dear to the hearts of philosophers of science: the limits of scientific explanation, the nature of physical laws, and different types of underdetermination, for example" (607). However, prior to the 1960s this discipline (like metaphysics for a time before 1980), was widely eschewed as unscientific and speculative at best. As Smeenk puts it, the challenge concerns how we can formulate a scientific theory of the "universe as a whole". He goes on to maintain that this field is a legitimate part of science despite it having its own unique challenges. Based on the great achievements cosmologists have made in discovering and confirming theories like big bang nucleosynthesis, Smeenk explains there "are no convincing, general no-go arguments showing the impossibility of secure knowledge in cosmology; there are instead specific problems that arise in attempting to gain observational and theoretical access to the universe" (608).

This restriction on our observational powers highlights the significance of what is known as our *selection effect*. A unique feature of cosmology as a science is its inherently speculative and philosophical nature, which results from our inability to probe the very early and distant reaches of our universe (Carr 2007, 3–4). Starting in the 1920s–1930s, data were enriched by a widening of our observational spectrum from optical to ultraviolet, infrared, X-ray, and so on. The *cosmic background radiation* observations of COBE and the WMAP satellites have provided crucial evidence for the global topology of our universe. Today, astronomers use as many means of observation as possible (the widest spectrum possible) because objects are often detectable at certain wavelengths but not at others.

As Ellis explains, "philosophical choice will to some degree shape the nature of cosmological theory, particularly when it moves beyond the purely descriptive to the explanatory role" (2007, 1184). He emphasizes that cosmology is far more than merely choosing or determining a number of physical parameters, as some hardnosed physicists maintain: cosmology *assumes* that the Universe is "expanding and evolving", and that "the laws of physics are the same everywhere, and underlie the evolution of the universe" (1185, emphasis omitted). Hence, the selection effect of cosmology highlights a lesson applicable to all the sciences: *our conception of reality depends upon our methods, tools, and disposition of observation.*

Livio (2008) cites the four cosmological observations that have bolstered this assumption and framed our investigation into the emergence of complexity in the universe:

1. The spectra of distant galaxies are red shifted.
2. To a precision of better than 10^{-4}, the cosmic microwave background (CMB) is thermal, with a temperature of 2.73 K.
3. Light elements like deuterium and helium, have been synthesized in a high temperature phase of the early universe.
4. Deep observations have shown that galaxies farther away appear younger (e.g. smaller with higher fraction of irregular morphologies) (114).

Taken together, these observations reveal good evidence that the universe is expanding and cooling; evidence he considers not proof of, but "consistent with the Hot Big Bang model" (115).

Indeed, the SM of cosmology extrapolates local laws to the universe as a whole; a move that implies our global-to-local constraints have not yet been adequately accounted for, and given the problem of induction, they may never be. Because our observations are made from a region of the universe that is hospitable to the existence of observers this results in the biasing of any global theories we construct (Carr 2007, 15). The important conclusion from this overview of cosmology has been anticipated by Harris's reasoning and eloquently captured by Gunn (2011): "as a whole", the Universe "is not an object of experience, nor can it ever become such" (258). Gunn goes on to argue that even if we could somehow "transcend" the Universe and "reengage it" our object would now only constitute part of the Universe proper. Hence, it remains "beyond all empirical determination" (ibid.).

Following Husserl's phenomenology, Thompson (2007) reveals that we find a similar notion concerning the structure of experience:

> Anything that comes forth, manifests, or emerges does so in an open clearing or expanse, delimited by a horizon. The horizon of every possible horizon is the world. Yet the world-horizon cannot be the synthesis, totality, or mereological sum of all these possible horizons because it is pregiven or *a priori* with respect to any of them and thus is *sui generis*. (35–36)

To say that the world-horizon cannot be reduced to any material or other formally constructed conception is to point out, with Harris, that Nature as a whole is an *a priori* metaphysical *synthesis*; one that is inherently incomplete. In the following section, I examine one implication of this conclusion that reads between the lines of Harris's original framework for the study of cosmos and mind.

3.3.1 Anthropic Reasoning

What can be understood about the nature of our universe from examining the presence of intelligent observers and the conditions of their cosmic neighbourhood? To address this question is to consider the *anthropic cosmological principle*, which was so named and then elaborated into strong and weak versions by Brandon Carter (1974). Carter later lamented that the term was improper because it really has nothing to do with human beings *per se*, but refers to the relationship between the physical conditions necessary and sufficient for the emergence of complex systems in a given space-time region: *observations of the cosmos must reveal physical parameters of a* universe *that are capable of supporting the emergence of observers.* Hence, the anthropic principle is used to constrain our physical theories by accounting for the presence of and necessary conditions for observers. Overlooking this distinction between observers and humans can lead to several unwarranted conclusions, like extending the principle to theological interpretations that grant humanity a privileged position in the Universe (Carr 2007, 85–86).

Fundamental to this principle, Robert Dicke (1961) was the first to notice the prevalence of *large number coincidences* with respect to the combination of fundamental physical constants. Interestingly, all of these dimensionless parameters—including the age/expansion rate of the universe, the gravitation/mass ratio of all stars, and the ratio of the Planck energy density to the (apparent) vacuum density—are some power of $N = 10^{40}$. For example, McMullin notes the following: "(1) the ratio of the gravitational to the electric force is ~N - 1; (2) the ratio of the mass of the universe to the mass of the proton is ~N2; and (3) the Hubble age of the universe in atomic units is ~N" (2008, 71).

What is further interesting about these respective parameters is that we appear to be in a unique location and epoch of cosmic history to observe them. Specifically, the time it takes for carbon to be cooked up in and dispersed by stars, the age of the universe, and the time it has taken light to travel to us since the Big Bang are all 10^{10}y (Carr 2007, 77). Further, the matter density and vacuum energy density run through a range of values over the history of the Universe and as Vilenkin (2007) explains, there is only one epoch at which they are roughly equivalent according to the SM—the one we are currently living in. This has resulted in what has been called the *cosmic coincidence problem*. Anthropic principles provide numerous ways of responding to this problem by positing how we have become a part of the universe at this most auspicious time in history.

Weinberg (2007) explains that *anthropic reasoning* makes sense for a given constant if its landscape of "possible" variation is large compared with its anthropically conducive subset of values (33). Nevertheless, he considers it understandable why some find anthropic reasoning distasteful because such theories "based on anthropic calculations certainly represent a retreat from what we had hoped for: the calculations of all fundamental parameters from first principles" (39). As discussed in Sect. 2.2, Harris maintained that the unavoidable use of metaphysics in *scientific philosophy* requires that one take as first principles both one's own existence and the Universe (the whole of which one is a part). The development of knowledge then proceeds as follows: "Self-awareness and reflection go hand in hand with an insistent demand for self-knowledge, for understanding of ourselves and our place in the world; and that demand carries with it the inevitable need to unify and systematize our experience of that world and of ourselves" (1988, 11). Hence, on Harris's account, to have self-knowledge requires a systematized worldview, one that relies upon empirical observation and is organized by metaphysics. He also maintained that the "anthropic principle" is "scientific" and "if respected, can give rise to significant observational predictions crucial to the acceptance of cosmological hypotheses" (1991, 1).

This is to say that recognizing inherent selection effects in what and how we conceive of the Universe is necessary for scientific progress. So, while our view of the world is inherently biased by our selection effect, critically assessing this bias results in ever clearer depictions of both cosmos and mind. Indeed, these ideas have been echoed by numerous cosmologists in recent years. Like Harris, Linde (2004) notes that our theory of physical reality has been so successful that we have forgotten that our starting point was our phenomenal experiences. Although the anthropic principle runs the risk of serving as a mere "painkiller" for our confusion if uncritically embraced, he considers many criticisms of anthropic reasoning to be unwarranted. He maintains this principle could "help us to understand that some of the most complicated and fundamental problems may become nearly trivial if one looks at them from a different perspective" (2004, 427). In a later (2007) work, Linde maintains that "in order to answer the question of why we live in our part of the Universe, we must first learn the answers to the questions 'What is life?' and 'What is consciousness?'" (140).

Likewise, following physicist John Wheeler's spirit, Gleiser (2004) points out that in asking "How come us?" by implication we are positing "the existence of (1) a universe capable of (2) harbouring life which, furthermore, is (3) intelligent enough to ask about its origins" (638). The question thus "encompasses all three origins, of cosmos, life, and mind; they may be (and often are) treated separately, but they are part of an indivisible whole (ibid.). Additionally—short of being realists about the mathematical objects comprising our theories—any sufficient theory addressing the nature of the Universe must include an explanation of how the theory fits into its own model, i.e. a theory of the structures of knowledge arising in minds. As Nesteruk (2013) claims, in the effort to maintain a naturalistic epistemology, through the search for its own facticity, scientists are led to a ***transcendental problematic***: "the complete picture of physical reality must include the conditions of its explicability and constitution" (2–3).

For Harris there is a two-way street (i.e. a biconditional) connecting ontological *theories* of mind and the Universe (the whole to which mind belongs). On the one hand, the conditions of mind are precisely what have resulted in our selection effect, which must be accounted for in order to establish an adequate cosmological theory. On the other hand, statements about the necessary conditions of mind must be grounded in some background theory about what is not mind, that is, a theory about how the cosmos (natural processes) are distinct from and permit consciousness. This background theory (cosmology) changes depending upon where and how we draw the boarder that distinguishes the conditions of mind from the rest of nature. Thus, having a complete theory of mind is possible iff we acquire a complete theory of the Universe.

Implicit in Harris's works is the notion that any adequate theory of either consciousness or the Universe will be grounded upon theories of the other, i.e. that these theories are *epistemically interdependent*. This *a priori* contention is evidently derived from his process-relational ontology and dialectical methodology (as outlined in Chap. 2), which posits that adequately speaking of a given phenomenon requires explaining its relationship to the system within which it occurs. Quite simply then, if we wish to have an ontology of consciousness that is scientifically grounded, we are obliged to fit this theory within the total paradigm of the natural sciences. This system is organized by cosmology, and hence such a theory of consciousness is adequate only when related within an overarching theory of the Universe. Likewise, an adequate theory of the Universe requires that we account for the conditions of consciousness—our selection effects. The Universe however, is an object that is beyond the scope of direct scientific investigation; a unique object more adequately understood as a metaphysical concept or *synthetic idea*. This establishes a bipolar structure for the dialectical holist epistemology (at this stage), but its ontology remains to be assessed.

3.4 Conclusion

In Sect. 3.1, I have shown that Harris's appeal to relativity theory to support his conception of the unifying principle has received support from numerous authors. While these theorists have taken different approaches to the ontological interpretation of space-time, they make common appeals to interdependence, self-organization, and irreducibility of global order, more recently articulated as *meta-laws*. Harris's optimism concerning the unity of particle physics is another story. The discussions of Sect. 3.2 have revealed that while such optimism is shared by advocates of the SM, this model aims for reduction, which is counter Harris's system. While the underdetermination of quantum theories remains a significant challenge, recent developments of synthetic unity (effective field theory) and relational physics support Harris's conception of the Concrete Universal as self-differentiated into irreducible scales—an issue to which I return in Chap. 5.

In Sect. 3.3, I have clarified the foundation of Harris's anthropic reasoning and supported his contention that cosmology is dependent upon philosophical *choices* or presuppositions that (in line with Ellis 2007, 1242–45) shape the details of our physical theories. Hence, metaphysics is inherently interwoven into scientific theory. One consequence of this reasoning is Harris's largely implicit argument concerning the interdependence of cosmology and consciousness studies. Although Harris has attempted to use this line of thought to arrive at ontic conclusions regarding mind and cosmos, at present the supporting evidence only permits a dialectical holist to endorse epistemic interdependence between *studies* of *mind* and *cosmos*: *implicating central roles for anthropic and unifying principles.*

For proponents of dialectical holism, the sort of internally related whole found at the quantum and relativistic scales is no surprise because it partially mirrors the simplest and most coherent dynamical form of phenomenology with which we began our investigations. Given the above, an invaluable means of evaluating Harris's metaphysics is to consider the extent to which the anthropic and unifying principles may provide insight into the hard problem of consciousness. Towards this end, in the following chapter I examine how Harris's *a priori* conception of wholeness led him to the stronger versions of the anthropic principle and evaluate his reasoning in light of the mainstream discussions.

Notes

1. In his 1965 work, Bohm likewise proposes there is an "irreducible participation of the observer" that arises in an analogous way to Heisenberg's *indeterminacy principle*, in which there is a limitation to what can be known at any instant. In special relativity he explains, due to the *principle of relativity* (laws are isotropic in the universe) and the impossibility to transmit signals faster than light, "the observer is part of the universe" and "does not stand outside of space and time and the laws of physics [...] he is trying to study" (137).

2. Barrow (2007) further echoes this interdependence: "The geometry of space and the rate of flow of time are no longer absolutely fixed and independent of the material content of space and time. The matter content and its motion determine the geometry and the rate of flow of time, and symbiotically this geometry dictates how matter is to move" (80).
3. http://map.gsfc.nasa.gov/universe/uni_shape.html
4. Diagrammatically, the aperiodic effects of *quasicrystal* structures described by Penrose (1989) capture the kind of non-local, dialectical unity Harris has in mind (1991, 33). For a technical overview of quasicrystals see Samavat et al. (2012); for a consideration of quasicrystals as a *paradigm shift*, see Ashkenazi and Lotker (2014).
5. The simplest possibility for a *grand unified theory* is based on the symmetry group SU(5): "In SU(5), we have 24 gauge bosons. In this scenario, SU(5) should be spontaneously broken to UY (1) Å ~ SUL (2) Å ~ SU(3) at 'low' energies. Below the SU(5) breaking scale, the masses of 12 gauge bosons would be of the order the SU(5) breaking scale, while UY (1) Å ~ SUL (2) Å ~ SU(3) would remain the residual symmetry above the electroweak scale with 12 massless gauge bosons" (Bambi and Dolgov 2016, 46).
6. Symmetry types include: (1) space-time/non-space-time; (2) continuous/discrete; (3) local (gauge)/global (ridged); (4) geometrical/dynamical; (5) discrete versus continuous symmetry, for example, square versus circle "invariance" (Bangu 2013, 4).
7. *Quantum logic* reveals is a logic of measurement-outcomes in closed subspaces of Hilbert space H forming an algebra called a *Hilbert lattice* L(H), where not all measurements are commutable (Wilce 2017). Consequently, the standard distributivity law of propositions does not hold and "the set of projectors on a Hilbert space do not generate a Boolean algebra, but a more complicated structure, called an orthomodular lattice" (David 2015, 81).
8. For a technical treatment of Bohm's argument for indivisibility in the EPR experiment see his (1980, ff. 117; Bohm and Hiley 1993, ff. 134). For further details regarding the formalism of *hidden variables* see David (2015, ff. 116).
9. Harris contends that "the theory of relativity gave precedence to the concept of the field over that of the particle" (2000, 98). Earman (2007) claims that in *quantum field theory* "fields are the fundamental entities and particles are epiphenomena of field behaviour" (1407).

Works Cited

Ashkenazi, D., & Lotker, Z. (2014). The Quasicrystals Discovery as a Resonance of the NoN Euclidean Geometry Revolution: Historical and Philosophical Perspective. (pp. 25–40) In: Philosophia. 42.

Bambi, C., & Dolgov, A. D. (2016). Introduction to Particle Cosmology: The Standard Model of Cosmology and its Open Problems. Heidelberg Germany: Springer Pub.

Bangu, Sorin. "Symmetry." In Robert W. Batterman, ed. *The Oxford Handbook of Philosophy of Physics*. 287–317. New York, NY: Oxford University Press, 2013.

Barad, Karen. *Meeting the Universe Halfway: Quantum Physics and the Entanglement of Matter and Meaning*. Duke University Press, 2007.

Barrow, John D. "Cosmology and Immutability." In John D. Barrow, Paul C.W. Davies, and Charles L. Harper, Jr., eds. *Science and Ultimate Reality: Quantum Theory, Cosmology and Complexity*. 402–425. Cambridge University Press, 2004.

Barrow, John D. (2007). New Theories of Everything: The Quest for Ultimate Explanation. Oxford University Press.

Belinski, Vladimir, and Verdaguer, Enric. *Gravitational Solitons. Cambridge Monographs on Mathematical Physics*. Cambridge University Press, 2001.

Berkovitz, Joseph. "Action at a Distance in Quantum Mechanics." Edward N. Zalta, ed. In *The Stanford Encyclopedia of Philosophy*, 2016.

Blanchard, Benjamin S. "Harris on Internal Relations." In Philip T. Grier, ed. *Dialectic and Contemporary Science: Essays in Honor of Errol E. Harris*. University Press of America, 1989.

Bohm, D. (1980). *Wholeness and the Implicate Order*. New York NY: Routledge.

Bohm, D., & Hiley B. J. (1993). *The Undivided Universe: an Ontological Interpretation of Quantum Theory*. New York NY: Routledge.

Capra, Fritjof. *The Tao of Physics: An Exploration of the Parallels between Modern Physics and Eastern Mysticism*. Boston, MA: Shambala Pub, 2010.

Carr, Bernard. "Cosmology and Religion." In Philip Clayton, ed. *The Oxford Handbook of Religion and Science*, 139–155. New York, NY: Oxford University Press, 2006.

———. "Introduction and Overview." In Bernard Carr, ed. *Universe or Multiverse?* New York: Cambridge University Press, 2007.

Carter, Brandon. "Large Number Coincidences and the Anthropic Principle in Cosmology." In *Confrontation of Cosmological Theories with Observational Data*. 291–298, 1974.

David, F. (2015). *The Formalism of Quantum Mechanics: An Introduction*. Switzerland, Springer Pub.

Davies, Paul C.W. "Algorithmic Compressibility, Fundamental and Phenomenological Laws." In Weinert, F. ed. *Laws of Nature: Essays on the Philosophical, Scientific and Historical Dimension*. 248–267. Berlin Germany: De Gruyter, 1995.

Dicke, Robert H. "Dirac's Cosmology and Mach's Principle." In *Nature*. 192, 440, 1961.

Earman, J. (2007). Aspects of Determinism in Modern Physics. (pp. 1370–1434). Butterfield, J. & Earman, J. (Eds.). Handbook for the Philosophy of Physics. The Netherlands, Amsterdam: Elsevier Pub.

Ellis, George. "Issues in the Philosophy of Cosmology." In Butterfield, J. and Earman, J., eds. *Handbook for the Philosophy of Physics*. 1183–1282. The Netherlands, Amsterdam: Elsevier Pub, 2007.

Gleiser, Marcelo. "The Three Origins: Cosmos, Life, and Mind." 637–653. In John D. Barrow, Davies, Paul C.W., and Cliff L. Harper, Jr., eds. *Science and Ultimate Reality: Quantum Theory, Cosmology and Complexity*. Cambridge University Press, 2004.

Gunn, David. *Pneumatology of Matter: A Philosophical Inquiry into the Origins and Meaning of Modern Physical Theory*. Winchester, UK: Iff Books, 2011.

Harris, Errol E. *The foundations of Metaphysics in Science*. London: Allen and Unwin Pub, 1965.

———. *Hypothesis and Perception: The Roots of Scientific Method*. London: Allen & Unwin Pub, 1970.

———. *Formal, Transcendental and Dialectical Thinking: Logic and Reality*. State University of New York Press, 1987.

———. *The Reality of Time*. Albany, NY: State University of New York Press, 1988.

———. Cosmos and Anthropos*: A philosophical Interpretation of the Anthropic Cosmological Principle*. Atlantic Highlands, NJ: Humanities Press International Inc, 1991.

———. The Spirit *of Hegel*. Atlantic Highlands, NJ: Humanities Press International, Inc, 1993.

———. The Restitution *of Metaphysics*. Amherst, New York: Humanity Books, an imprint of Prometheus Books, 2000.

Healey, Richard. "Holism and Nonseparability in Physics." In Edward N. Zalta, ed., *The Stanford Encyclopedia of Philosophy*, 2016

Heil, John. "Relations." In Robin Le Poidevin, Peter Simons, Andrew McGonigal, and Ross P. Cameron, eds. *Routledge Companion to Metaphysics*. 310–321. New York, NY: Routledge Press, 2009.

Jenann, Ismael. "Quantum Mechanics." In Edward N. Zalta ed., *The Stanford Encyclopedia of Philosophy*, 2015.

Lange, Marc. "Meta-laws of Nature and the Best System Account." In *Analysis*. 216–222. 71 (2), 2011.

Laudisa, Federico, and Rovelli, Carlo. "Relational Quantum Mechanics." In Edward N. Zalta, ed., *The Stanford Encyclopedia of Philosophy*, 2013.

Linde, Andrei. "Inflation, Quantum Cosmology, and the Anthropic Principle." In Barrow, J.D., Paul C.W. Davies, and Charles L. Harper, Jr., ed. *Science and Ultimate Reality: Quantum Theory, Cosmology and Complexity*. 426–458. Cambridge University Press, 2004.

Linde, A. (2007). *The Inflationary Multiverse.* (pp. 127–149) In: Car (Ed). Universe or Multiverse? New York: Cambridge University Press.

Livio, Mario. "The Interconnections Between Cosmology and Life." In John D. Barrow, Simon C. Morris, Stephen J. Freeland, and Charles L. Harper, Jr., eds. *Fitness of the Cosmos for Life: Biochemistry and Fine-Tuning.* 114–131. New York: Cambridge University Press, 2008.

Manchak, John Byron. Global space time structure. In Oxford Handbook of Philosophy of Physics, 2013.

Maudlin, Tim. "Distilling Metaphysics from Quantum Physics." In Michael J. Loux, Dean W. Zimmerman, eds. *The Oxford Handbook of Metaphysics.* 461–487. New York: Oxford University Press, 2003.

———. *Quantum Non-Locality and Relativity: Metaphysical Intimations of Modern Physics.* Third Ed. Malden, MA: Wiley-Blackwell, 2011.

McMullin, Ernan. "Tuning Fine-Tuning." In John D. Barrow, Simon C. Morris, Stephen J. Freeland, and Charles L. Harper, Jr., eds. *Fitness of the Cosmos for Life: Biochemistry and Fine-Tuning.* 70–94. New York: Cambridge University Press, 2008.

Morrison, Margret. "The New Aspect: Symmetries as Meta-Laws – Structural Metaphysics". In Weinert, Friedel, ed. *Laws of Nature: Essays on the Philosophical, Scientific and Historical Dimension.* 157–188. Berlin, Germany: De Gruyter, 1995.

———. "Unification in Physics." In Robert Batterman, ed. *The Oxford Handbook of Philosophy of Physics.* 381–414. New York, NY: Oxford University Press, 2013.

NASA.gov (2014): http://map.gsfc.nasa.gov/universe/uni_shape.html

Nesteruk, Alexei V. "A 'Participatory Universe' of J. A. Wheeler as an Intentional Correlate of Embodied Subjects and an Example of Purposiveness in Physics." In *Journal of Siberian Federal University.* 415–437. Humanities and Social Sciences, vol. 6, n. 3, 2013.

Penrose, R. (1989). *The Emperor's New Mind: Concerning Computers, Minds, and the Laws of Physics.* Penguin Books.

Pike, S. M. (2009). *A Model of Mind from the Perspective of Temporal Structuralism.* NZ: University of Canterbury.

Pooley, Oliver. "Substantivalist and Relationalist Approaches to Spacetime." In Robert Batterman, ed. *The Oxford Handbook of Philosophy of Physics.* New York, NY: Oxford University Press, 2013.

Rovelli, Carlo. "Quantum Gravity." In Jeremy Butterfield, and John Earman, eds. *Handbook for the Philosophy of Physics.* 1288–1329. The Netherlands, Amsterdam: Elsevier Pub, 2007a.

———, and Smerlak, Matteo. *Relational EPR.* In *Foundations of Physics*, 427–445. Vol 37, 2007b.

Samavat, F., Tavakoli, M. H., Habibi, S., Jaleh, B., & Ahmad P. T. (2012). *Quasicrystals.* Open Journal of Physical Chemistry, 2012, 2, 7–14 http://dx.doi.org/10.4236/ojpc.2012.21002 Published Online February 2012.

Serway, A., Jewett, J., Wilson, K., &Wilson, A. (2013). *Physics Vol. 2. Asia-Pacific Edition.* Melbourne: Cengage Learning Australia.

Smeenk, Christopher. "Philosophy of Cosmology." In Batterman, R. ed. *The Oxford Handbook of Philosophy of Physics.* 607–652. New York, NY: Oxford University Press, 2013.

Smolin, L. (1997). *The Life of the Cosmos.* New York, NY. Oxford Pub.

Teller, Paul. "Relational Holism and Quantum Mechanics." In *British Journal for the Philosophy of Science.* Vol: 37, No. 1. Oxford University Press, 1986.

Thompson, Evan. *Mind in life: Biology, Phenomenology, and the Sciences of Mind.* Cambridge, MA: Belknap Press of Harvard University Press, 2007.

Tomizawa, Shinya, and Mishima, Takashi. "New Cylindrical Gravitational Soliton Waves and Gravitational Faraday Rotation." In *Physics Review D: Covering Particles, Fields, Gravitation, and Cosmology,* 2014.

Unger, Roberto M., and Smolin, Lee. *The Singular Universe and the Reality of Time: A Proposal in Natural Philosophy.* Cambridge, UK: Cambridge University Press, 2015.

Vilenkin, Alex. "Anthropic Predictions: The Case of the Cosmological Constant." In Bernard Carr, ed. *Universe or Multiverse?* 163–179. New York: Cambridge University Press, 2007.

Weinberg, Steven. "Living in the Multiverse." In Bernard Carr, ed. *Universe or Multiverse?* 29–42. New York: Cambridge University Press, 2007.

Wilce, A. (2017). Quantum Logic and Probability Theory. The Stanford Encyclopedia of Philosophy. Edward N. Zalta (ed.)

Yudell, Zenja. "Lange's Challenge: Accounting for Meta-Laws." In *British Journal for the Philosophy of Science.* September. 1–23, 2012.

CHAPTER 4

The Anthropic Principles of Dialectical Holism

4.1 The Weak Anthropic Principle

According to the Newtonian worldview, the Universe is a machine guided by the laws of physics that can be understood independently of observers. According to Carter's original (1974) assessment, our region of the universe, the planet Earth within its stable solar system, is epistemically privileged. The *weak anthropic principle* (WAP) is thus modern cosmology's (Copernican) revision of this worldview, stating that we are inherently *biased* to observe a universe from a space-time location that permits the existence of observers. Unlike stronger versions, the WAP makes no ontic claim about the nature of observers within the Universe and today; it has been widely accepted as a basic epistemic principle for the logic of modern cosmology (e.g. Carr 2007; Tegmark 2014, 111; Ellis 2007b, 1254). Appealing to phenomenological reasoning however, Harris has used the WAP as a launchpad for stronger, more contentious anthropic principles.

According to Vilenkin, the WAP recognizes that there are certain values of some cosmological variable(s) (X) whose variance influences the chances of intelligent observers evolving: "$X_{min} < X < X_{max}$" (2007, 164). Both the lower and higher values of X will not be observed because they are inconsistent with the development of observers. The WAP thus constrains our theories about where we are in space and time and the predictions we can make about certain measurable parameters. In discussions of AR,

J. Schofield, *A Phenomenological Revision of E. E. Harris's Dialectical Holism*,
https://doi.org/10.1007/978-3-030-65029-2_4

multiverse theories arise simply by authors remaining realist about the unseen and logically possible values of these cosmological parameters. This move is captured by Barrow's final pillar of scientific advance, *transfiguration*: "the discovery that our supposed constants are a small part of a vastly more exotic structure" (2004, 407). He explains the constants of nature "may not be specified by the self-consistency of a 'theory of everything' and so they may be permitted to take on any (or a wide range of) values" (411). Consequently, the possibility transfiguration permits multiverse theories, to which appeals have been made both for and against anthropic reasoning. In this vein, Vilenkin's "principle of mediocrity" is the flip side of the WAP. Whereas the WAP says we should expect our cosmic location to be "fine-tuned" such that it is consistent with the existence of "observers", the principle of mediocrity says that we should not expect our location to be "over-tuned", or more privileged than necessary to explain our existence within a multiverse.

Harris claims that the discoveries in contemporary physics supportive of the WAP depict a universe in which life can now be "understood as a development continuous with the non-living; and the world is observed as providing conditions for the emergence of intelligent beings" (1991, 4). Harris maintains, "we exist because the universe is the way we observe it to be, so we could not observe it otherwise" (ibid.). This is to claim that our existence in some sense constrains and is constrained by the world-as-it-is-observed. *A fortiori* he maintains, "[w]e can only imagine the universe—any universe—as it would appear to our observation, even if it were different from what we actually observe" (ibid.). While this may *prima facie* appear to be a mere truism, Harris clarifies that "[w]hat we observe is conditioned not only by the fact of our existence, but also by the nature and capacities of our perceptive and intellectual faculties" (5).

In this vein, I propose three types of *selection effects* largely implicit in Harris's system and invaluable for evaluating his AR:

1. *Epistemological selection effect* (ESE)—denotes the limitations of *a posteriori* knowledge resulting from the method of investigation and *choice of conceptual relevance*. A common *metaphor* is the mesh of a net used for fishing, which permits fish smaller than the mesh size to escape and therefore avoid being observed.
2. *Physical selection effect* (PSE)—involves any force acting on material entities such that certain phenomena are restricted while others are permitted. The above metaphor is here taken literally as the *physical*

constraint *and sensitivity* some entity (e.g. a net) has over a system (e.g. a population of fish). The PSE necessarily contributes to but does not exhaust the ESE.

3. *Metaphysical selection effect* (MSE)—considers the impact that a priori presuppositions have on the form of the observer's physical theories, for example, rules of logic, the rules for mapping abstract terms onto identifiable objects, and choice of methodology. Here the constraint arises from a teleonomic trajectory (e.g. a paradigm), which limits how accurately physical entities can be individuated. The MSE gives rise to the ESE.

Harris does not try to deny a priori that some multiverse might turn out to be a viable theory but maintains such theories must be meet certain standards of theoretical coherence. He claims the "Universe is everything that is and ever has or will be; there can be only one. To speak of many Universes is therefore a misuse of the term" (1991, 11). This is a fair and apparently benign etymological reminder, but he goes on to explain a crucial implication. Each region (or *u*niverse) "must somehow, and in some sense, be mutually related; otherwise they could not be distinguished, or counted, or regarded as many" (ibid.). Likewise, the totality must "constitute a single complex, within which there may be many distinguishable regions or epochs, but these would not strictly be Universes, even if between them no communication of information could pass" (ibid.). He argues that if "we cannot observe them, for whatever reason, we must at least be able to infer to their existence from what we do observe, or else their postulation becomes scientifically otiose" (ibid.).[1] Ultimately, Harris concludes on metaphysical grounds that multiverse theories are of little consequence in the face of his teleological anthropic reasoning. To clarify Harris's approach, in the following section I introduce two multiverse theories and exemplify the three selection effects listed above.

4.1.1 *Levels I and II Multiverses*

According to Tegmark (2007), Level I multiverse is totally uncontroversial, requiring only that *space and matter are homogenous beyond our cosmic horizon*. This kind of multiverse has the same effective and fundamental laws throughout, but random quantum density fluctuations give rise to an infinite *ergodic* space containing Hubble volumes of *all possible initial conditions*. This means that the probability distributions of events in any given volume permit that everything that could have happened here has or will happen somewhere else at some calculable distance from Earth.[2] Interestingly, Tegmark argues that this level dissolves the debate between free will and determinism:

> If there are indeed many copies of "you" with identical past lives and memories, you would not be able to compute your own future even if you had complete knowledge of the entire cosmos! The reason is that there is no way for you to determine which of these copies is "you." Yet their lives will necessarily differ eventually, so the best you can do is predict probabilities for what you will experience from now on. (104)

Tegmark contends that we can accept the existence of things that are at present unobservable on the basis that they are in principle observable under certain conditions (like a ship beyond the horizon) (2014, ff. 100). In this way, our inference to the existence of this multiverse is by no means costly or ungrounded. Although Harris did not address this version of the multiverse directly, it certainly satisfies his requirement that such theories must be phenomenologically grounded.

Concerning AR, Tegmark and Carr agree that if quantum density fluctuations of the early Universe are to blame for the conditions that give rise to living observers, then they are contingent on accidental features of symmetry breaking arising from the initial conditions of our universe. In this case the WAP would be necessary to discern which initial conditions (the presumed PSE) have given rise to our Hubble volume out of a range of *possibilities* (derived from some MSE).[3] On the other hand, if a future ToE will determine all universes uniquely then there would be no room for AR, since it would be possible to derive the parameters of our universe from this theory (Carr, 83). Since this multiverse theory is relatively non-contentious, it is the later versions that will reveal much more about Harris's anthropic reasoning.

Level II multiverse posits an infinite set of Level I multiverses as a result of *chaotic inflation*. Level II bubbles arise from the same *fundamental laws* but contain different constants, particles (due to symmetry breaking), dimensionality, and starting conditions. Level II multiverse implies an endless Universe in space and time with no ultimate Big Bang to get it started, that is, the *big bang* is always occurring *somewhere*. In further contrast to Level I multiverse, we could *never* observe or get from any one of the worlds in Level II multiverse to another because of ongoing inflation. Though more controversial, Tegmark explains that we have come to many theoretical insights that go beyond what we can *immediately observe* (ESE). Arguing from analogy he says that if a fish never left the ocean it might "mistakenly conclude that the properties of water are universal, not realizing that there is also ice and steam" (2007, 100). He claims likewise, "Space can 'freeze', transitioning between different phases in a landscape of possibilities, just like water can be solid, liquid or gas. In different phases, the effective laws of physics (particles, symmetries, etc.) could vary" (ibid.).

Linde (2007) holds that inflation theory explains local homogeneity by suggesting that at distances beyond our light cone the Universe is completely inhomogeneous. Based on chaotic inflation, Level II multiverse appears as an intricate fractal that expands eternally, the structure being predicted by *mathematical laws* (MSE). "The different parts of this fractal are enormous and may have dramatically different properties. They are connected to each other, but the distance between them is so large that for all practical purposes they look like separate universes" (128). In a rapidly expanding universe such as the chaotic inflationary model, it is argued that a scalar field goes down slowly "like a ball in a viscous liquid, with the viscosity being proportional to the speed of expansion" (129–30). He goes on to say, when the field value becomes small, inflation ends and the field oscillates near the minimum and like

> any rapidly oscillating classical field, it loses its energy by creating pairs of elementary particles. These particles interact with each other and come to a state of thermal equilibrium with some Temperature T. From this time on, the Universe can be described by the standard hot big bang theory. (130)

As previously discussed, such a theory will by its very nature be beyond empirical proof. For Harris such theories are unavoidable, and our only available strategy is to make sure that the theory is parsimonious and congruent with observations that are available. Likewise, Rees (2007) warns, "consistency is not enough; there must be grounds for confidence that such a theory is not a mere mathematical construct, but applies to external reality" (64). He proposes that such confidence can be achieved by a theory of the Universe or Multiverse structure "accounting for things we can observe that are otherwise unexplained" (ibid.). He argues that such worlds are analogous to many galaxies that are beyond our current technological observability: they are included in our cosmological discourse on the basis of what we already know about the observable universe. Hence, space-times completely disjoint from ours and perhaps even governed by different "laws" are no less "real" legitimate objects of scientific consideration than regions (e.g. with redshift 10) that never come within our horizon.

What evidence is available for Level II? Tegmark (2007) finds that since Level II multiverse is a direct result of inflationary cosmology, evidence in favour of inflation should do the job of supporting this multiverse. He claims, "cosmological measurements have confirmed two key predictions of inflation: that space has negligible curvature and that the clumpiness in the cosmic matter distribution was approximately scale-invariant" (106). In agreement with Linde, he claims that Level II necessitates anthropic reasoning because in a multiverse with different constants, "physicists will never be able to determine the values of all physical constants from first principles. All they can do is compute probability distributions for what they should expect to find, taking selection effects into account" (109).

One of the most important implications of chaotic inflation is *universality*, according to which, everything that is possible inevitably comes into existence at some point. Ellis (2007b) concedes that this scenario would validate the WAP but points out that the physics for this position relies upon the undiscovered inflation field(s) (PSE). Importantly, these fields arise from the universe itself at a fundamental level, acting as order parameters for the *phase transitions* (i.e. evolution) of the respective worlds. In this regard, Ellis asks, "What determines what is possible?" (1257). Importantly, what is *physically possible* in this sense (PSE) may or may not be identical with what is considered to be *logically possible* (MSE). In either case, Harris's *Concrete Universal* may be retained. Hence, Levels I and II multiverses do not conflict with Harris's system but much more remains to be discussed.[4]

4.1.2 Weak Anthropic Reasoning

By my reading, Harris (1991, ff. 2) attempts to demonstrate both mischaracterizations (e.g. Hawking 1988, 124) and what he considers to be the fruits of the WAP. Harris first considers whether taking our selection effects seriously implies that "our theories reveal more about their authors than about their subject-matter" (1991, 5). He admits that theorizing about the world is possible even in cases where observation is "confined within irremovable horizons" (e.g. black holes), via extrapolations from known laws (ibid.). He recognizes that "[t]o speak of the selection effect is to presume a known (or at least ascertainable) range of facts beyond what is selected" (6). This he maintains must be true of both our (present) sense faculties and our theoretical models.

Aiming to avoid a common pitfall, Harris argues that if we consider our sense faculties to be "incorrigibly" afflicted by selection effects, we are destined for "subjectivism" and "epistemological disaster", because science would no longer provide knowledge of the real world (ibid.). The WAP is then "scientifically important because it indicates a significant selection effect must be treated with the utmost caution and restraint" (ibid.). So, for Harris, *selection effects* not only reveal that our observation is *consistent* with some presumed cosmological PSE, but also emphasize our limitation in individuating the parameters of our physical world as a result of our ESE.

Towards an illumination of Harris's AR, Carr, Barrow and Hawking's explanations of the WAP provide important contrasts. Taking a conventional approach, Carr holds this principle "accepts the constants of nature as given and then shows that our existence imposes a selection effect on when (and where) we observe the Universe" (2007, 78). More specifically, Barrow maintains that our observations "must not be viewed as having been taken from some unconstrained ensemble of possibilities but from some subset conditioned by the necessary conditions for carbon-based observers like ourselves to have evolved before the stars die" (2007, 193). Going further, Hawking (2007) takes a "top down" approach, in which the history of the Universe is given by the *Feynman path integral* that "allows every possible history for the Universe" and is constrained by our present measurement surface. Hawking reasons, the "histories of the Universe depend on what is being measured, contrary to the usual idea that the Universe has an objective, observer-independent, history" (97).

Consequently, they each endorse at least one of the following three assumptions:

(i) The constants of nature are sufficiently individuated within our current models.
(ii) The unseen parameters correspond to a space of physical universes.
(iii) Our measurements impose a physical influence upon the history of our universe.

Evidently, Barrow, Carr, and Hawking are all committed to (i). For Carr and Barrow, taking the ***relevant*** physical constants as given implies that their logical manipulations reveal a material possibility space (ii). Hawking further endorses (iii) that the history of our cosmos is adequately individuated by the path integral, which is itself influenced by observers.

The concern is that many arrangements of physical constants that are *logically possible* (MSE) might never *physically obtain* (PSE). Arguments are needed to justify either that the two conceptually distinct possibility spaces are not being conflated, or that they are coincident. Barrow (2011) recognizes this concern with a sense of humility. He cautions that we do not know what it truly means to consider the counterfactual alteration of a given constant:

> At present we can imagine the effects of changing rates of radioactive decay without worrying if there are consequences for gravity or atomic structure. A fully unified theory would reveal all those interconnections and ensure that a small change in one part of physics would have extra consequences elsewhere. (231)

More recently, Harris's abandon of anthropic reasoning that invokes multiverse models has been supported by Smeenk (2013). Despite the purported explanatory power of chaotic inflation, Smeenk questions the theoretical integrity of Level II multiverse. His first worry is that "the detailed accounts of how the multiverse arises are typically beyond theoretical control" (645). This he claims creates a risk that "the claimed multiverse explanations are just so stories where the mechanisms of generating the multiverse is contrived to do the job" (ibid.). He goes on to say that it is not at all clear why it is helpful to shift our attention from explanations of the apparent value of some parameter(s) X, which is contingent upon historic events of our universe, into questions about the location of our universe within a much larger ontological landscape where the parameter varies. "Analogously, the success of historical explanations in evolutionary biology does not imply the existence of other worlds where pandas have more elegant thumbs" (ibid.). Emphasizing parsimony, Smeenk finds no reason why in explaining X, we must posit the existence of multiple universes to house its alternative values. Such a multiverse may in this sense be an excessive ontological commitment.

While a great deal of scientific progress proceeds in an inductive manner, it would appear to become problematic in cosmology because in this case we assume that the Universe may be characterized by the parameters we observe, despite our *present* (insurmountable) inability to examine these constants from *alternative perspectives*, let alone a *view-from-nowhere* (i.e. first principles). In this respect, Ellis (2007b) poses a further word of caution:

> Reality is not fully reflected in either observational or theoretical models. Problems arise from confusion of epistemology (the theory of knowledge) with ontology (the nature of existence): existence is not always manifest clearly in the available evidence. The theories and models of reality we use as our basis for understanding are necessarily partial and incomplete reflections of the true nature of reality, helpful in many ways but also inevitably misleading in others. They should not be confused with reality itself! (1247, emphasis omitted)

Two of the most important examples of this confusion that he offers are *computer simulations of reality* and *the laws of physics themselves* being *identified with our mathematical formalism.* Ellis concludes that because the supposed causal (or pre-causal) processes of multiverse theories are either unproven or untestable, they are not physics but rather metaphysics (1263).

The upshot of the present approach to the WAP, I maintain, can be understood in much the same way as Bohm's implicate order is applied to quantum mechanics: As appeals to probability in quantum theory are for Bohm due to our ignorance of an implicate order guiding quantum particles, appeals to multiverse cosmology can be used *as a crutch* for our inability to sufficiently individuate the necessary parameter(s) (of a super-implicate order) that instantiates observers. Consequently, appealing to a possibility space that *explains* the occurrence of observers commits a *hysteron proteron* by presuming precisely what we do not and may never have, that is, sufficient *empirical* knowledge of nature-as-a-whole that permits *proper derivation (individuation)* of X parameters and their *relation to observers* ($\exists^R$). In lieu of empirical experimentation on the universe itself, appeals to multiverse cosmology may therefore conflate two selection effects—the coherence of mathematical systems and the physical organization of the Universe—resulting in an *MSE-PSE fallacy* (MPF). I maintain that this (rather ironically) reveals the true philosophical fruit of the WAP.

So, while the WAP may not appear impressive, merely proclaiming that we will find ourselves in a galactic region suitable for (consistent with) observation, it also reveals the antiquity of an objective worldview. Additionally, on the dialectical holist account, the WAP reveals that theories about the Universe must confront the same challenges as efforts to establish a foundation in metaphysics. Central to Harris's system is the importance of accounting for the biases inherent in our individuation (diffractive/levative) practices in our efforts to establish such a foundation.

Towards this end, Harris's claim that "we could not observe the universe besides how it is revealed to us" requires further elaboration and investigation into the *strong anthropic principle*. This principle was originally conceived by Carter as saying that "the Universe (and hence the fundamental parameters on which it depends) must be such as to admit the creation of observers within it at some stage" (1974, 294). Beginning with Barrow and Tipler (1986), arguments surrounding this principle concerning vitalism and design have been significantly elaborated. Though widely discussed in subsequent years, the strong anthropic principle has been mostly criticized by scientists and invoked by theists. In the following sections I consider three means by which authors have maintained the *necessity* of observers despite our inability to acquire knowledge of the trajectory and initial conditions of the cosmos. These are the *participatory, final,* and teleological *anthropic principles* respectively. I maintain that arguments for the participatory and final anthropic principles are weak at best, but that each may be subsumed by Harris's teleological anthropic principle, which has not yet received an adequate evaluation.

4.2 The Participatory Anthropic Principle

The *participatory anthropic principle* (PAP), first articulated by Wheeler and Zurek (1983), proposes that observation is required for the universe to come into being (Wheeler and Zurek, ff. 182). Harris aims to defend the PAP in an extreme but particularly nuanced form that makes use of both orthodox and Bohmian quantum theories. In this section I orient Harris's PAP among some alternatives and evaluate its likely implications for dialectical holism moving forward.

Harris first introduces the PAP by appealing to the orthodox conception of uncertainty, from which it was originally developed. He maintains the uncertainty imposed by Planck-level influence upon measurements of quantum systems is not just a feature of our knowledge, but of nature itself:[5]

> The Psi-function describing the state of the system defines only a probability amplitude for such quantities, so that the system has no determinate properties until a measurement is made, when the probability amplitude collapses, and only then can the property disclosed be assigned to the system as determinate and actual. Observation and measurement thus become prerequisite to the actuality of all determinate properties of quantum entities, without which they do not exist in reality but are merely potential [...] It then follows that intelligent beings, through their observation and measurements, must participate in the actualization of the universe at large. (1991, 6–7)

Experimentation is by this reasoning necessary to bring about a switch from an abstract possibility of quantum formalism to physicality. Hence, by appealing to the collapse of the wave function, the PAP begins with a phenomenological qualification about how we can speak about quantum phenomena and attempts to distil an ontology about how observing beings relate to nature ($\exists^R$).

Harris recognizes that this interpretation may result in a circularity given that it took billions of years for observers to arrive on the scene. He considers it incoherent to maintain a mutual dependence, such that "[i]f physical reality depends on the existence of mind, and mind depends on the prior existence of physical reality, neither can exist unless both can come into being simultaneously" (1991, 7). Harris also rejects Berkley's subjective idealist interpretation, criticizing that it will inevitably lead to solipsism. On the one hand he endorses the Copenhagen interpretation for purportedly maintaining that particles are "latent, or potential, before they are observed" (1991, 146). He maintains that the measuring instrument and what it measures constitute "a single indivisible complex, within which the measured property comes into being. What the measuring instrument registers, however, has meaning as a value only when (and as) read by an investigator" (2000, 100).

On the other hand, Harris's appeal to Bohmian hidden variables is consistent with his emphasis on indivisible wholeness and (qualified) realism. According to Harris, Bohm's theory shows we can "no longer insist upon strict individuation of particles and must therefore acquiesce in their dependence upon the energy system for their properties and behaviour" (1965, 138). He goes on to say,

> The proper interpretation of quantum mechanics, therefore, may depend upon the recognition that the physical real to which it refers is a whole or structured totality in which individuality is constituted by form, and particu-

late existence is only provisional, approximate, temporary and adjectival to the dynamic pattern in which it occurs. (139)

Harris appreciated Bohm's implicate order because he claims it "maintains the holistic conception of a physical reality that requires dialectical explication [...] the system only becomes clearly explicit when cognized and expressed in scientific terms, so that the Participatory Anthropic Principle still holds" (1991, 147).

Despite his apparent sympathies for Bohmian holism, Harris attempts to remain agnostic claiming that in one form or another quantum theory can justify both claims that (1) observer and measurement are "inseparable from the very actuality of elementary particles", and (2) "reflective awareness, in the guise of observation and interpretation, is constitutive of the very being of the universe" (1991, 145). While Harris recognized that the two systems are incompatible, he made no attempt at synthesis nor did he decide between orthodox and Bohmian accounts. Crucially, to establish a PAP of Bohm's theory is a task that neither Harris, nor (to my knowledge) anyone since has explicitly undertaken. Hence, questions remain concerning what such a "dialectic explanation" entails and which quantum theory is most accommodating of Harris's metaphysics. Towards a clarification of these issues, I next highlight Bohm and Harris's mutual dismissal of the Everettian *many-worlds interpretation* (MWI).

4.2.1 Level III Multiverse

According to Tegmark, Level III multiverse is a result of Everett's (1957) *many-worlds interpretation* (MWI). This interpretation grew out of dissatisfaction with Bohr's *Copenhagen interpretation*. Advocates for Everett's view such as Mukhanov (2007), complain that the timing for the collapse of the wave-function is not derivable from the equations of orthodox quantum mechanics. Moreover, orthodox equations are so overly complicated that they appear laughable when applied to the Universe. To avoid having to contend with "schizophrenic" superpositions, Everett argued that there is *no collapse of the wave-function* and at these points the universe splits into respective probabilistic outcomes or *worlds*.[6]

Counter the PAP, Mukhanov maintains that the MWI can preserve the classical conception that there is a "one-to-one correspondence between reality and the mathematical symbols which are used to describe it. We also return to the idea that physics describes not only 'our knowledge and perceptions' but also the world 'out there,' which *existed* and *will exist* without any observers" (270). As Vaidman (2014) explains, while quantum events appear random, according to MWI, the wave-function evolves *deterministically* in a 3D Hilbert space, whose state vector branches into all possible outcomes. The *temporal evolution* of this structure thereby provides a foundational ontology. Hence Hogan (2007) maintains "the cosmic wave-function never collapses, but only appears to collapse from the point of view of observers who are part of the wave-function" (227).

Importantly, Tegmark (2007) notes, Level III branching effects may be found in Level I, with the key difference that in Level I, the "branches" are far away in space-time, whereas in Level III, they exist in a Hilbert space. Moreover, according to Wallace (2013) within this Hilbert space, the terms (coefficients) in each branch can interact, reinforce, and cancel out. He goes on to say, instead of indefiniteness, in MWI microscopic quantum superpositions are cases in which one or more definite thing is occurring at the same time. Wallace explains this plurality consists of two parts:

> [A] contingent physical postulate, that the state of the Universe is faithfully represented by a unitarily evolving quantum state; and an *a priori* claim about that quantum state, that if it is interpreted realistically it must be understood as describing a multiplicity of approximately classical, approximately noninteracting regions that look very much like the "classical" world. (465)

Tegmark argues that the same reasoning also applies to Level II, except that at Level III the "process of symmetry–breaking did not produce a unique outcome but rather a superposition of all outcomes, which rapidly went their separate ways" (2007, 113).

Hogan maintains anthropic reasoning is built into MWI at a basic level: "Viewed from a disinterested perspective outside the Universe, it is as though living beings swim like salmon up their favourite branches of the wave-function, chasing their favourite places" (227). In this weak sense it is relevant to consider our selection effect to establish what branch we inhabit. However, supporters of MWI would not claim that observation serves as some sort of "higher level" causal force as in the PAP. Indeed, a proponent of MWI claim that the conditions required for the existence of intelligent observers emerge from quantum fluctuations and since every possible fluctuation is actualized, observation plays no special role in the process.

Harris contends that MWI "violates wildly Occam's principle of parsimony, *entia non multiplicand sunt praeter necessitate*" (1991, 13). Though he did not recognize that MWI retains one of the simplest quantum formalisms on offer, Harris finds the theory metaphysically excessive. He argues, to the extent that these parallel worlds are "inaccessible to inspection" and incapable of "mutual communication", they appear to be "cosmological lumber of no scientific use" (ibid.). He thus concludes:

> At best (if "best" is an appropriate word), they are things-in-themselves, whose existence is postulated without means of verification. They are, in the worst sense, "metaphysical entities." Much less can the theory remove the objections to an anthropic selection effect, or to any of the four versions of the Anthropic Principle. (Ibid.)

Here, Harris maintains that MWI cannot surpass our selection effects via a view-from-nowhere and that the theory cannot avoid his stronger anthropic reasoning. It has been widely recognized that while Bohmian mechanics "philosophically" achieves more than the MWI, it does so "at the price of adding the non-local dynamics of Bohmian particle positions" (Vaidman 2014). This means Harris's appeal to Bohm's *realist* holism of the EPR experiment (Sect. 3.3) marks a metaphysical separation from MWI; but this is a choice based on consistency and taste, not parsimony. Nevertheless, his claim that MWI cannot escape his stronger anthropic reasoning may yet be vindicated.

For his part, Bohm and Hiley (1993) recognizes that some may believe that the greatest advantage of MWI is that "everything can be put in some correspondence with Hilbert space" (316). It follows that while we are directly aware of the explicate particle aspect of nature via the senses, the subtler wave function aspect is inferred. So Bohm's approach does not assume that all we are aware of are particles, as in MWI, but maintains "at least in science as we know it thus far, all information comes ultimately through the manifest level (recalling that it comes not only through the particles, but also through the classical limit of electromagnetic field beablels)" (314).

So, his alternative interpretation of quantum mechanics is arguably superior to the MWI not for its parsimony, but for more technical reasons. He claims his view provides an explanation of why quantum theory "implies a relatively autonomous domain in which the quantum potential can be neglected so that an approximately classical behaviour will result" (314). Additionally, he claims his approach provides an "intuitively understandable account of how quantum processes take place" (315). In agreement with Harris, Bohm maintains it "defies the imagination to grasp intuitively" how both world and awareness "could split into many parts that are not aware of each other, and even more so how awareness could be, in effect identified with some part of a vector in Hilbert space" (ibid.).

4.2.2 *Participatory Anthropic Reasoning*

On an Orthodox account, quantum probability consists of preparation, transformation, and measurement, in which the past can only be in the form of classical particle physics whereas the future is conceived as a probability or wave-function (Ψ). Hence, quantum theory pertains only to the future. According to Dyson (2004), the role of the observer is to "draw the line" between the two by changing "the point of reference from before the event to after it" (84). As Zeilinger (2004) elaborates, the experimentalist's "choice of apparatus" can determine what quality becomes reality in the experiment, and so "is constitutive of reality", but he warns not to take a "subjective interpretation" of the role of the observer because in contradiction to a widespread misunderstanding, "the consciousness of the observer does not influence the particle at all" (209).

Concerning the origins of the PAP, Davies (2004) agrees that observation entails "meaningful information" and that "[m]easurement implies a transformation from the realm of mindless material stuff to the realm of knowledge" (8). It was this idea that led Wheeler to conceive of the participatory universe:

> [O]ne that makes full sense only when observers are implicated; one that is less than fully real until observed. He envisaged a meaning circuit, in which atomic events are amplified and recorded and delivered to the minds of humans—events transformed into meaningful knowledge—and then conjectured a return portion of that meaning circuit, in which the community of observers somehow loops back into the atomic realm. (Ibid.)

One way this circuit can be completed is with the wave/particle dual nature that is effectively "decided" by an experimental arrangement. Davies points out however that the PAP does not claim that "the universe doesn't exist unless it is observed", but rather that past states do not possess "a full set of physical attributes, such as all particles having a definite position, motion, etc." (10). Consistent with Harris's reading, observers in the present are thus believed to determine the actuality of the past and transform Nature, not bring it into being.

With these strictures in mind, if the dialectical holist were to invoke orthodox quantum theory in defending the PAP, Stapp's (2011) proposal for the *Mindful Universe* appears a likely refuge. Stapp argues that the conscious choice of an observer is an irreducible and necessary component of physical reality, however (contra Harris), Stapp recognizes that the orthodox account is "intrinsically dualistic" (49; ff. 79). Following von Neumann's interpretation, three processes are required:

- *Process 1*—the act of intervention that partitions a system by experimental arrangement and entails an "acquisition of knowledge".
- *Process 2*—the evolutionary process governed by Schrodinger's equation that unfolds as "a representation of this acquired knowledge".
- *Process 3*—the result of 1 and 2, in which we receive some form of phenomenological *data*, for example, a Geiger counter reading (24).

Stapp maintains that "process zero" is the free choice of the experimenter that ties together physical and psychological aspects of reality into one "whole" quantum event.

Following what Stapp calls "Whiteheadian quantum ontology" he considers real what Bohr accepted only pragmatically, namely that our conscious intentions are the cause of our actions and the psychological correlate of a process 1 action:

> Within orthodox thinking, the physical process 1 action results from, as von Neumann's words emphasize, an intervention from outside the physically described domain. This process has, according to contemporary quantum theory, no sufficient causal roots in the physical alone. The experimenter's "free choice" participates in the selection of the needed partition that physical processes alone are unable to achieve. (108)[7]

Note that Stapp's position amounts to an argument from ignorance for free will (i.e. orthodox account cannot account for experiment choices), but does not detail any ontological relationship between observation and the cosmos: (1) *the* meaning *made by us may be an essential part of reality*, but this requires its own argument and even if it were true, in no way does it imply (2) *that the* meaning *we make influences reality beyond the empirical results of our experimental arrangements.* Presently neither view is justified. Hence, insofar as they reject dualism, a dialectical holist PAP remains unsupported by an orthodox interpretation.

Expanding upon the implications of quantum theory for the nature of information, Zeilinger (2004) argues that any theory about reality must be based on "information we receive" and so "the concept of a reality without at least the ability in principle to […] obtain information about its features is devoid of any possibility of confirmation or proof. This implies that the distinction between information that is knowledge, and reality is devoid of any meaning" (218–19). Zeilinger here contends that it is meaningless to speak of a world that we are in principle incapable of observing and likewise, what we can take to be "real" is knowledge based upon our worldly engagements. The aim of these points is to deflate the Orthodox PAP and render meaningless the distinction between phenomenal and noumenal domains. Admirable as they are, I maintain these moves are achieved at the expense of also glossing over any prospect of *theoretically* accounting for present or future *selection effects.* That is, we cannot both maintain that the world-as-scientifically-observed establishes knowledge (i.e. is real) and admit that any one of our theories may be wrong in the face of future discoveries *accounting* for present selection effects because we have no means of determining what is *real* at any particular time.

Initially in agreement with Zeilinger, Bohm (1980) maintains, we risk rendering our explanandum meaningless if the observing apparatus is divided from what is observed "because the content of the observed fact cannot coherently be regarded as separate from modes of observation and instrumentation and modes of theoretical understanding" (124). Following Bohm, Pylkkänen (2007) notes, "we do not check a preexisting property of the observed system without influencing it" (196). As science progresses, Bohm holds a "new whole" is comprehended with a new range of relevant aspects that will itself doubtlessly "be revealed as an aspect in yet another new whole" (1980, 137). Consequently, he finds, "the observing apparatus and the observed system […] each participates in the other to such an extent that it is not possible to attribute the observed result of their interaction unambiguously to the observed system alone" (1990, 275).

Just two years after Harris published his *Cosmos and Anthropos*, Bohm and Hiley wrote the following passage without ever mentioning AR, but nevertheless provided some indication of what their response to the issue would be:

> In a certain sense we could say that the overall quantum world measures and observes *itself*. For the classical "sub-world" that contains the apparatus is inseparably contained within the subtle quantum world, especially through those nonlocal interactions that bring about the classical behaviour. In no sense is the "observing instrument" really separate from what is observed. The relative autonomy of the classical level [...] is then what makes it possible for the total quantum world to manifest and reveal itself within itself in a measurement. Thus in contrast to the classical notion of measurement we should regard a quantum measurement as a manifesting process. (Bohm and Hiley 1993, 179)

While Harris agrees that reality exists prior to observation, with Bohm he further contends that the very conception of "particle" is (dialectically) inseparable from an experimental arrangement that includes observer *participation*. As Bohm's analysis reveals however, we are continually accounting for and thus moving beyond any given observation-manifestation enactment. Additionally, Bohm, like Harris, maintains that we will never achieve an exhaustive account of how respective means of participation *explicates* the world, that is, *thereby transcending our selection effects*.

Perhaps the closest orthodox account to Harris's position has come from Nesteruk (2013), who maintains the PAP may be maintained on "*phenomenological*" grounds. He explains, the physics discoveries of the twentieth century led Wheeler towards "a phenomenological stance" that situated the human observer at "the centre of disclosure and manifestation of the world" (2). He goes on to say that Wheeler

> attempted to approach physical reality not as something "out there," which is passively described by observers, but to see it as a genesis through conscious dialogue between observers-participants and physical reality, so that the universe emerges as a special articulation of the relationship between human intelligence and physical reality. (Ibid.)

Reminiscent of Dewey's conception of the "situation", Nesteruk contends that in the PAP, meaning emerges through a "dialectic relationship between man and the world" and that it is a mistake to say that meaning of one "first" comes from the other (6). "Embodiment", he claims, must be the starting point of any discussion on corporeity, space, time, and so on. This is "living being in relation to other beings and to the world, in whom this relation is announced and articulated by the way of its sense reaction and its comportment, or its action in situation" (11). Phenomenologically speaking, knowledge about objects in the world arises in and through relations to this very embodiment.

Though she never mentions AR, Barad's (2007) *agential realist* account appears in agreement with Nesteruk's phenomenological PAP. Barad's point is not merely that knowledge practices have material consequences, but that "practices of knowing are specific material engagements that participate in (re)configuring the world. Which practices we enact matter—in both senses of the word" (91). The construction of knowledge is more than establishing facts, it is about "making specific worldly configurations—not in the sense of making them up *ex nihilo*, or out of language, beliefs, or ideas, but in the sense of materially engaging as part of the world in giving it specific material form" (ibid.). Knowledge then is generated as part of the world, meaning it is subjective, but does not presume "the preexisting distinction between object and subject that feeds representationalist thinking). At the same time, objectivity cannot be about producing undistorted representations from afar; rather, objectivity is about being accountable to the specific materializations of which we are a part" (ibid.). As a result, "matter" is reconsidered as a "process of materialization" that is neither a social construction, nor a term corresponding with something outside of the subject (210).

The above sympathetic reading of Harris's PAP reveals that his (1) claim, that the observer is inseparable from the very nature of elementary particles has received sufficient support on *epistemic* grounds. This is to say we are restricted from positing *objective* elements external to a whole system partially constituted by observer participation because without which, the supposed information in such a system would no longer be grounded upon *meaningful relations*. This PAP line of reasoning reiterates the findings of Sect. 3.4.1 concerning observer's inability to transcend their selection effects and conceive of an independent world but adds an *ontological* feedback loop from the ESE to PSE. This loop runs from the constrained information we receive from the world (ESE), to our unavoidable *participation* in physically partitioning the world to make this information possible (PSE). Crucially, such partitioning occurs in and through our embodied being as a prerequisite to our experimental choices. I maintain that the nuanced phenomenological and agential realist contentions fill out Harris's notion of a "dialectical relation" between observer and world, while (given Chap. 2) Bohm's ontological commitments remain preferred.

The resulting PAP does not support the subjective idealistic conclusion that *only* what obtains in and as mentality is real, and neither does it maintain with Kant that the *world-in-itself* is unknowable. Rather, this principle maintains that the world we discover is grounded upon and *inseparable* from embodied and *participatory relations*—not representations or mere constructions, but biological, social, and technological enactments of the world. Bearing in mind the discussion of Sect. 2.5, this interpretation once again establishes an isomorphism between Harris's appeal to Bohm and AE conceptions of *reflexive meaning-making*, dynamic co-emergence, and now what Varela et al. have called the *groundlessness of experience* (1991, ff. 217).

Consequently, for the dialectical holist, counterfactual alteration of the relevant constants (X) *as they are presently individuated* becomes *inconceivable* insofar as observations of the resulting uninhabited universes would not be possible *even in principle*. This implies that in agreement with Harris, on the present phenomenological and Bohmian account, multiverse theories are thus far irrelevant in the face of the dialectical PAP. Nevertheless, Harris has not yet found contemporary support for his (2) claim, concerning how (if at all) *observation* itself alters the Universe; that is, there is currently no justification for granting observation such ontological status (MSE). To this issue I return in Sect. 8.3. Remaining with the task of elucidating Harris's AR, I next consider prospects for a *post-humanist future*.

4.3 The Final Anthropic Principle

Whether appealing to our as yet unlimited technological potential (Hoefflinger 2012), our intellectual development, or human-computer interfacing (Graimann et al. 2010) the *final anthropic principle* (FAP) claims: "Intelligent information-processing must come into existence in the Universe, and, once it comes into existence, it will never die out" (Barrow and Tipler, 23). This is to say that the lineage following present observers, or their technology will evolve to such a degree that observation and information processing are maintained indefinitely. In this section I clarify Harris's appeal to the FAP only to set it aside for reasons that will become clear anon.

Typically, those who invoke the FAP rely upon a *singularity* (Vernor 1993) of some kind that will presumably permit humanity to transcend our biological (i.e. mortal) constraints (More and Vita-More 2013, 361). Such views are maintained most rigorously by authors such as Shanahan (2015) and Kurzweil (2013), who consider the different forms that such a singularity might take as well as the moral, socio-political, and technological complications they entail. Some of the most popular means by which we are suspected to reach the singularity are via the potential development(s) of *artificial intelligence* (AI), *artificial life* (AL), and the *transhumanist* dream of transferring a human mind into software and thus achieving *immortality.* Of all the anthropic principles, the FAP has certainly received the most scientific attention, perhaps because it has the most immediate and tangible implications for contemporary research: the transcendence of nature and immortality are apparently desirable. Thus, we find the FAP appealing to our sense of anthropocentrism as well as our fear of death, in postulating that irrespective of the evolution of the universe—likely heading towards *heat death*—our collective ingenuity will equip us with the means to persevere indefinitely.

For proponents of this stance, appealing to the FAP depends upon a fundamental assumption that *information processing* is a necessary function of the Universe (i.e. it must come into being and proliferate). To defend this particularly strong move, some consider the universe as a kind of computer (Zenil 2012). Without regard for the FAP, Lloyd (2013) lends support to this move. He argues that as long as the laws of a given universe support *universal quantum computation* (PSE) all that is needed is access to *free energy* and a range of increasingly complex, information processing systems will be generated. In a similar vein, Carr maintains that careful study of the Big Bang theory can support a cosmic history of increasing organization in which "Heat death is avoided because local pockets of order can be purchased at the expense of a global increase in entropy, and, if the Universe continues to expand forever, intelligent beings may be able to delay their disintegration indefinitely" (2007, 86).

Susskind (2007) explains another well-known means by which heat death can be overcome is that even after thermal equilibrium, over enough time (T_r = expS t_p), *Poincare recurrence* shows that in any closed system large fluctuations will occur such that "the phase point will return over and over to the neighbourhood of any point in phase space, including the original starting point" (254). Faced with this scenario of having no way of favouring one vacuum over another, he claims anthropic reasoning combined with String theory "provides a framework in which this can be studied in a rigorous way" (262).

Futurists like Kurzweil (2013) take the exponential increase of information processing as given and predict that within a few generations, humans will have fused with technology to transcend our biological constraints. Tipler (1996), on the other hand, takes a pseudo-religious approach in appealing to an *Omega Point*. He hypothesizes that beings of the far future could have infinite computing power acquired through the final stages of a Big Crunch. Consequently, he maintains these beings would have "unlimited subjective time" and the ability to (re)create conscious beings of their own.

Initially, Harris expresses caution against enthusiasm for AI manifesting in "von Neumann-probes", or self-replicating biotechnology that could live on indefinitely. Such a claim, he says, "seems rather speculatively ambitious and smacks of over-confidence" (1991, 8). Harris considers it a real possibility that due to our "greed, short-sightedness, and imprudence" technological achievements may bring about the extinction of humanity far sooner than its "transcendence". He considered it an unjustified equivocation to associate "the manipulation of 'bits' of information by mechanical means" with "self-reflective reason" of "rational thinking" (156). He claims that there is no evidence that AI can achieve the latter even if it one day writes its own programs, corrects errors, and reproduces. For Harris, unless such technology achieves self-reflectivity, (conceiving oneself as part of the universe) it could not be considered a success of the FAP.

Numerous recent authors have raised similar criticisms. Proudfoot and Copeland (2012a, 2012b) argue that futurists such as Kurzweil tacitly assume that the mind is *Turing computable*. This is the pivotal step that would permit the kind of psycho-technological interface required to transcend our biophysical constraints. Unfortunately, they point out, there is currently no evidence that downloading a mind is or will ever be possible. Similarly, Bostrom (2002) holds that if reframed as a hypothesis about the future of intelligent observers, there may be some value to the postulate, but as a principle he agrees with Gardner's (1986) criticism that it is pure speculation and more accurately dubbed the "*completely ridiculous anthropic principle*" (CRAP).

In defence of the FAP however, Harris maintained that the generation of complexity (or scale of forms) could not continue forever, but that it would inevitably reach a final culmination. In at least partial agreement with Tipler, Harris advanced his own view of the *Omega Point* (1988) and later fleshed out what he thought were the theological and moral implications of this end-directed explicative process (1992). These works however were not based upon any speculations about the future abilities of consciousness or technology but came more elegantly from his a priori conception of nature-as-a-whole. Barring AI, Harris reasons that under certain conditions the FAP would be inescapable:

> If the universe as a systematic whole can only achieve completion in reflective consciousness, so that such self-awareness is, as it were, built into its implicate order, the emergence of intelligent life must be intrinsic to its very nature. In that case, the extinction of humanity in this planetary system will simply stimulate the development of intelligent creatures in some other, as the plucking of dead heads from a rose bush encourages the blossoming of new roses. (1991, 158)

Although Harris is appealing to metaphor, this passage does further illustrate the kind of relationship between consciousness and the Universe that he has in mind. Harris is at this point in line with some of the more extreme futurists in so far as he considers self-reflective awareness to be an irremovable function of the universe. Unfortunately, however, Harris's defence of the FAP relies upon the collapse of the wave function brought about by an observer, a move that has already been shown to be incompatible with dialectical holism. Moreover, since he makes no further use of "information processing", something else would be needed to validate a dialectical holist appeal to the FAP.

Putting the above speculations aside, despite Harris's stern and repeated assertion that there must be a culmination to $\mathfrak{E}$, recent observations paint a much more desolate view of the universe's evolutionary trajectory:

> At t≈1 second, the SM describes the universe as filled with matter and radiation, where the latter initially has much higher energy density. Because the energy density of radiation dilutes more rapidly than that of matter, the initial radiation-dominated phase is followed by a matter-dominated phase that extends until the present. Current observations indicate the presence of "dark energy" [...] with properties like a Λ term. Supposing these are correct, in the future the universe will eventually transition to a dark-energy-dominated phase of exponential expansion, given that the energy density of a Λ does not dilute at all with expansion. (Smeenk 2013, 612)

Although this principle was important to Harris, when we consider his faulty reasoning, I maintain that beyond this point it adds nothing to an understanding of his system that cannot be gained by an analysis of teleology. Hence, I set the FAP aside and move on to an analysis of the latter topic that will take centre stage in the discussions to follow.

4.4 The Teleological Anthropic Principle

The *teleological anthropic principle* (TAP) states that "[t]here exists one possible Universe 'designed' with the goal of generating and sustaining 'observers'" (Barrow and Tipler, 22). Following Aristotle (1936), the telos or "final cause" is that "for the sake of which a thing is done" (*Physics* 2.3). This means that an end goal directs a series of events as their superordinate purpose. Here, the means is not necessarily important, only the result or end product. Though the Aristotelian view of nature held sway for nearly two thousand years, with the rise of Newtonian physics in the early eighteenth century, nature was reduced to a chain of efficient causation with no room for *purpose*. Today, the TAP is either immediately dismissed for its *unwarranted appeal to "design"* or invoked to support such design. Inherent in Harris's conception of the dialectical whole is a specific brand of teleology that does not require Godly intervention, and if adequately understood, provides complete insight into the structure of his metaphysics.

From those who address the TAP, a number of objections have recently arisen:

(i) Teleology is unscientific on the grounds that it posits some future event that "causes" a present or past event.
(ii) Teleology depends upon an external force that guides material processes to a desired end.
(iii) Positing purpose to inorganic and non-conscious processes is anthropocentric and ignores firmly established explanations of efficient causation in modern science.

If Harris (or a subsequent dialectical holist) is to justify teleology, adequate responses to the above concerns must be provided. Invoking a brand of teleology derived from Kant and Hegel, Harris clearly dismisses the first two concerns: "Explanation would be 'teleological' if it made the parts of a whole intelligible in terms of the organizing principle that constituted them a totality, or processes understandable in terms of the dynamic system to which they belonged" (1965, 262).

Denying (i), he maintains that in his teleology, the principles that direct the activity "are not to be identified wholly with the final event in any series of effects" (ibid.). With regard for (ii), Harris disputes super-natural determination by holding, "[i]t is the immanence of the principle of order in the parts of a structured whole that constitutes its teleology" (1991,

168). Importantly, when Harris says the parts are explained "in terms of the whole", he means a "teleological process is one in which the whole being generated determines the stages by which it comes to maturity. It is a process directed by the organizing principle of the whole" (1991, 27). Possible forms, relations, and so on are considered implicit throughout the evolutionary process, which renders the whole "inadequate" at each stage until its *possible* forms become actual. Like the completion of a painting, the development of a seed into a tree, an embryo into an adult, or the enactment of a musical score, for Harris the aim is not the final state, "but the symphonic whole" (1991, 168).

What these metaphors purportedly exemplify is that "the activity is the goal in the making, and the goal is what the activity is all the time generating" (2000, 135). He later adds that if "a process is teleological in the sense described, no explanation will be complete that omits the principle regulating its variations and its course as a whole, or that overlooks the structural totality it is generating" (2000, 138). Harris's TAP therefore maintains that some set of later (or lower-level) laws and their corresponding material systems (e.g. life) are determined by the Concrete Universal.

It is important to notice that extending Bohm's ontology into cosmology almost completely mirrors Harris's TAP. Bohm holds that a "super-implicate order" unfolds into "an infinite number of levels" that "objectively and self-actively differentiates and organizes itself into independent sub-wholes, while determining how these are interrelated to make up the whole" (Bohm 1987, 46). Bohm (1980) grants that "life" (246) and the "individual human being" (264) more specifically, are such sub-wholes enfolded in the totality. Harris goes further however, claiming that the Universe "must tend to explicate itself in and as an intelligent experience" (1991, 28). In this way Harris's teleological explanation is invoked not in the sense of causation by a final state, but rather, by progressing through a scale there is a "satisfaction" in having made the process complete. As Harris's response to (iii) is the most complex however, consideration of this concern will have to wait until Chap. 5. I next introduce Tegmark's (2014) *mathematical universe hypothesis* (MUH), to bring to the fore a number of issues latent in Harris's TAP.

4.4.1 Level IV Multiverse

With Level IV multiverse we must confront one of the oldest debates in philosophy, between the Aristotelian and Platonic interpretations of reality. According to the Aristotelian (nominalist), mathematical formalisms are merely abstract approximations that do not obtain in reality. On the other hand, the Platonist (realist) views formal systems such as hypothetical laws, functions, sets, spaces, and operators, as real. Whereas Tegmark has defended the attempted realist position, Harris has attempted to transcend the debate. Arguing from the mathematical notion of *equivalence* Tegmark proposes the MUH:

> if some mathematical equations completely describe both our external physical reality and a mathematical structure, then our external physical reality and the mathematical structure are one and the same, and then the Mathematical Universe Hypothesis is true: our external physical reality is a mathematical structure. (2014, 217)

In addition to endorsing this hypothesis, Tegmark maintains that every sufficiently coherent mathematical system is identical to a universe. The resulting set of universes is Level IV multiverse.

Importantly, by appealing to Level IV multiverse, Tegmark maintains there is no distinction between mathematical (MSE) and physical (PSE) structures. Level IV multiverse subsumes the previous multiverses (Levels I-III), since it entails that everything exists as mathematical structures. By implication, if our statements are mathematically consistent, we are describing structures of reality; the question is whether we are describing our world or another because across Level IV universes, even the fundamental laws may be different. Level IV multiverses contain alternative mathematical structures, what is *possible* is no longer limited to what we *presently* deem logically consistent. In this way, Level IV multiverse subsumes and goes considerably beyond David Lewis's (1986) modal realism, the main difference being that Lewis's worlds are all material, whereas Tegmark's need not be.

Here, Ellis (2007a) poses a valuable criticism that in discussing the Universe, any mention of probability becomes deeply problematic (1218). He claims that in order to discuss a multiverse we first need to define the *possibility space*, which "means making some kind of assumptions about physics and geometry that will then apply across the whole family of models considered possible in the multiverse, and excluding all other possibilities" (1260). He contends, we would need "a distribution function *f*(m) of actually realized universes" (1261, emphasis in original) and it is currently unclear how or if we will ever have access to such information.

Consequently, he considers the multiverse an untestable philosophical proposal and not a scientific theory:

> Direct observations cannot prove or disprove that a multiverse exists, for the necessary causal relations allowing observation or testing of their existence are absent. Their existence cannot be predicted from known physics, because the supposed causal or pre-causal processes are either unproven or indeed untestable. However, some self-consistency conditions for specific multiverse models can be tested. (1263, emphasis omitted)

While Ellis's contention would be welcomed by Harris and appears well warranted with regard for Levels II and III, at Level IV this criticism loses traction. One powerful but counterintuitive argument in support of Level IV also replies to Harris's requirement of parsimony. Tegmark contends an entire ensemble of universes is simpler than any one of its members:

> The algorithmic information content in a number is, roughly speaking, the length of the shortest computer program that will produce that number as output [...] the set can be generated by a trivial computer program, whereas a single number can be hugely long. Therefore the whole set is actually simpler. (2007, 123)

He goes on to say that the "lesson is that complexity increases when we restrict our attention to one particular element in an ensemble, thereby losing the symmetry and simplicity that were inherent in the totality of all the elements taken together" (ibid.). Tegmark maintains that Level I removes our need to specify initial conditions, Level II need not specify physical constants, and Level IV requires no specification at all; that is, there is no difference between mathematical possibility and true existence.

Davies (2008) cautions against the typical method of positing a multiverse that contains some biofriendly subspace, within which our constants obtain. He reasons that while a future theory might render all parameters "fixed", so they have no freedom to vary independently,

> [A] final theory that contained no free parameters would be a curve in the M-dimensional space that would pass through the biofriendly volume. This curve would define a particular multiverse model, being the ensemble of universes corresponding to the points on the curve. A different curve would define a different multiverse model characterized by a different final theory. (110–11)

Requiring an explanation for why *our* curve takes on biofriendly values it does, Davies claims we are liable to posit a series of such curves, thus shifting "the problem up one level, changing the question from Why this universe? to Why this multiverse?" (111). Interestingly, the MUH is immune to this problem because there should exist some mathematical law that simultaneously specifies the spaces of mathematical and physical possibility.

Level IV multiverse hereby provides a means of bypassing Ellis's concern that multiverse theories are philosophy rather than physics since mathematical possibility is taken to be reality. Moreover, Tegmark avoids what I have previously dubbed the MPF—concerning the unjustified conflation of metaphysical and physical structures—by maintaining that our task is to identify the maximally stable set of mathematical structures obtaining within our universe, that is, our "*cosmic postal address*" (2014, 253). Although this would clearly hold an important role for the WAP as outlined above, Tegmark's response to the TAP requires further elucidation.

4.4.2 Teleological Anthropic Reasoning

What has been key (albeit implicit) throughout the present and previous chapters is the need to posit a synthetic form beyond what is observable that accounts for the regularity of (and our relation to) all observable phenomena. In discussing the balance between necessary and excessive metaphysical postulation, Bangu's (2013) advocates for Wigner's *superprinciple* (or "principles of invariance"): "That is, just as the laws constrain what events might take place in the world, such a superprinciple would constrain or determine the laws—in other words, it would have the role of a meta-law which would somehow explain the laws" (311). In his own words, Nobel laureate, Eugene Wigner asserted:

> it is hard, if not impossible, to believe that the laws of nature should have such complexity as implied by four or five different types of interactions between which no connection, no analogy, can be discovered.
>
> It is natural, therefore, to ask for a superprinciple which is in a similar relation to the laws of nature as these are to the events. The laws of nature permit us to foresee events on the basis of the knowledge of other events; the principles of invariance should permit us to establish new correlations between events, on the basis of the knowledge of established correlations between events. (1963, 10)

It is from Wigner's proposal that contemporary discussions of meta-laws have developed. Harris (1965) appears to have independently posited its analogue in the form of his Concrete Universal, the philosophical implications of which guide his entire metaphysics. Harris holds that however conceived, the overarching form of Nature and its component worlds "can be imagined only on the basis of what we already know. If they can be imagined at all, and if they can be legitimately postulated, the Anthropic Principle is relevant to them in the same way as it is relevant to the region that we inhabit and the epoch in which we live" (1991, 11). As a result, he claims this makes "any such Anthropic Principle applicable to our world automatically applicable to the Universe of which it is inevitably a part" (ibid.). At first glance this means that just as we expect to observe our world from a life-habitable space-time location, so too we should expect to find ourselves in a life-habitable Universe. More importantly however, Harris is not maintaining that if there are many, all universes will necessarily give rise to life. Indeed, citing *prevalence of life* (whether high or low) is a *red herring* concerning Harris's TAP. Harris means to argue that even if we admit such a multiverse mostly devoid of intelligent observers, what matters is identifying the unifying principles therein, those that have been necessitated by the presumed Concrete Universal.

As Ellis (2007b) points out, in discussing the multiverse we make a set of assumptions, the most important of which is the (either explicit or implicit) posit of a meta-law that provides a space of possibilities within which our observed universe obtains: "The very description of the space *M* of possibilities is based on an assumed set of [...] meta-laws that determine the laws of physics" (390). In another work Ellis (2007a) adds however that the concept of a unique law of physics that applies only to one object is "questionable" (1217). This is because unlike effective laws, this law cannot be deduced from anything else, meaning the Universe is not a *class of entity* that belongs to a larger super-set. So, Ellis concludes, "we cannot establish higher-level effective laws that apply to all universes and determine their structure, as we can at all other levels of the hierarchy of complexes" (2007a, 1217). Nevertheless, Ellis argues some such super-principle or meta-law is required to establish "regularities of properties across the class of universes", provide a common "generating mechanism", and causal connection (even if it functions across distributed space-times), otherwise we could not even describe the Universe, let alone calculate a distribution function (2007b, 398).

Tegmark (2014) recognizes that while the different mathematical structures in Level IV multiverse are not connected physically, "at the meta-level" they bear a range of important relations. "For example [...] one can be the combination of others [...] one structure can in a sense describe another: the elements in the first structure can correspond to the relations in the second, and relations in the first can describe what happens when you combine relations in the second" (254). Conveniently, "initial conditions" are rejected because a ToE positing that everything just "started out" in an unspecified state is considered an incomplete description (262). Responding to Ellis's worry regarding our requirement of a distribution function, according to the MUH, "physical reality is a mathematical structure that is completely specified" by its place within what Tegmark calls a "master list" of all possible (computable) structures (ibid.). This implies the Universe could "be simulated by quite a short computer program", and if our world is such a computable structure, then "these computations don't evolve our Universe, but describe it by evaluating its relations" (269). Evidently this atemporal conception of mathematical realism conflicts with Harris's process ontology, but the issue of teleology remains to be seen.

Towards this end, it is instructive to briefly consider Lange's (2009) eloquent metaphor for discussing laws of nature. The laws, he maintains, fail to supervene on the *Humean mosaic* (local matters of fact), just as

> the rules of a game fail to supervene on the moves that are actually made on one occasion (or every occasion) when the game is played. If neither player ever castles in the course of a game like chess, for instance, then the moves they make leave undetermined whether castling was permitted. (52)

Lange maintains that laws can be externally related such that some may never be instantiated, while others could have been counterfactually absent without influencing the Universe.[8] The dialectical holist would immediately take issue with both claims of unrealized possibilities on the one hand and counter-factuality of laws on the other. It is important to realize that the rules of a chess game supervene upon the superordinate social system to which they belong. It is possible therefore to claim that the rook and king *could have* castled after what might have been the only game ever played *only* because we can define such rules through media *beyond* the bounds of the game; for example, we can write down *possible* moves.

In the case of the Concrete Universal, there is nothing "external" to which counterfactual appeals can be made, thus dialectical holism rejects realism about subjunctive (abstract) possibilities. For Harris, it is nonsense to consider the counterfactual alteration of laws because laws are dialectically related to one another and to the material evolution of the Universe. In addressing teleology and counterfactuals Harris argues,

> there are no causal chains but only causal networks and proliferations [...] events are linked not in successive pairs but in interlacing complexes. In every case the determining factor is the structure of the system, and necessity arises from the impossibility of altering a single nexus without subverting the entire system. Not two of the distinguishable relata depend upon themselves alone for their relational web. Within this web, and given the law of its construction, it is never possible that any nexus could be other than it is and the rest remain the same. (1965, 475)

Here, Harris assumes (for now) that the Universe is the total space of dialectical relations among its unifying principles or "nexes" (chess pieces *and* their moves), each dictated by and expressions of the Concrete Universal.

Tegmark evidently has no problem maintaining that unrealized *moves* are possible, because for him, there is no difference between possibility and reality—as mathematical structures, *everything that is possible just is.* This however does not in itself respond to Harris's TAP. Getting to the heart of the matter, one of the most interesting implications of Tegmark's MUH is an implicit appeal to dialectical relations, for what amounts to the same reasoning as Harris provides (see Sect. 2.3):

> The MUH solves the infamous infinite regress problem where the properties of nature can only be explained from the properties of its parts, which require further explanation, *ad infinitum*: the properties of nature stem not from properties of its ultimate building blocks (which have no properties at all), but from the relations between these building blocks. (2014, 210)

Tegmark further claims that according to his MUH, "[r]egardless of whether anything seems random to an observer, it must ultimately be an illusion, not existing at the fundamental level, because there's nothing random about a mathematical structure" (2014, 262). If this is to say that such structures are necessitated given the *form* of the system within which they obtain, then Tegmark's MUH may inevitably support Harris's notion of teleology.

In sum, four important points should be remembered: for Harris's TAP, it is irrelevant whether (i) intelligent observers are *described* as a natural kind, mathematical structure, and so on; or if (ii) any, all, or none of the respective multiverses obtain; (iii) what matters is whether observers may be characterized as unifying principles and thereby bear a dialectical relationship to the (heretofore *presumed*) Concrete Universal. What remains to be examined concerning Tegmark's system is whether he does indeed posit consciousness as one such "structure" in a manner similar to Harris's "unifying principles"—an issue to which I return in Chap. 8. (iv) Nevertheless, dialectical holism takes issue with Tegmark's claim that it is possible to establish our *cosmic postal address*, because the former camp cannot endorse the metaphysical assumption that mathematics may be *identified* with a universe or any *part* of it.

4.5 Conclusion

The above overview of Harris's appeal to anthropic reasoning have highlighted a variety of issues with import for the philosophies of science and mind. In spite of his apparent disregard for the multiverse, Sect. 4.2 reveals that Harris's conception of the Concrete Universal is consistent with at least Level II multiverse theory, in that each bubble of chaotic inflation may be governed by their own principle of order. While Levels I and II multiverses might be unavoidable if chaotic inflation is an accurate theory, these levels nonetheless imply our reliance upon the WAP. Following Harris, as our ESE confounds our ability to make large-scale claims regarding the nature of our universe, unverifiable projections of theoretical structure (MSE) onto nature-as-a-whole become necessary on both scientific and philosophical grounds. Working from this foundation however entails a great limitation on how we individuate the constants of nature, meaning their counterfactual multiplication and alteration on a theoretical basis cannot be justified on a phenomenological basis—lest we commit the MPF.

Although Harris's reasoning concerning the *orthodox* PAP is demonstrably at odds with the rest of his system, in Sect. 4.3 I have attempted to show that his reasoning supports a weaker *phenomenological* version of the PAP. I have shown that Level III multiverse is discarded on Harris's account and indeed the majority of his system supports Bohmian ontology. Accordingly, the observers' inability to exhaustively account for their selection effect(s) makes it impossible to coherently conceive of the

Universe independent of our *biased* relation to it ($\exists^R$). Positing a universe in which this relation is void undermines the coherence of at least AE epistemology, that is, such a universe is *meaningless*. The resulting *reflexivity* of mind and Universe thus justifies the PAP on epistemic grounds. The question remains open however as to whether this reflexivity can be ontologically interpreted to provide a full TAP.

The above analysis has shown that none of the multiverse theories pose a significant challenge to Harris's AR: regardless of whether a particular multiverse obtains, it is irrelevant in the face of his TAP. In this vein, it no longer matters whether we describe the essence of the world to be mathematical or material, what matters is rather what unifying principles can be identified. For Harris, the Concrete Universal is the full space of forms that are instantiated and (as will be discussed below) he argues consciousness is one such form. Concerning Level IV multiverse, I have shown that despite his Platonic sympathies, Tegmark's implicit appeal to dialectical relations can be interpreted such that Harris's TAP may be inevitable.

Given the above, Harris's TAP may be the simplest and most fundamental (foundation for) what Barrow has called a cosmological *transfiguration*. As Harris aims to establish his theory in ontological terms however, his reasoning for how the development of unifying principles may be empirically observed and theoretically justified must be further assessed. The remainder of this text is devoted to this task. Towards this end, in Part III I consider a range of contemporary theories concerning cosmological and biological evolution that both support and further elucidate Harris's system.

Notes

1. Ellis (2007b) agrees, "the idea of a completely disconnected multiverse—with regular properties but no common causal mechanism—is not viable. Some pre-realization causal mechanism must necessarily determine the properties of the universe in the ensemble. What are claimed to be totally disjoint universes must in some sense be causally connected, albeit in some prephysical or meta-physical domain" (398).
2. The Hubble constant measures the expansion rate of the observable universe and its value is denoted H0. Today h0 ≈ 0. 70. The distance that light has travelled in 13.8 billion years (the oldest observable stars) composes a sphere whose radius defines our observable universe, or "Hubble volume" and denotes our observable horizon. "Above the Virgo Supercluster, we

find the visible Universe: it contains about 108 superclusters, which all together count something like 1011 galaxies, its total mass is 1023 M◉, and its radius is about 15 Gpc" (Bambi and Dolgov 2016, 15).

3. Weinberg concludes that the multiverse provides for cosmology something analogous to what natural selection provided for biology: "Just as Darwin and Wallace explained how the wonderful adaptations of living forms could arise without supernatural intervention, so the string landscape may explain how the constants of nature that we observe can take values suitable for life without being fine-tuned by a benevolent creator" (2007, 39).
4. Although Harris devoted Chap. 4 of *Cosmos and Anthropos* (1991) to discussing evidence for "fine-tuning", I have found this material to add nothing to the central vein of his anthropic reasoning and so have neglected its discussion in the present argument.
5. Heisenberg's *indeterminacy principle* states "If the position of a particle is known with an uncertainty Δx, its momentum can only be known with an uncertainty Δp_x such that the product of the two uncertainties is never smaller than $h/2$: $\Delta x \Delta p_x \geq \hbar/2$ where $\hbar = h/(2\pi)$. That is, it is physically impossible to measure or know simultaneously the exact position and exact momentum of a particle. The uncertainties Δx and Δp_x do not arise from imperfections in practical measuring instruments. Rather, these inescapable uncertainties arise from the quantum mechanical nature of matter" (Serway et al. 2013, 1089–90).
6. Though maintaining determinism, in Carter's (2007) revision of the Everettian view (called the micro-anthropic principle of quantum mechanics) he concludes that his view is in accordance with Ockham's principle and he claims (unspecified) *Eastern philosophies* within Hinduism and Buddhism (308). In this theory "the number of distinct perceptors which can most economically be postulated" is "just one". While he admits that this conclusion is devoid of scientific implications it carries a strong ethical impact in that "the injunction to 'love one's neighbor as oneself' acquires a new significance when one recognizes that the 'neighbor' may be another incarnation of 'oneself'" (318). As will be discussed in greater detail in Chap. 8, such a view is in striking accord with Harris's ontology of mind. Unfortunately, however the details of Carter's theory are too technical for a rigorous comparison with Harris's system within the present text.
7. Stapp argues that repeating process 1 intervention in rapid succession, for example, answering "yes" to raising one's arm, the mixed quantum state evolves into a discrete state. In this way, mental effort increases a sequence of "identical intentional acts", which then causes what he calls the "template for action" to be sustained. This template for action is said to produce the brain activity that permits a given intended feedback (36). He calls this the

quantum Zeno effect: the mechanism by which the mind is believed to have a downward causation to the brain and body (114).

8. For example, Lange argues, "the fundamental dynamical law [F=ma] would still have held, had the world been populated by different kinds of forces or different kinds of fundamental particles, or had the strengths of those forces or the characteristic properties of those particles been different" (Lange 2009, 39). For the dialectical holist this demonstrates a clear MPF as per the reasoning of Sect. 4.2.2.

Works Cited

Aristotle. *Physics*. William D. Ross, ed. Oxford, Clarendon Press, 1936.

Bambi, C., & Dolgov, A. D. (2016). *Introduction to Particle Cosmology: The Standard Model of Cosmology and its Open Problems*. Heidelberg Germany: Springer Pub.

Bangu, Sorin. "Symmetry." In Robert W. Batterman ed. *The Oxford Handbook of Philosophy of Physics*. 287–317. New York NY: Oxford University Press, 2013.

Barad, Karen. *Meeting the Universe Halfway: Quantum Physics and the Entanglement of Matter and Meaning*. Duke University Press, 2007.

Barrow, John D., and Frank J. Tipler. *The Anthropic Cosmological Principle*. Oxford University Press, 1986.

Barrow, John D. "Cosmology and Immutability." In John D. Barrow, Paul C. W. Davies., and Charles L. Harper Jr. eds. *Science and Ultimate Reality: Quantum Theory, Cosmology and Complexity*. 402–425. Cambridge University Press, 2004.

Barrow, John D. *The Book of Universes: Exploring the Limits of the Cosmos*. Bodley Head Inc, 2011.

Bohm, David. "The Enfolding-Unfolding Universe and Consciousness." In Lee Nichol ed. *The Essential David Bohm*. 78–120. New York, NY: Routledge, 2003, 1980.

Bohm, D. (1987). *Hidden Variables and the Implicate Order*. (pp. 35–48). In: Hiley, B. J., & Peat, D. D. (Eds). *Quantum Implications: Essays in Honour of David Bohm*. New York, NY: Routledge.

Bohm, David. "A New Theory of the Relationship of Mind and Matter." In *Philosophical Psychology*. 271–286. 3: 2, 1990.

Bohm, D., & Hiley B. J. (1993). *The Undivided Universe: an Ontological Interpretation of Quantum Theory*. New York NY: Routledge.

Bostrom, Nick. *Anthropic Bias: Observation Selection Effects in Science and Philosophy*. New York, NY: Routledge, 2002.

Carr, Bernard. "Introduction and Overview." In Bernard Carr, ed. *Universe or Multiverse?* New York: Cambridge University Press, 2007.

Carter, Brandon. "Large Number Coincidences and the Anthropic Principle in Cosmology." In *Confrontation of Cosmological Theories with Observational Data,* 291–298, 1974.

Carter, Brandon. "Micro-Anthropic Principle for Quantum Theory." In Bernard Carr ed. *Universe or Multiverse?* 285–319. New York: Cambridge University Press, 2007.

Davies, Paul C.W. "John Archibald Wheeler and the Clash of Ideas." In John D. Barrow, and Charles L. Harper, Jr. eds. *Science and Ultimate Reality: Quantum Theory, Cosmology and Complexity.* 3–24. New York, NY: Cambridge University Press, 2004.

Davies, Paul C.W. "How Biofriendly is the Universe?" In Simon Conway Morris., Stephen J. Freeland., Charles L. Harper Jr eds. *Fitness of the Cosmos For Life: Biochemistry and Fine-Tuning.* 169–196. New York: Cambridge University Press, 2008.

Dyson, Freeman J. "Thought-Experiments in Honor of John Archibald Wheeler." In John D. Barrow, and Charles L. Harper, Jr. eds. *Science and Ultimate Reality: Quantum Theory, Cosmology and Complexity.* 72–89. Cambridge University Press, 2004.

Ellis, George. "Issues in the Philosophy of Cosmology." In Jeremy Butterfield., and John Earman eds. *Handbook for the Philosophy of Physics.* 1183–1282. The Netherlands, Amsterdam: Elsevier Pub, 2007a.

Ellis, George. "Multiverse: Description, Uniqueness and Testing." In Bernard Carr, ed. *Universe or Multiverse?* 387–409. New York: Cambridge University Press, 2007b.

Everett, Hugh. *On the Foundations of Quantum Mechanics.* Ph.D. thesis, Princeton University, Department of Physics, 1957.

Gardner, Martin. "WAP, SAP, FAP & PAP." In New York Review of Books. 22–25. 33 (May 8), 1986

Graimann, Bernhard, Allison Brendan, Z., and Pfurtscheller Gert. eds. *Brain-Computer Interfaces: Revolutionizing Human-Computer Interaction.* Springer Pub, 2010.

Harris, Errol E. *The foundations of Metaphysics in Science.* London: Allen and Unwin Pub, 1965.

Harris, Errol E. *Cosmos and Anthropos: A philosophical Interpretation of the Anthropic Cosmological Principle.* Atlantic Highlands, NJ: Humanities Press International Inc, 1991.

Harris, Errol E. *Cosmos and Theos: Ethical and Theological Implications of the Anthropic Cosmological Principle.* Amherst New York: Humanity Books, an imprint of Prometheus Books, 1992.

Harris, Errol E. *The Restitution of Metaphysics.* Amherst New York: Humanity Books, an imprint of Prometheus Books, 2000.

Hawking, Steven. *W. Brief History of Time: From the Big Bang to Black Holes.* Toronto, New York, London: Bantam Books, 1988.

Hawking, Steven. "Cosmology from the top down." In Bernard Carr, ed. *Universe or Multiverse?* 91–98. New York: Cambridge University Press, 2007.
Hoefflinger, Bernd. Ed. *Chips 2020: A Guide to the Future of Nanoelectronics.* Springer Pub, 2012.
Hogan, Craig J. "Quarks, Electrons and Atoms in Closely Related Universes." In Bernard Carr ed. *Universe or Multiverse?* 221–230. New York: NY Cambridge University Press, 2007.
Kurzweil, Ray. "Progress and Relinquishment." In Max More., and Natasha Vita-More eds. *The Transhumanist Reader: Essays on the Science, Technology, and Philosophy of the Human Future.* 451–453. Wiley-Blackwell. John Wiley & Sons, Inc, 2013.
Lange, Marc. *Laws & Lawmakers: Science, Metaphysics, and the Laws of Nature.* New York: Oxford University Press, 2009.
Lewis, David. *On the Plurality of Worlds*, Cambridge, MA: Blackwell, 1986.
Linde, Andrei. "The Inflationary Multiverse." In Bernard Car ed. *Universe or Multiverse?* 127–149. New York: Cambridge University Press, 2007.
Lloyd, Seth. "On the Spontaneous Generation of Complexity in the Universe." In Charles H. Lineweaver., Paul C. W. Davies., and Michael Ruse, eds. *Complexity and the Arrow of Time.* 80–112. New York: Cambridge University Press, 2013.
More, Max, and Natasha Vita-More. eds. *The Transhumanist Reader: Essays on the Science, Technology, and Philosophy of the Human Future.* Wiley-Blackwell. John Wiley & Sons, Inc, 2013.
Mukhanov, Viatcheslav. "Cosmology and the Many Worlds Interpretation of Quantum Mechanics." In: Bernard Carr, ed. *Universe or Multiverse?* 267–272. New York: Cambridge University, 2007.
Nesteruk, Alexei. V. *A* "'Participatory Universe' of J. A. Wheeler as an Intentional Correlate of Embodied Subjects and an Example of Purposiveness in Physics." In Journal of Siberian Federal University. Humanities & Social Sciences. 415–437. vol. 6, n. 3, 2013.
Proudfoot, D., & Copeland, J. (2012a). *Our Posthumanist Future.* (pp. 73–78) In: *The Philosopher's Magazine.* Issue 57, 2nd Quarter.
Proudfoot, D., & Copeland, J. (2012b). *Aritificial Intelligence.* In: Margolis, E., Samuels. R., & Stich, S. P. *The Oxford Handbook of Philosophy of Cognitive Science.* New York, Oxford University Press.
Pylkkänen, Paavo T. I. *Mind, Matter and the Implicate Order.* (The Frontiers Collection). Heidelberg, Germany: Springer Pub, 2007.
Rees, Martin. "Cosmology and the Multiverse." In Bernard Carr, ed. *Universe or Multiverse?* 57–75. New York: Cambridge University Press, 2007.
Serway, Raymond A., Jewett, John W., Wilson, Kate, and Wilson, Anna. *Physics Vol. 2. Asia-Pacific Edition.* Melbourne: Cengage Learning Australia, 2013.
Shanahan, Murray. *The Technological Singularity.* MA: MIT Press, 2015.
Smeenk, Christopher. *Philosophy of Cosmology.* In Robert Batterman, ed. *The Oxford Handbook of Philosophy of Physics.* 607–652. New York NY: Oxford University Press, 2013.

Stapp, Henry. *Mindful Universe: Quantum Mechanics and the Participating Observer* (The Frontiers Collection). Heidelberg, Germany: Springer Pub, 2011.
Susskind, L. (2007). The Anthropic Landscape of String Theory. (pp. 247–266) In: Carr, B. (Ed). Universe or Multiverse? New York: Cambridge University Press.
Tegmark, Max. "The Multiverse Hierarchy." In Bernard Carr, ed. *Universe or Multiverse?* 99–125. New York: Cambridge University Press, 2007.
Tegmark, Max. *Our Mathematical Universe.* New York: NY. Knopf, Borzoi Books, 2014.
Tipler, Frank J. *The Physics of Immortality: Modern Cosmology, God and the Resurrection of the Dead.* Pan Books, 1996.
Vaidman, Lev. "Many-Worlds Interpretation of Quantum Mechanics." In Edward N. Zalta, ed. *The Stanford Encyclopedia of Philosophy*, 2014.
Vernor, Vinge. "The Coming Technological Singularity: How to Survive in the Post-Human Era." NASA Lewis Research Center and the Ohio Aerospace Institute, March 30–31, 1993.
Vilenkin, Alexander. "Anthropic Predictions: The Case of the Cosmological Constant." In Bernard Carr, ed. *Universe or Multiverse?* 163–179. New York: Cambridge University Press, 2007.
Wallace, David. "The Everett Interpretation." In Robert Batterman, ed. *The Oxford Handbook of Philosophy of Physics.* 460–488. New York NY: Oxford University Press, 2013.
Weinberg, Steven. (2007). *Living in the Multiverse.* (pp. 29–42). In: Bernard Carr. ed. *Universe or Multiverse?* New York: Cambridge University Press.
Wheeler, John A., Zurek, Wojciech H. "Quantum Theory and Measurement." Princeton, NJ. Princeton University Press, 1983.
Wigner, Eugene. "Events, Laws of Nature, and Invariance Principles." Nobel Lecture, December 12, 1963. Reprinted in *Science.* 995–999. 145(3636), 1963.
Zeilinger, Anton. "Why the Quantum? 'It' from 'bit'? A Participatory Universe? Three far reaching challenges from John Archibald Wheeler and their relation to experiment." In John D. Barrow, Paul C.W. Davies, and Charles L. Harper, Jr. eds. *Science and Ultimate Reality: Quantum Theory, Cosmology and Complexity.* 201–220. Cambridge University Press, 2004.
Zenil, Hector. ed. *A Computable Universe.* World Scientific Press, 2012.

PART III

Towards a Dialectical Holist Theory of Evolution

CHAPTER 5

Levels of Complexity Within the Cosmos

5.1 Cosmogenesis (𝕮)

In opposition to the classical paradigm that considers both knowledge and nature to be ultimately composed of externally related basic entities, Harris proposes a framework of process, dialectical relations, and a self-directed series of irreducible unifying principles. These principles characterize a metaphysics that he claims to *most* adequately account for self-organization in contemporary sciences. The primary questions to be considered in these sections are as follows:

(a) What current theories support Harris's conception of 𝕮?
(b) What are the necessary and sufficient conditions for something to be a scale in Harris's system?
(c) To what extent do Harris's depictions of 𝕮 and the scale of forms provide an adequate conceptual bridge from cosmological to biological evolution?

As discussed above, Harris begins his depiction of cosmic evolution by positing a single "field", or "matrix". He claims that as the field expands, it differentiates into a series of contrasting forces that further manifest as a range of waves and particles, energy, and matter (1991, 42). Harris claims that atomic physics "shows individuation to be secondary to system". In

J. Schofield, *A Phenomenological Revision of E. E. Harris's Dialectical Holism*,
https://doi.org/10.1007/978-3-030-65029-2_5

this case, what Harris means by "individuation" is that the differentiation of a system into respective forms is always the result of some pre-existing order. He goes on to argue that "if the principle of organization is prior to the discrimination of units, what produces order cannot possibly be the random shuffling of those units" (1965, 148-49). In a later work he elaborates this contention:

> The whole, with its principle of structure is, therefore, prior to the parts. It is the universal or pervasive influence of this structural principle that makes the distinguishable and diverse elements what they are, that determines their relations to one another, and adjusts one of them reciprocally to every other. It is the universal principle which the differentiations are manifestations or (as we shall shortly see more convincingly) exemplifications. (1987, 143)

By *reductio* he reasons, "a possible world without physical laws could be conceived [...] even if it were total chaos", but to do so would "strain the meaning of possibility intolerably; for [...] pure chaos is a self-destructive notion, chaos being always parasitic on order of some kind" (1991, 50). In earlier works Harris develops these ideas more fully, arguing that the more frequently order is reversed, the more random or chaotic the system: "as the frequency increases to infinity so the continuum approaches homogeneity" (1970, 325). He goes on to claim that complete homogeneity is impossible and order is thus "marked by its degree of departure from complete homogeneity" (ibid.).[1] Identity is by this reasoning based upon "recurrences" and "periodicities" within a "manifold" of relations that "serve as clues to the principle of order fundamental to the structure" (ibid.). Consequently, order involves change and difference but also requires "some element of sameness, similarity or recurrence" (326). Harris argues that if we take abstract homogeneity to obtain in nature and thereby consider it *real*, we become victims of what he calls "the fallacy of spurious homogeneity" (1965, 462). Ultimately then, Harris is identifying total chaos with homogeneity and considers it contradictory to attribute such a state to Nature because any conception of Nature must involve some principle of order.

Harris models cosmogenesis as a process by which the *prior* organization of the whole directs the evolution of its respective phases. By this reasoning, being and *non-being* are empty concepts when considered independently. "Being" must thereby be reconsidered as "a dynamic and not a static concept, and in its dynamic opposition and interchange with nonbeing it issues as Becoming" (1987, 174). By "Becoming" Harris means "subsumation" (1965, 283), "sublation" (1987, 180), or "*Aufheben*" (1993, ff. 50): a synthetic activity leading to a genesis of increasingly complex systems. This process is extended over cosmic history and guided by a "nisus" akin to "Spinoza's doctrine of the conatus *in suo esse perseverare*" (1965, 154):

> [T]he advance seems constantly to result from a kind of self-enfoldment of the simpler form: if a field of force (energy) is curvature in space, matter is a higher degree of such curvature—a superposition of curvatures [...]—and the atom is a convolution of fields—the bending in upon itself of the wave-packet which constituted the electron around the complication of nuclear fields. (Ibid.)

To Harris's mind, the series from energy to matter, from the simplest to the most complicated molecules, demonstrates a continuous differentiation and merging of opposites: "each develops from and incorporates its predecessor [...] energy (waves) opposed to matter (particles), which nevertheless overlap as wave packets, proton (positive) opposed to electron (negative), each excluding yet complementary to the other" (1988, 58-59). In this way, considering the numerous kinds of atom, molecules, and chemical compounds, he holds that each step of the series "is not only more complex but also more comprehensive in content than its predecessor" (1965, 154).

An example that Harris finds in support of his case is the versatility of carbon, second only to hydrogen, which forms an astonishing number of compounds. This is significant, Harris claims, because carbon is capable of storing more information than any other element and "the generation and reproduction of information is the essential characteristic of life" (1991, 56). For Harris, all events within the Universe are to be understood as "the infinitely various specific manifestations of the principles of order implicit in this whole—that is, of the laws of physics—and the temporal process is simply the self-differentiation of the whole" (1988, 58). These passages reiterate that for Harris, rather than existing eternally, laws emerge through a teleological process of cosmogenesis.

Although contemporary science had not yet identified this supposed *nisus*, Harris claims that recent discoveries depict "a gamut of evolving physical forms, that manifestly grow out of one another in increasing complexity in a manner which demands explanation in terms of some persistent tendency toward self-enfoldment, self-differentiation and self-elaboration" (1965, 155). In other words, Harris argues the genesis of diverse forms and processes within the Universe requires teleological explanation because each is the consequence of the whole preforming some function upon its constituents. At the time, Harris suspected the "proper science" to study the underlying "teleological mechanisms" that drives the explication of scales "would be cybernetics—the science of regulation or governance" (1965, 265).

By this reasoning, Harris attempts to establish a continuum, a common language from particle differentiation to social complexity. On this account, there ought to be identifiable dynamics underlying evolutionary processes at every spatiotemporal scale in virtue of their common source, namely the function of the Concrete Universal (ℭ). If Harris is correct, then cosmic evolution will demonstrate a propensity for self-organization and enfoldment of one element or system into the next, on a trajectory of increasing complexity towards life and mind. This however remains to be seen. In the following sections I consider responses to Harris's thesis from the contemporary philosophy of complex systems (Fig. 5.1).

5.1.1 *Metaphysical Implications of GST*

To what extent is Harris's posit of ℭ supported by contemporary accounts of cosmic evolution? A brief examination of the metaphysical underpinnings of AE may reveal further clarification of Harris's approach. Opposed to the classical view that finds low-level efficient causation sufficient to provide exhaustive explanations of all empirical phenomena, an increasing number of *general systems theorists* have maintained that such reductivism has serious limitations. As will be outlined in this chapter, proponents instead find it necessary to appeal to formal and final causation in the analysis of complex and non-linear systems. I am here interested to elucidate the sympathetic theoretical accounts, while the following Sect. 5.2.2, endeavours to identify empirical support in defence of Harris's TAP.

In establishing the ontological basis of AE, Thompson (2007) appeals to complex systems and contends that emergence through collective self-organization has two aspects: "One is local-to-global determination, as a result of which novel macro-level structures and processes emerge. The other is global-to-local determination whereby global structures and processes constrain local interactions" (61). Hence, AE is founded upon terms of "emergent process" rather than "emergent properties", because

Fig. 5.1 Aperiodic quasicrystal. To understand this account, it is helpful to set aside the well-known model of the universe exploding outwards and, instead, to imagine one system undergoing a kind of self-differentiation as an ever-widening range of structure-processes manifest upon its topology. In line with Harris's reference to Escher's plane division (1987, ff. 140) and his appeal to these same structures (Sect. 3.3), *quasicrystals* provide a useful means of visualizing ℭ. In *aperiodic* quasicrystals, identifiable 2 and 3D patterns (like the above) are projections (shadows) from a higher 6D structure (Steurer and Deloudi, 2009 ff. 61). For Harris (in line with Bohm), the totality of Nature's higher-dimensional process remains *hidden* from view. Penrose tiling recreated with permission from Jos Leys: http://www.josleys.com/show_image.php?galid=238&imageid=7197.

> it does not make sense to say that a property emerges, but only that it comes to be realized, instantiated, or exemplified in a process or entity that emerges in time. Emergence is a temporal process, but properties (whether considered as universal or as linguistic abstractions) are atemporal). (418)

Further, he claims emergent processes are part of networks whose "coupled elements have nonlinear interactions" (e.g. *attractors*) that are "nonadditive or nonproportional", producing "systems whose activities cannot be derived aggregately from the properties of their components"

(419). So eschewing abstracta, relational processes across global and local domains are awarded primacy.

The term *downward* causation is here considered a misnomer. As introduced in Sect. 2.3.1, rather than *lower-level entities* being affected by *higher*, there is what Thompson calls "*dynamic* co-emergence", in which "part and whole co-emerge and mutually specify each other" (2007, 431). In complex systems there is "reflexive global-to-local influence that happens in a system that has dynamic global coherence in and through collective self-organization" (434). Hence the emphasis is on self-organization as a global *function* that both emerges from and *constrains* the spatio*temporal* relations of its composition.

Examining the metaphysical implications of GST, Bishop (2011) clearly endorses a metaphysical view akin to Thompson's, arguing that the "analysis of determinism in complex systems is complicated by the fact that there are additional forms of causation arising in such systems that must be taken into account" (125). To this end he explains that "in complex systems the formation of control hierarchies often comes about when a new form of dynamics arises that exhibits downward constraint on system constituents and is self-sustaining" (127). Echoing Harris, Bishop further laments the tendency of analytic philosophers to completely ignore the formal in preference for efficient causation. Such ignorance he adds, misses many of the dynamical relations relevant to complex systems analysis. For example, concerning laws of nature, he argues we should not appeal to either efficient or formal causation as primary in our explanations: "Sound explanations of complex systems are likely to involve both appeals to causal mechanisms and ordering/constraining structure via the interrelationships among the lower-level and higher-level dynamics" (130). He maintains "[h]ierarchies and wholes in complex systems act to constrain or direct the possibilities made available by lower-level laws as opposed to somehow violating those laws" (ibid.).

Given the above, it seems reasonable to consider Silberstein's (2009) contention that AE relies upon some kind of *ontic structural realism* (OSR). Bringing this into focus, Wolkenhauer et al. (2011) have found that "[i]n systems theory objects and relations between objects have identical ontological status" (355). More specifically still, McKenzie's (2014) revision of OSR, has found that *fundamentality* in the philosophy of physics is granted neither to structures (symmetry relations) nor to their relata (kinds of physical entity): "If we conceptualize priority in terms of ontological dependence, then, rather than symmetries being more fundamental than particles the two seem to be on an ontological par" (1101). Certainly recommending her account to the dialectical holist, she concludes that this position should be agreeable to those whose "hunch is that physical symmetries cannot exist independently of physical objects—any more than that there can be laws of nature without anything to behave in accordance with them" (ibid.).

While Silberstein's embodied neutral monism (ENM) agrees that neither relata, nor their relations can exist without the other, both categories of "particle types" and of "symmetry relations" are too abstract to be granted ontological primacy. For ENM, materiality must be rephrased as a *processes of relation* because such a conception remains consistent with our embodied phenomenology, or how the world is disclosed. Importantly, this addition of process also preserves *causal efficacy* within ontology and thereby avoids the common criticism that proponents of OSR are just masquerading Platonists. Both Thompson and Harris, however, base their respective positions upon axioms of *dialectical relations* and process ontology. Hence, they would agree with Capra's contention that in systems theory neither structures nor relations correspond to independent units in the world, rather "every structure is seen as the manifestation of underlying processes" (2014, 81). Moreover, according to Thompson,

> there is no bottom level of basic particulars with intrinsic properties that upwardly determines everything else. Everything is process all the way "down" and all the way "up," and processes are irreducibly relational—they exist only in patterns, networks, organizations, configurations, or webs. (2007, 440)

Here Thompson is claiming that according to systems theory, nature is composed of co-emerging structures and irreducible relational processes with no fundamental level. Importantly, McKenzie's final point, implying a theory of laws that co-evolve with Nature, points to a thesis unrecognized by any enactivists, but one that already receives support from Harris and within the philosophy of complex systems.

Bickhard (2011) has provided one of the most explicit arguments for how complex systems theory relies upon *process* metaphysics. In agreement with Harris's contention that chaos presupposes order, Bickhard argues "process has whatever causal powers that it does in part in virtue of its organization [...] organization cannot be delegitimated as a potential locus of causality without eliminating causality from the universe" (95).

Consequently, Bickhard maintains that in physics fields are granted primacy over particles and so any "return to a substance or particle framework is precluded by the empirical confirmation of multiple non-localities, and dynamic space-time and vacuum effects. Such phenomena are not consistent with the local independence and fixedness of particles and substances" (97). By turning to a process metaphysic, he maintains, many of our traditional assumptions concerning boundaries and individuation are brought into question. Informed by the analysis of complex systems, positing boundaries requires explanations in "terms of their natures, origins, and forms of maintenance" (98). Extrapolating this view to the cosmos would therefore mean endorsing his contention that a "process metaphysic" is not only central to systems theory but "is arguably the only framework that offers a viable orientation for the scientific future" (102).

If this is right, a plausible way of vindicating Harris's TAP would be to argue that if such criteria of complex systems (e.g. Bishop, 112; Yin, et al. 2011, 404) can be extrapolated to the Universe, it would be reasonable to posit the kind of governance upon the evolution of its *constituent forms* that Harris had in mind.

The Universe as a Complex System

1. *Process/structural ontology*—process ontology in combination with OSR is a viable (if not preferable) foundation for understanding complex systems. Accordingly, relational processes constitute entities at all scales.
2. *Emergent constraints*—forms arise through self-organization that both restrict and enable future states, dynamics, structures, and relations.
3. *Historicity*—complex systems are *path-dependent* meaning small events can become locked-in to constrain future dynamics and emergent structures.
4. *Deterministic function*—the components of a complex system are interrelated or coupled such that if one of the constituents is absent, the function (or evolutionary trajectory) is incomplete.
5. *Hierarchical levels*—nested structures obtain within complex systems, each requiring different (irreducible) descriptions.
6. *Situated dynamics*—the constituent states and structures depend upon the environment within which they are embedded.

7. *Intricate and cycling* behaviour—complex systems obtain between simple order and total chaos, which instantiates periodic behaviour that does not exactly repeat.

C. If the Universe is a complex system, its function is identified with a space of determinant forms (states) that obtain throughout its evolution (ℭ).

This brief overview of the metaphysical commitments of systems theory has provided further illumination of the commonalities between enactivists and Harris. In this framework, the emphasis is on *formal-governance* and self-organization (or constraint and co-emergence). The upshot is that—in opposition to classical conceptions—complexity reveals that nature does not consist of basic particulars, but interdependent relational processes obtaining across irreducible levels. By generalizing the above philosophical claims, the cosmos as a whole may be regarded as a self-organizing system such that each form that develops is determined by and is a reflection of the whole. Yet this would not conflict with the efficient causation of local interactions.[2] I maintain that this "constraint" on efficient causation may be identified with what Harris calls "nisus" and would entail the evolution of laws of nature. What remains to be seen however, is support from cosmologists and physicists that the evolution of the universe can and should be understood as such a self-determining complex system.

5.1.2 Empirical Evidence for ℭ: Towards an Enactivist Cosmology

According to Harris's metaphysics it should be possible to identify common principles guiding the evolution of complexity in both living and non-living systems. In this section I consider two theories that may serve this end: Kurakin's (2011) *self-organizing fractal theory* (SOFT) and Brender's (2013) appeal to symmetry breaking. I show: (i) these works provide empirical support for Harris's original conception of cosmic evolution; and (ii) these works reveal a number of cosmological consequences implied by *AE* metaphysics as outlined in the previous section.

Kurakin begins by considering an apparent conflict between the second law of thermodynamics and the evolution of complex systems. He maintains that this conflict only exists if we assume that

> the energy/matter comprising the Universe is near equilibrium and that it evolves toward an equilibrium state via disorganization and disordering, obeying the laws of equilibrium thermodynamics. The conflict disappears, however, if we postulate that the energy/matter making up the Universe is far from equilibrium, that it exists as an evolving flow, and that the energy/matter flowing through and comprising the Universe evolves from simplicity and disorder to complexity and order via self-organization, in accordance with the empirical laws of nonequilibrium thermodynamics. (Kurakin, 2)

Providing a theoretical foundation for Kurakin's project, Brender (2013) argues for a paradigmatic use of symmetry breaking as a means of explaining the emergence of complexity. Here symmetry is considered invariance under transformation, which means the greater the transformations that maintain invariance, the greater the symmetry.[3] Contrary to intuition but in line with Kurakin, greater symmetry does not imply greater order or structure but rather a *structureless uniformity*. Crucially then, form only arises due to a *symmetry breaking* that introduces differences. Disorder, Brender explains, is more symmetrical than order: "Thus the question of the genesis of form is not how symmetry arises out of disorder, but rather how the symmetry of disorder gets *broken* in determinate ways to produce the characteristic *a*symmetries of the forms we find in nature" (267).

Contrary to the classical view that attempts to conceptualize the emergence of form by appeal to identity, where each difference is caused by a prior equally complex effect, Brander finds this approach inadequate in its linearity and addresses the issue via nonlinear systems theory. Here he claims, we can have "the emergence of behaviours that are less symmetrical than their causes" (268). The patterns of convection rolls, for example, are said to arise "for free", not with the addition of some imposed order, but through the "*subtraction*" of symmetries that results in the apparent form.

In perfect alignment with Brender's paradigmatic appeal to symmetry breaking, Kurakin describes an example wherein a temperature gradient gives rise to self-organization:

> When the temperature gradient is relatively weak, heat propagates from the bottom to the top by conduction. Molecules move in a seemingly uncorrelated fashion, and no macro-order is discernible. However, once the imposed temperature gradient reaches a certain threshold value, an abrupt organizational transition takes place within the liquid layer, leading to the emergence of a metastable macro-organization of molecular motion. Molecules start moving coherently, forming hexagonal convection cells of a characteristic size. As a result of the organizational transition, conduction is replaced by convection, and the rate of energy/matter transfer through the layer increases in a stepwise manner. (Kurakin, 4)

Importantly, a flow of energy/matter through an open physicochemical system of interacting components will "spontaneously" result in the coupling of gradients, which in turn results in a self-organizing, non-equilibrium systems (e.g. *Belousov-Zhabotinsky reaction* and *Benard cells*). The resulting *emergence* of global order is due to a self-organized phase transition when the system passes a critical threshold of energy/matter flow rate. Macrostructure processes that emerge in far-from-equilibrium systems "display both configurational dynamics and flow dynamics", constraints that connect everything within the system, mediating its evolution, transformation, and the organization of its internal structures (ibid.).

Following this line of reasoning, both Kurakin (2011) and Chaisson (2013) have independently made the same point about the evolution of complexity. Chaisson aims to defend the thesis that "specific energy flow" can help us to understand the nature of evolution in general. His working hypothesis is that the energy rate density (ϕ_m), "the amount of energy flowing through a system per unity time and per unit mass" is the most fundamental process "capable of building structures, evolving systems, and creating complexity in the universe" (71, emphasis omitted). Anticipating Chaisson's thesis, Kurakin argues that what guides the evolution of complex systems is the relationship between the system's level of complexity and *rate of energy flow*:

> A relatively higher degree of complexity and order requires and, at the same time, supports a relatively higher rate of energy/matter flow. Increasing the rate of energy/matter flow normally leads to a stepwise increase in relative complexity and order within an evolving nonequilibrium system. Conversely, decreasing the rate of energy/matter flow results in organizational relaxation via a stepwise decrease in relative complexity and order. (4)

By Chaisson's analysis, "energy flow as a universal process helps suppress entropy within increasingly ordered localized systems evolving amidst increasingly disordered, surrounding environments" (73).[4] While earlier stars and galaxies demonstrate the lowest energy rate densities, our galaxy, the biosphere, and human societies he claims, have the highest known. However, despite their lower complexity he finds galaxies have nearly the same degree of "adaptability" and "metabolic" capacity as life forms. Chaisson proposes that this theory depicts evolution as a "universal phenomenon" and stands as "a unifying principle throughout natural science". He goes on to conclude:

> Energy is a common currency; energy rate density (ϕ_m) generally correlates with system complexity and may drive, at least in part, the process of evolution itself.
>
> Selection and adaptation are ubiquitous in Nature; the emergence, maintenance, and fate of all complex systems are often determined, again partly, by their ability to utilize energy. (Chaisson, 78)

Kurakin further develops this line of thought in a way that renders Harris's anticipation painfully evident. Kurakin argues that if the "Universe is far from equilibrium", then the evolving flow of energy/matter through the "Universe spontaneously self-organizes into self-similar (fractal) structures-processes on all scales" (5).[5] Moreover, these "scale-specific" forms are "interconnected and co-evolve as a nested set of self-organizing and interdependent structures-processes" (ibid.). Providing a concrete example of what undergoes symmetry breaking in cosmological evolution, Kurakin proposes a general nonequilibrium model of electron transfer. He maintains electron transfer chains can provide "scale-invariant" and "universal principles" applicable to any structurally adaptive, living or nonliving medium:

1. Different environments will favour different electron transfer chains. As a result, environmental and internal changes in energy/matter flow will drive the emergence, evolution, and adaptation of the electron transfer chains.
2. In instances of local relaxation, individual chain components will compete for the momentarily viable pathways of the existing chains, which will lead to encouraging and/or dissolving the chains that are available.

3. Following global relaxation, alternative electron transfer chains will compete until some of the chains are optimized for efficient electron transport under the environmental circumstances.
4. The system of electron transfer as a whole will *co-evolve* with the constraints of its environment and thus, the electron transfer chains that are most efficient, adaptive, and stable will persist.
5. Consequently, if the universe is a complex open thermodynamic system, it is expected to give rise to new and increasingly complex systems over time (Kurakin, 16).

According to Brender, following symmetry breaking to its wider metaphysical conclusions provides a means of reconciling our dichotomous conceptions of "being and non-being" as well as "form and matter". He goes on to say, "Being is no longer defined by self-identity, but rather by self-differentiation. The opposite of being is not non-being or negation, but rather the absence of negation; uniformity or indifference" (269). If we apply the concept of symmetry breaking to the history of the universe we find one system with high symmetry (low order and low entropy) in its early phase, lowest symmetry (highest order and medium entropy) in its middle phase, and a return to high symmetry (with low order and high entropy) in its late phase.[6]

In combination with Kurakin's electron transfer model, Brender's conception of symmetry-breaking seems to have reconciled the apparently paradoxical contention that whole and part arise from each other, while the whole is nonetheless prior to the parts. Brender has thus offered a tremendous insight into conceptions of Harris's 𝕰 and AE's *dynamic* co-emergence. However, Brender seems to have simply not noticed Harris's warning: it is anti-naturalistic to posit that our *physical* Universe under some condition could have such homogeneity as to be form*less* or *absolutely disordered* since these terms are abstractions at best and contradictions at worst (from Sects. 2.2.2 and 5.2). In the following section, I examine this issue more closely with respect to one way *natural order* can be conceived in dialectical holism.

5.1.3 *Implications for the "Law(s) of Nature"*

Although Harris at no point in his works explicitly posits a theory of laws of nature, I argue that some very clear implications for how laws ought to be conceived in dialectical holism can be derived from what has been said above concerning cosmic evolution. The purpose of this section is to assemble some recent works that demonstrate roughly how this line of reasoning can run.

Beginning with the above conception of meta-laws (Sects. 3.2 and 4.5), the dialectical holist can clearly appeal to Ellis's contention that "laws may depend on the boundary conditions of the universe" and the cosmos itself may "influence the *nature* of local physical laws, rather than just their initial conditions" (2007, 1239, emphasis in original).[7] Contrary to Descartes's story of an orderly world created out of disorder (one that would by definition have *no initial constraints*), whatever phenomena arises, requires some kind of prerequisite order and must be consistent with the parameters of our universe throughout its history. Ellis calls this prerequisite order a *possibility space*. As novel physical phenomena emerge over the course of cosmic history, their very existence

> is allowed by the boundary conditions provided by the universe for local systems, together with the possibility space generated by the underlying physics. While their physical existence is novel, every new thing that comes into being is foreshadowed in possibility structures that precede their existence. (2007, 1242, emphasis omitted)

Interestingly, the Universe is the only system that necessarily sets its own boundary conditions with no prior constraints. On the other hand, excluding the presupposition of an ordered physical superprinciple or Concrete Universal has permitted physicists like Laurence Krouss to conceive *A Universe from Nothing* (2012). As Vilenkin (2006) has previously warned however, the very process of a universe tunnelling into existence from *nothing* would be "governed by the same fundamental laws that describe the subsequent evolution of the universe" (2006, 205). It follows, he argues, that the laws must be understood to exist "prior to the universe itself". He goes on to say that without space, time, and matter, the laws are expressed in the form of mathematical equations, but the medium of mathematics is the mind, and thus we are left with the (unwanted) question of whether mind precedes the universe (ibid.).

Though Vilenkin may have gone too far in believing that this line of reasoning necessarily presupposes mind before Nature, it's clear that such speculation must face either some form of idealism or an infinite regress. While each may be acceptable by those who invoke nomic necessity about laws (e.g. Lange 2009), for Harris and AEs alike, such an explanation will not provide scientifically or philosophically satisfying answers concerning the origins or nature of complex systems. The reason for this dissatisfaction is that nomic theories reduce physical laws to abstract rules that *just are the case*, independently of both observers and matter. Here we come to Harris's warning regarding the *fallacy of spurious homogeneity*.

Argument from Absolute Chaos

- P1. In a condition of absolute disorder, no principles, patterns or laws can obtain.
- P2. To posit a state of Nature is to distinguish an instance of order.
- P3. A state of perfect symmetry is one that lacks all contrasts, patterns, and regularity, that is, order.
- P4. To posit a state of nature in perfect symmetry is tantamount to saying there is a state of matter that is *no-thing*, that is, abstract homogeneity.

- C1. Therefore, nature can never be in, nor arise from a state of absolute disorder, randomness, or symmetry.
- C2. Any disorder or randomness of the Universe presupposes self-differentiated order.

Harris's posit of a Concrete Universal is meant to account for this presupposed order. By a sympathetic account of Harris's reasoning, even the highest level of symmetry in nature must also be governed by an overarching order that specifies the (concrete) possibility space and constrains the process of symmetry breaking that follows. Harris has consistently maintained that the

> influence of the universal is transmitted uninterrupted, through forms of growing complication and self-enfoldment, along a scale of increasing degrees of adequacy in its exemplification, which guarantees that life is the fruition of what is already potentially present in the physical. Its emergence is simply the continuation of an already-evident tendency to build more integral, more versatile, and more self-maintaining wholes. (1991, 61)

To my mind, the monistic picture of cosmogenesis that has been laid out thus far leads directly to a dialectical holist theory of laws. Supportive of Kurakin's SOFT and Harris's TAP, Gleiser (2004) proposes his own principle that characterizes just such a theory:

> [T]he set of physical laws is also an emergent property of the cosmos, together with the cosmos itself [...] it may be meaningless to talk about physical laws before the existence of a physical reality where these laws are enacted: I propose that the laws and the material reality they describe can only exist together. The emergence of cosmic order may be the result of an optimization process rooted on the evolution of complex material structures, resulting from random experimentation between form and functionality. The laws we use to describe nature are a consequence of this optimization process, not its cause. That is, the only truly fundamental principle in nature is that of economy of performance, what could be called a *cosmic optimization principle*, or COP. (2004, 641)

In apparent agreement with Harris, Gleiser suggests that a result of this principle is that to some (unspecified) degree, the universe could not have been different. For Harris, the reason the universe could not have been different is strictly because what obtains ultimately consists of material relations—thus it is nonsensical to claim that either the Concrete Universal or its constituent material forms can be altered without altering the other. Nevertheless, there remains a possible discrepancy depending on what exactly Gleiser meant by "random". As mentioned above, Harris considered randomness to be something that appears within and between the *kinds* of phenomena that arise, but not among them.

In a number of sympathetic works Davies has argued that the principle that underlies the evolution of complexity in the universe "can be traced to symmetry breaking" (2013, 27) overarched by a *Cosmic Blueprint*, and specified into a series of "organizing principles" across a range spatial scales (2004a ff. 152). Davies argues for the *predisposition thesis*: "the laws of nature are such that matter will inevitably be led along the road of increasing complexity toward life. In the same vein, the existence of intelligence and conscious beings is also regarded as part of a natural progression that is somehow built into the laws" (201). Davies likewise claims that this theory "receives support from those experiments that show how complexity and organization arise spontaneously and naturally under a wide range of conditions" (202). In agreement with Harris, Davies claims that in a model of this kind, the general pattern of development is "predestined", but the details are not (ibid.).

Davies's teleological conclusion depends upon an argument for the non-duality of laws and matter. As he has persistently argued "[i]f one accepts the idea of evolving laws then the distinction between fundamental and phenomenological laws fades away. As new and ever more elaborate physical systems come into existence with time, so might the laws that apply to them" (1995, 262). His thesis of evolving laws comes from a dissatisfaction with (i) the common dualist tendencies to depict laws as the software of physical events; (ii) any objectivist picture of nature; and (iii) intuitions that there is a fundamental level to laws of nature. "If one were to pursue this as far as cosmology, it would imply that the universe is a sort of gigantic feedback system where the laws that apply today depend on the details of cosmic history" (263). In this vein, laws are inseparable from the self-organizing systems whose regularities they enact.[8]

Why should we believe that the non-duality of laws and matter result in what Davies (2007) describes as an *unavoidable* teleology *of* life *and mind* (499)? Appealing to analogy from cellular automata, Davies argues that the state of a given cell depends not only on the state of its neighbour, but also upon the "global state". He explains that in this scheme, "physics and biology co-evolve under the action of a (precise) principle operating at the multiverse level, in such a manner that teleological behaviour emerges" (500). A key difference between this model and Darwinian evolution is that there is a causal link between laws and product states: "Thus life is neither a statistical fluke in an indifferently random set of laws/universes, nor is the Universe designed in an *ad hoc* way for life. Instead, life and mind, laws and universe, are common products of an overarching principle" (503). Hence, in accordance with the conception of the complex system noted above, he claims there must be a "self-consistent, self-supporting loop" between the composite systems and their "overarching principle" (ibid.). In line with Gleiser's COP and Harris's TAP, in a later work Davies concludes, it is possible that the state of the world evolves in a way that reflects "law-like optimality rather than any optimality in the intrinsic property of the states themselves" (2013, 39). It is in this sense he considers the universe "fine-tuned for life, mind, and comprehension" (ibid.).

Extrapolating the *developmentalist perspective* (of GST) to its limit, Coffman (2011) has articulated a view of laws and causation that is naturalistic, yet anti-reductionistic:

> [T]he ultimate arbiters of causality, the "laws of physics," are themselves no more than organizational constraints produced by (and contingent upon) the early development of the universe. The causal relationships that define chemistry and biology are more highly specified organizational constraints produced by later development. (287)[9]

More specifically, he claims, "systems develop within systems that evolve over time, an iterative, bootstrapping process wherein the organization that emerges developmentally within a given system is constrained by the information (regulatory networks) that developed previously in a larger, more general system" (300). These "organizational constraints" appear to be analogous to Harris's unifying principles or scales (to be further discussed in the following section).

Coffman finds that given the evidence, it can be argued that a model of this kind "may even characterize the birth of our universe" (301, emphasis omitted). Consequently, *far from equilibrium complex systems* create autocatalytic cycles that create their own evolutionary trajectory by constraining future dynamics at each stage of development. Coffman claims, they first

> create information by selectively organizing a system, which in turn establishes attractors that constrain the flow of energy toward producing specific outcomes [...] this process ultimately leads to a senescent state that lays the groundwork for the emergence of new systems that develop even higher levels of specification. (302, emphasis omitted)

In line with Harris and Brender's symmetry breaking, "evolution" Coffman maintains, "leads to individuation" and he thus finds it appropriate to invoke Aristotle in summarizing the developmentalist perspective of causality: "stochastic events constrained by the organizational context of a specified flow network provide circumstances (formal causes) that generate specific developmental attractors (final causes) which entrain the effects of physical chemical mechanisms (material and efficient causes)" (302). Again, if the universe can be depicted as an open thermodynamic system as Kurakin and others have argued, then the developmentalist model supports Harris's 𝕰.[10]

At this point I only hope to have revealed: (i) the extent to which Harris anticipated what are today a number of interrelated (though mutually unrecognized) theses concerning cosmological evolution premised on self-organization; and (ii) that when brought together, this picture of cosmic evolution follows directly from some of the founding principles of AE—namely, neutral monism, process ontology, dialectical relations, and symmetry breaking. Nevertheless, Harris's notion of teleological scales or unifying principles requires further clarification. In what follows, I critically assess and develop Harris's theory of the scale, which will serve to ground his later discussions of life and mind.

5.2 Updating Harris's Scale of Forms

In the interest of evaluating Harris's TAP, the previous sections of the present chapter have been devoted to revealing the extent to which his 𝕮 can be identified in modern theories of cosmological evolution. The remainder of this chapter more carefully examines Harris's conception of the Levels of Nature. Harris maintained that chaos theory provides a non-reductive paradigm, through which to view natural phenomena across a wide range of sciences and which became integral to his defence of *unifying principles* (henceforth ϕ). For example, in one of his later works he argues as follows:

> The same conviction of holism has emerged from the study of turbulence in what has come to be known as chaos theory, which has revealed the delicate sensitivity of complex, dynamic systems to initial conditions and the unexpectedly intricate and beautiful patterns of order underlying what seem at first sight to be chaotic processes. The discovery of what came to be called "strange attractors" operative in turbulence led investigators to seek patterns amid apparent randomness rather than local mechanisms [...] So the study of complex, dynamic systems has persuaded its pursuers that no reductionist program in science can produce genuine explanations, and that only a grasp of the entire system can render the details intelligible. (2000, 100–101)

Here he makes explicit appeal to the notion of patterns arising from apparent randomness (e.g. the butterfly effect), to provide evidence for emergence and ontological hierarchy. Specifically, Harris believed that chaos theory and fractal geometry could exhibit the antiquity of reductionism and he appealed to this research to make five central claims in defence of his scale of forms:

(1) Harris argues the *self-specification* found in fractal geometry can be connected to Bohm's articulation of the implicate order, which "is enfolded in the 'holomovement' of physical activity, such that an organizing principle is implicit in every point in space-time, as in a hologram—the whole immanent in each part" (1991, 35). He later writes, the "concrete universal [...] is a self-representative system, dialectically structured, like a fractal curve in complex dynamic systems theory" (1993, 238). Bearing in mind the discussion of Sect. 5.2, Harris appealed to fractal geometry as empirical evidence for a self-specifying Concrete Universal that is differentiated into a series of levels, each reflective of the whole.

(2) Harris maintained ϕs are irreducible; for example, "physical and chemical properties" are "consequences of the structure as a whole, and are determined by its organizing principle of unity and not by a simple summation of the properties of its several constituents or parts" (1965, 166). Additionally, he found, "Problems of phase transitions (e.g., between solid and liquid, liquid and gaseous states, the magnetizing of metals [...]) proved tractable when treated in terms of fractals, revealing that they all obey the same rules" (1991, 35). In these instances (and others), Harris believed phase transitions could provide a way of not only identifying, but also *systematizing* what he considered the *irreducible scales of* Nature.

(3) By the terms of Sect. 2.2, Harris maintains that ***attractors*** are proper wholes and so ϕs should involve attractors. He claims there is "a periodic repetition of similar patterns on different scales", which he takes to be indicative of "a universal law governing 'strange attractors,' which turn out to be, or at least to involve, fractal curves" (1991, 35). This is to say ***attractors*** are ***indicative*** of ϕ in virtue of being "a dynamic structure toward which all motions in a complex dynamic system tend at specific degrees of energy flow" (ibid.). Harris maintained that the strange attractor demonstrated that "the whole" has become of utmost importance for science, "apart from which it is futile to examine and try to understand the parts" (36).

(4) With regard for the relationship between scales, Harris goes on to say that "[t]hresholds governing phase transitions are the boundaries between the pulls of strange attractors, and these boundaries are themselves fractal forms, which (as it turns out) correspond to the shapes and dispositions of natural objects" (1991, 36). With this claim, Harris attempted to supply evidence that each ϕ will be dialectically related (i.e. they should appear co-emergent or interdependent) and that ***their boundaries will appear fractal in*** nature.

Towards an elaboration of ϕ, it will be instructive to consider one instance in which Harris explicitly discussed emergence. Harris considers attempts by analytic philosophers such as Kim (1984) to preserve downward causation "by giving it a conceptual interpretation: that is, as referring to concepts rather than phenomena or properties in the real world; by describing them in different languages" (2006, 167). Harris argues that such epiphenomenal or conceptual distinction amounts to paying epistemic lip service to different Levels of Nature. With this complaint, Harris is making it clear that mere conventional or epistemic emergence is too weak.

Harris suspects that a particular hindrance to typical analyses of emergence is the presumption that such phenomena are externally related to their background. In accord with Thompson's comments above, Harris claims this mistake arises directly from the philosopher's tendency to abstract natural phenomena until what remains are a-temporal *properties*.[11] Harris claims that proper examination of nature reveals a series of self-organizing forms, each dialectically related to the rest, so what the analytician neglects is "the central condition for the emergence of new qualities [...] the holism of the configuration from which they emerge" (2006, 168). Appealing to dialectical relations, he claims, means that "downward causation never occurs except in the sense of *final* causation, which, properly understood, is the regulation and determination of the parts or elements of a complex structure and the process of their development by the configuration of the whole" (2006, 169).

(5) By this reasoning, Harris is again (albeit implicitly) led to the conclusion that laws evolve and are characterized by respective kinds of self-organization (ϕ) via the "interlock" and "cooperation" of components:

> Kim and his Analyst followers, therefore, have simply missed the main point of the Emergentists' doctrine: that special configurations of entities at one level of development give rise to new wholes on a higher level, in which the interlock of the constituents modifies them so as to cooperate to produce a new entity with new properties inexplicable by the laws that govern those at the lower level [...] The lower level constituents are necessarily involved, but in the new configuration are transformed so as to cooperate according to higher level laws to which they are not subject in isolation, or in causal collation, at their own original stage of development. (2006, 169)

What is *ontologically emergent* then are the lawful regularities enacted by ϕ. Note however that just as a different set of unifying principles could not be identified with the same Concrete Universal, each ϕ or mode of self-organization, likewise, cannot be instantiated by totally different underlying ϕs. This does not however mean that Harris is poised to argue against multiple-realizability (e.g. of identifiable relations or functions), only that a given ϕ is to be defined by some necessary and sufficient "cooperation" of composite ϕs.

Harris's argument for the nature of ϕs (which have been discussed above as meta-laws or unifying principles) can be considered on epistemic, ontological, and empirical terms:

1. Ontologically speaking, each instantiate unique (meta-)laws characteristic of particular spatiotemporal scales.
2. In epistemic terms, each is expected to resist reduction to any other level.
3. The boundary or *limit* of each scale should be marked by phase transitions appearing fractal in their interrelations with one another.

In consideration of Harris's proposal for a scale of forms, the questions we must ask are: (a) Does modern physics provide necessary *and* sufficient conditions of disparate scales in accord with or in contrast to Harris? (b) What if any bearing does Harris's view of scales provide in conceiving of typical spatiotemporal *scales*? (c) What bearing does this thesis have on the current debate concerning emergence? (d) To what extent can the respective scales be rendered necessary with respect for Harris's TAP? I now examine what replies to Harris's claims may be found in current literature concerning scales and emergence.

5.2.1 What Is a Scale?

Towards establishing a more adequate understanding of scale and emergence, an invaluable first step is to dispel some long-held misconceptions about both of these concepts. To do so, an examination of the *scale* of classical mechanics will serve as an invaluable starting point. Wilson (2013) points out that our epistemic state of affairs is far more tumultuous than many might expect:

> [T]he thesis emerges that physics cannot begin its descriptive task until it has first indulged in a preliminary degree of *essential idealization*: smallish portions of materials must be credited with patently incorrect characteristics. After we reach a complete modelling, we can throw away the idealized ladder we have climbed, for our final equations will describe materials that behave identically at every scale. (101, emphasis in original)

He later elaborates, a result of this "idealization" is that if one were to look for the "foundation" of classical physics "we are likely to feel as if we have become trapped in a novel by Kafka, with particular branches of a vast bureaucracy claiming greater authorities than they truly possess and, when challenged, shunting us off to other departments that assist us no further in our quest" (104). This however is just what one would expect of a dialectically related system of the sort articulated by Harris, Bohm, and AE, in which each level depends upon its relations within the system-as-a-whole.

Contrary to the orthodox interpretation, a purported advantage of Bohm's (1993) *ontological interpretation of* quantum mechanics is that it implies "an approximately classical behaviour for objects containing a sufficiently large number of particles (which means that there is a classical level that is essentially independent of measuring instruments)" (160). *Prima facie* his account appears reductive because "autonomy" arises "wherever the quantum potential can be neglected", meaning the classical domain "is actually an abstraction from the subtle quantum world" (177). However, Bohm argues the action of a quantum potential depends on its form and not on its magnitude, which implies it can have dominant effects even when its intensity is small. As a result, there is

> a strong nonlocal connection of distant particles and a strong dependence of the particle on its general environmental context. The forces between particles depend on the wave function of the whole system, so that we have what we may call "indivisible wholeness." This means that for different wave functions we can have radically different connections between particles (not expressible, for example, in terms of a predetermined interaction potential). (Ibid.)

This is to argue that the classical world is irreducible to the quantum and the quantum world actually "contains an approximately classical 'sub-world' that gradually emerges" from its own "nonlocality and undivided wholeness" (178).[12] As previously discussed, Bohm goes further to maintain that the quantum is itself an abstraction contained within the *super* implicate *order*: "Evidently we could go on indefinitely to higher levels of implicate order. Since each is related to the one below, as the one below is related to the one still further below [...] we have an order of implicate orders" (380). At this point the reader should remember the discussion of Sect. 2.4.1, in which I noted Bohm's resistance to *fundamentality in physics*. With this in mind, his present contentions reveal that the influence of "hidden variables" may come from either "above" or "below" the scale of a given phenomenon. So, for Bohm, it is a mistake to seek a *first scale* at some microscopic level, rather, higher and lower distinctions are abstractions from a unified holomovement whose self-differentiation gives rise to quantum and classical domains—in agreement with symmetry breaking (𝕮).

Addressing reductionism and scales, Batterman (2013) holds a view that properties at one scale indeed depend upon "dynamical changes" that take place *between* others. He explains that the macro properties of steel, for instance, cannot be reduced to the micro alone (258). He contends that it is even "naïve" to believe that "one can, in any straightforward way derive or deduce from atomic facts what are the phenomenological parameters required for a continuum model of a given material" (272). The result is that neither the quantum, nor the classical scales can be derived from the other. Instead, as Batterman maintains, we need to recognize the interdependence of numerous levels of description:

> Bottom-up modelling of systems that exist across a large range of scales is not sufficient to yield observed properties of those systems at higher scales. Neither is top-down modelling. After all, we know that the parameters appearing in continuum models must depend upon details at lower scale levels. The interplay between the two strategies—a kind of mutual adjustment in which lower scale physics informs upper scale models and upper scale physics corrects lower scale models—is complex, fascinating, and unavoidable. (283)

So, when carefully examined, the very notion of an objective chain of efficient causation at the classical and continuum levels becomes an idealization. With this in mind, the notion of ***intrinsic properties*** identified on any particular level of nature may at least be called into question and could even become antiquated given further investigation. Instead (when we retain ***phenomenological honesty***), identifiable properties that characterize a given scale appear only in virtue of relations ***across spatiotemporal levels***. In the dialectical holist spirit, Batterman concludes a later work by noting that there has been too much focus in the reduction/emergence debate on "what is the actual fundamental level and whether, if there is a fundamental level, non-fundamental (idealized) models are dispensable. I am arguing here that the focus on the 'fundamental' is just misguided" (2013, 133). Agreeably, Meyer-Ortmanns (2015) concludes her argument concerning reduction by holding that "whenever reductionism is pushed to its extreme, it will exact a price from us. It is as if hidden subtleties pop up and take revenge for the decomposition into simple constituents, so that the effort required of us is kept constant" (33). She argues (in part) that it is not a matter of computational power to exhaustively account for higher scale phenomena with fundamental laws, but that such a final theory may be impossible *because* of our reliance upon *idealization* (ff. 36).

In order to better understand how one might distinguish between respective scales it is instructive to consider emergence. According to Bishop (2011) *emergence* may be characterized as follows:

> When the behaviours of the constituents of a system are highly coherent and correlated, the system cannot be treated even approximately as a collection of uncoupled individual parts ("the whole is *different* than the sum of its parts"). Rather, some particular global or nonlocal description is required taking into account that individual constituents cannot be fully characterized without reference to larger-scale structures of the system. (114)

How, the reader might ask, is it helpful to appeal to emergence to distinguish scales when the very definition of emergence relies upon a distinction of scale? This question brings us closer to realizing the point at issue: the very notion of scale is only intelligible with reference to *significant* and identifiable spatiotemporal differences, efficacious relations, and dynamics of material systems, that is, emergence.

5.2.2 What Kind of Emergence?

In the interest of elucidating Harris's conception of "scale", in the following sections I orient dialectical holism among four orders of emergence respectively. I argue that Hans Jenny's (2001) *triadic phenomenology*—originally devised as a theoretical framework for studying *cymatics*—can sufficiently characterize the first three orders of emergence (i.e. *structures, kinetics,* and *periodicity*) in agreement with Harris's reasoning, while the fourth must be rejected.

First-Order Emergence

This level of emergence is typically invoked with the expressed intention of maintaining ontic reductionism, while preserving respective levels of description. As Davies (2006) notes, for many working scientists the motivation for invoking this type of emergence is to permit a final ToE that explains the fundamental constituents of matter and thus, the natural world in toto (35). This *weak* emergence is characterized by "systems for which the micro-level laws in principle capture the entire physics of the system, but for which nothing less than inspection of the real system, or simulation, would reveal its behaviour" (37).[13] In other words, X bears predictive or explanatory emergence with respect to Y, iff Y cannot predict or explain X without appealing to properties of X.

Here, the literature will almost invariably invoke properties and structures. Deacon (2006) describes this level as non-recurrent and instantiated by the *relational* qualities of higher-order (but nevertheless *simple*) thermodynamic phenomena. Weakly emergent phenomena occur precisely because of the way a group of entities fit together to give rise to relations at a higher spatial scale. Ellis (2006) further explains, at this level we find bottom-up action plus boundary conditions leading to higher-level generic structures not directly implied by the boundary conditions, but no "higher-level complex structures or functions" (99, emphasis omitted). A property X, is thus weakly emergent if (i) X is microphysically realized; (ii) the laws, theories and, explanations true of X cannot be derived from those of the microphysical properties which realize X; and (iii) all microphysical events are nevertheless determined by the microphysical laws. In this way, first-order emergent phenomena *supervene* upon the lower level, which means a change in the former depends upon a change in the latter.

This kind of emergence is also exemplified by chaotic systems and fractal geometry in virtual models. In this case, emergence results from *sensitive dependence to initial conditions* (SDIC) that restrict our ability to make accurate predictions about a system's behaviour despite knowing that the system is determined by some underlying mechanism(s), for example, equations or laws. Importantly, this implies that one cannot derive a later state from an earlier state, but must compute each and every composite state.[14] For a concrete example, Wallace (2013) claims that this kind of weak emergence can be seen in cases where "there are structural facts about many microphysical systems which although perfectly real and objective [...] simply cannot be seen if we persist in analysing those systems in purely microphysical terms" (472). For example, he claims this may occur in statistical mechanics:

> the temperature of bulk matter is an emergent property, salient because of its explanatory role in behaviour of that matter. (It is a common error in textbooks to suppose that statistical-mechanical methods are used only because in practice we cannot calculate what each atom is doing separately:

even if we could do so, we would be missing important objective properties of the system in question ...). (473)

In sum, instances of first-order emergence are additive and *synchronic* in nature, present in cases that require that our descriptions of phenomena are distinguished by boundaries and shapes, but these "higher levels" remain *supervenient* upon and *reducible* to their microphysical base. I propose that at the threshold of first-order emergence ESEs may be identified due to such systems imposing limitations on our knowledge. In such cases of chaotic models and global thermodynamic phenomena, what emerges at this level can be characterized by Jenny's first parameter of static *structure*, which distinguishes respective *spatial* levels of description. Clearly Harris's claims imply a version of emergence that is non-supervenient, prospects for which I now consider in the second-order.

Second-Order Emergence

Second-order emergence is believed to occur when simple thermodynamic dispositions (e.g. the tendency for entropy of a system to increase) are overridden by structural regularities *as a function of time*. Deacon (2006) eloquently describes this order of emergence as *simple-recurrent* and claims that it introduces *morphodynamics*. By these terms he means we find "a deviation-amplifying dynamic that propagates throughout the system" (130). At this order, it is argued, chaotic and self-organized dynamics can introduce regularities and unpredictable global relations typically initiated by *phase transitions* and symmetry breaking.

In the case of crystallization for example, we discover shape-determined dynamics in which microscopic interactions cascade throughout a system to instantiate an organized lattice structure. Likewise, bearing symmetry breaking in mind, in Benard instability the "global system 'harnesses' the local forces, but at no stage is there a need for an extra type of force to act on an individual molecule to make it comply with a 'convective master plan'" (Davies 2006, 38). The resulting hexagonal structure arises and is sustained because it most uniformly distributes the dissipation of heat in liquid; that is, all the other forms were selected against by entropy (Deacon, ff. 131). At this order however, reference to the global features and environmental context (such as boundary conditions) is required in order to provide a satisfactory account of local dynamics, but there is no need to invoke two (local and global) causes respectively.

Crucially, this order of emergence is characterized by *feedback loops*: a circular causal process in which some portion of a system's output is fed back into the system's input, creating autocatalytic (self-generating) phenomena. According to Thompson (2007), feedback is captured by collective variables known as *order parameters*, which

> constrain or govern the behaviour of the individual components, entraining them so that they no longer have the same behavioural alternatives open to them as they would if they were not interdependently woven into the coherent and ordered global pattern. At the same time, the behaviour of the components generates and sustains the global order. (62)

More specifically, Snyder et al. (2011) define two types of feedback as follows: "Negative feedbacks act to stabilize the system, decreasing a new output change. In contrast, positive feedbacks act to destabilize the system, increasing or amplifying a new output change" (470). They go on to explain that, one implication of feedback in the case of weather systems is that what emerges cannot be predicted in terms of the constituent components one level below the target phenomenon (ff. 473).[15]

Feedback loops of second-order emergence are particularly important for permitting a new categorical distinction of natural phenomena. In his introduction to the *philosophy of complex systems* Hooker (2011) maintains it is "most useful to consider self-organisation to occur where (and only where) a system bifurcates, sufficiently under its own dynamics, so as to bring to bear an additional system-wide constraint" (27). Such a macro constraint he claims establishes "a new level proper in the system" (ibid.). Hence, associated with this level of emergence is a non-linear, non-additive, and "intrinsically global coherence" that in addition to limiting dynamical trajectories of the system's state space, also enables certain states that were otherwise unavailable (28-29). Hooker reasons, "since self-organisation involves a new dynamical form, it is reasonable to say that it obeys new dynamical laws characteristic of that form" (38). In apparent agreement with Harris, Hooker finds that this emergence of laws includes and surpasses first-order or *epistemic emergence*:

> [T]rying to simplify the problem by leaving out globally organized constraints, when they are central to functional characterization, much less by ignoring organization entirely, is to aim at capturing a physics-centered class of cases at the expense of missing much of the special importance and challenge of complexity. (40)

Among others, Bickhard (2011) concludes that supervenience does not apply to such *diachronic* emergent systems because "local versions of supervenience cannot handle relational phenomena" (101). For example, he notes that far from equilibrium processes of organization, such as a candle flame, requires persistent exchanges with its environment in order to maintain their existence. In these instances, there is no fixed set of particles that exhaust the phenomenon, but can only be understood as a holistic organizational process (102).

Important for later discussions, second-order emergence also appears to be manifest in the phenomena of *soft matter*. This class of material includes classical and chemically active systems composed of polymers (e.g. plastics, proteins, and DNA), fluids (e.g. solvents, paints, and soaps), gels, elastomers, colloids, foams, and membranes. Here, unique symmetry breaks result in *liquid crystal* (anisotropic materials) forming layers of molecules such as polymers aligned in a *translational order*. Specifically, while a crystal is fixed in space along *X*, *Y*, and *Z*, axes, a liquid crystal is only constrained along some coordinates (e.g. *X* and *Y*) involving layers and vertical orientation.[16] Consequently, soft matter can undergo reversible topological distortions of several hundred percent and in contrast to the broader category of *condensed matter*, soft matter gives rise to *dynamics* that cannot be easily predicted from the underlying atomic or molecular constituents because of their propensity for self-organization (Bouligand 2011, ff. 56).

Taking the above sentiments together, at this order of emergence new laws can be identified as a result of feedback loops across underlying constituents of a system. This line of reasoning can be traced back to Hans Jenny's (2001) *triadic phenomenology*, in which what emerges can be characterized by structures combined with the additional parameter of *kinetics* and has inspired Deacon's label of "*morphodynamic emergence*" (136). Indeed, numerous arguments from philosophy and physics maintain that at this level of emergence we find self-organization and symmetry breaking, by which the lower-level laws are not violated but are overarched by *diachronic* forms of governance, that is, what will be considered *formal governance* henceforth. In line with the findings of Sect. 5.2.1, second-order emergence appears to minimally characterize the metaphysical underpinnings of AE, but issues remain concerning how higher-order phenomena are to be classified.

Third-Order Emergence

Purportedly exemplified by living, evolving, and conscious systems, this level is characterized by causal influences that are not wholly determined by their microphysics. Many have argued that this level of emergence has been unjustly dismissed throughout much of the twentieth century due to unrecognized metaphysical assumptions within scientific theory. This reductivist paradigm posits the "*completeness of physics*" in the hope of discovering a ToE that shows all microphysical events are determined by prior microphysical events and the laws of physics.

In the face of this reductive explanation, Davies captures the motivation of many who invoke third-order emergence, contending that "a final theory would not in practice provide a very useful account of much that we observe in the world [… it] would not explain the origin of life, nor have much to say about the nature of consciousness" (2006, 35). I maintain third-order emergence is a paradigmatic corner stone of both AE's and Harris's metaphysics alike. As Silberstein (2006) notes, at this order of emergence the following three theses are disputed:

- *Completeness of Physics* (CoP): all physical events are determined by prior physical events, the physical laws that govern them, and efficient causation at each time step.
- *Physicalism* (PHY): all phenomena are *exhaustively* constituted by and identical to their microphysical parts.
- *Causal Inheritance Principle* (CIP): the causal powers of all higher-level phenomena are identical with, or determined by, the causal powers of their lower-level base (217).

Third-order emergent systems are comprised of at least two interconnected morphodynamic processes, autocatalysis and self-assembly. Whereas autocatalytic cycles concern the cyclic processes of chemical reactions, self-assembly involves the crystallization of molecules that assemble into geometric structures. In addition to the emergence of feedback constraints noted above, when occurring within the same system, these processes involve some form of information storage and "memory" structures such as DNA in the context of a cell, adaptivity, and evolutionary learning. Causality is consequently extended across time and space in virtue of memory, which renders the system self-referential and self-organizing.

In systems that are third-order emergent there is an "additional loop of recursive causality that transcends and encloses the second-order recursive causality of self-organized systems" (Deacon, 137). Like morphodynamics of the second-order, such systems provide biases of cancellation, amplification, and selection of the system's own influences. Along these lines, Yin and Herfel (2011) define complex adaptive systems by three criteria:

1. diversity and individuality of components,
2. localized interactions among those components, and
3. an autonomous process that uses the outcomes of those interactions to *select* a subset of those components for replication or enhancement (394).

Such amplification selectively reintegrates structures and processes back into the system (one non-equilibrium process that converges to a stable pattern and is then reinforced), resulting in a "higher-order" stochastic process over time. As Deacon claims, descriptions of such systems rely upon a "combination of multi-scale, historical, and semiotic analyses (analyses based on the relationships between signs)" (139). This is to say that third-order emergent systems essentially *produce* and *use* first-order structures and second-order processes *towards some end.*

In sum, third-order emergence denotes systems whose self-organization, and self-reference creates new *purposeful* laws (MSE), which provide unique *contexts and constraints for the further emergence of novel structures and functions.* With this feature in mind, I claim that the third-order of emergence can again be traced to the third level of Jenny's triadic phenomenology, being *periodicity.* Thus, analysis of these phenomena requires reference to concepts of representation, adaptation, information, and function. In agreement with Ellis (2006, ff. 100) and Silberstein's defence of emergence in AE (2006, 218), Deacon further claims that these features of third-order emergence are to be understood as *teleolodynamic* (143). As AEs and Harris appeal to this kind of emergence in their respective arguments for life, evolution, and consciousness, I now end this brief overview only to return to these issues in later sections. What remains for any proponent of dialectical holism is to defend the irreducibility of third-order emergence and (my next task) to explain its metaphysical limits.

Fourth-Order Emergence

Just as third-order emergence, this thesis entails that there are in principle no *bridge laws* connecting the higher and lower levels and that for certain systems supervenience fails. Also known as *ontological emergence*, fourth-order emergence refers to high-level causes that fundamentally change the system from which they arise and whose components are insufficient for the resultant force.

In contrast with the previous orders, Gillett (2006) holds that in ontological emergence "A property instance X, in an individual s, is O-emergent only if (i) X is instantiated in s, where this individual is either constituted by microphysical individuals or is a non-physical individual; (ii) X is an unrealized property; and (iii) X is causally efficacious" (809). What he means by X's being "unrealized" is that X is not found in the compositional hierarchy of the natural world but is granted causal efficacy. Similarly, Ellis finds that the structures of human creativity and culture indicate the existence of free conscious agents on the one hand (2006, 100) and a Platonic realm of rational ideas on the other (2007, 1272). Neither of these, he maintains, can be reduced to any other level of explanation. They instead exist as *independent* and *efficacious* possibility structures in their own right (2015, 422).

This kind of top-down efficient causation may more fittingly be described as *emergentism of the gaps*. That is, this thesis has been the go-to for various authors—for example, Bergson (1911), Alexander (1929), Clayton (2004), and Peacocke (2006)—hoping to insert a soul, God, *elan vital*, and so on into the gaps of scientific understanding. Such *ontological emergence* is then an explicit means of positing or has an unrecognized implication for dualism. Hence, rather than enriching science or philosophical understanding, appeals of this sort draw a veil of mystery over prospects of naturalizing knowledge and mind. Given the discussions of Sects. 3.2, 3.3, and 5.2, in which I have maintained that Harris supports relational holism and nomological holism, but not ontological holism; insofar as a (neutral) monistic position is to be maintained, fourth-order emergence could not be endorsed by a proponent of dialectical holism. Because all processes are presumed to be continuous with a unified $\mathfrak{E}$, there is no room for anything, whether Platonic or divine, outside of Nature.

To keep these issues straight the reader should bear in mind that for Harris, nature is to be subdivided into respective forms or unifying principles, each reflective of the Concrete Universal via characteristic *laws*. As I have argued above, third-order emergence is the paradigmatic cornerstone of enactivism and Harris's holism alike. This means that the soundness of *dialectical holism* in the context of cosmology rests upon the naturalization of this type of emergence. Alternatively, if supposed instances of third-order emergence (e.g. life and mind) can be reductively interpreted, then extrapolation to the Universe (as per Sect. 5.2.1) becomes increasingly problematic. The consequence for my evaluation of Harris is that his appeal to teleology would potentially fail and the scale of forms (while possibly remaining irreducible for epistemic reasons) could not be considered ontologic and necessary by any empirical means. Thus, a remaining task for anyone in the dialectical holist camp is to show that life and mind are indeed third-order emergentist phenomena. Towards this end, in the following section I exemplify ϕ in contemporary terms before examining Harris's application of this theory to life and mind in later chapters.

5.2.3 A Clarification of Harris's Unifying Principle

As previously discussed (Sect. 3.2), Harris considers the space-time field—distinguishing the scale of our universe—as an example of what now appears to be second-order emergence. Likewise, the quantum scale, he claims can be distinguished by *Pauli's exclusion principle* (PEP), which prohibits any two like particles in an atomic system from having the same state (1965, 131-32; 1991, 31).[17] As Harris explains, due to the PEP:

> Under appropriate conditions, molecules coalesce to form crystals, within the so-called leptocosm of which they are arranged in accordance with the same laws and principles, exemplifying the original universal at a higher level. More complex molecules combine into polymers, and the scale proceeds to organic substances, such as proteins and nucleic acids, necessary for the emergence of life. (1991, 43)

For Harris, crystalline structures are another example of one such level of nature, as their form depends upon an underlying constraint, involving the "interlock" (e.g. chemical bonds and the *exclusion principle*) of atoms and may obtain within a range of phases depending upon the parameters (e.g. temperature and pressure) of their environment. Consequently, each unifying principle is not considered a *basic unit* of nature, but is described as a "polyphasic unity" (1965, ff. 166). This means each ϕ both contains and depends upon such constraints from larger and smaller *scales*.

Towards an argument for the necessary and sufficient conditions of ϕ, Thompson's (2007) criteria for AE emergence provides an invaluable point of development. He maintains a *network* (*N*) of interrelated components in nonlinear *dynamics* (*D*) demonstrates an *emergent process* (*E*) with *emergent properties* (*P*) iff:

1. *E* is a global process that instantiates *P*, and arises from the coupling of *N*'s components and the nonlinear dynamics, *D*, of their local interactions.
2. *E* and *P* have a global-to-local ("downward") determinative influence on the dynamics *D* of the components of *N*.

And possibly,

3. *E* and *P* are not exhaustively determined by the intrinsic properties of the components of *N*, that is, they exhibit "relational holism" (418).

In this passage (1) specifies what kind of component dynamics will be necessary for an emergent whole, whereas (2) characterizes the (at least second-order) formal governance itself. Concerning the transition to (3) however, it can be argued that the identification of a dynamical nonlinear *system* is *synonymous* with that of an emergent whole. As Davies (2004b) maintains, "In a non-linear system the whole is much more than the sum of its parts, and it cannot be reduced or analysed in terms of simple subunits acting together. The resulting properties can often be unexpected, complicated and mathematically intractable" (25). Additionally, with respect for dynamic co-emergence and relational holism, it appears more appropriate to replace *properties* (*P*) with *relational process across spatio-temporal scales* (*R*). Hence, given 1 and 2, *R* will not be exhaustively determined by the endogenous relations of *N* (for further argument concerning dynamic co-emergence see Wu 2016).

Taking a sympathetic reading of Harris's line of reasoning and building on AE co-emergence implies that when theorists and scientists appropriately identify a scale of nature, they are always distinguishing characteristic space-time dynamics that obtain through global constraint(s) imposed by a self-organizing material system as a whole. Importantly, each ϕ is what it is in virtue of *R* and is hypothesized to "reflect" a more general whole to which it belongs. In the remainder of this section, I consider two empirical research projects concerning *phase transitions* and soliton *waves* respectively, which may provide support for the present conception of ϕ.[18]

A number of theorists have recently recognized the philosophical utility of *effective field theories* and symmetry breaking for not only establishing a theory of emergence, but also redefining unity in physics. The common practice in physics has been to consider the appearance of singularities to be an indication of modelling failure and that in order to learn about the physical system we must rid the theory of such information *sinks*. Batterman (2011) argues that a potentially finite theory that is devoid of these infinities (typically found in quantum field theory and condensed matter physics), cannot explain the existence of what he calls "emergent protected states of matter" (1032). Likewise, Morrison (2013) maintains the most significant hindrance to reductive unification is the modelling of *emergent phenomena* seen in condensed matter physics (including superconductivity and superfluidity). Such emergent phenomena mark what she calls a "decoupling of physics at different energy levels" and to be described, require choosing a relevant length scale while ignoring others (383). This is because the "defining feature of these phenomena [...] cannot be explained, predicted, or reduced to their micro constituents and the laws that govern them" (384).

Appealing to *renormalizable group theory*, Batterman argues, when a critical temperature is approached "there is a breaking of symmetry", which causes the order parameter of the system to undergo "a discontinuous jump from zero to nonzero values" (1038). Such a phase transition can be seen when at temperatures above a superfluid transition, a substance will assume random values (the orientation of a paramagnet), but if it is cooled through the superfluid transition phase the arrows align (the ferromagnet). At these critical points we find singularities in our models that are key signatures of "*the emergence at different scales of distinct and apparently 'protected' states of matter*" (1039). These "protectorates" are considered effectively "decoupled and largely independent of physics at shorter length/higher energy scales" (1040). He explains the correlation length (ξ) "is the typical distance over which the behaviour of one microscopic variable or degree of freedom can be correlated with the behaviour of another" (1042). At the critical temperature of phase transition (T_c), the correlation length "diverges" to infinity: $\xi(T\ ) \rightarrow \infty$ as $T \rightarrow Tc$, at which point many degrees of freedom become *coupled*, that is, thereby instantiating a global constraint.

Batterman concludes, if such decoupling of scales is related to the existence of emergent protected states of matter, it is natural to expect infinities and divergences to be involved in our understanding of how those protectorates are possible. He holds that the identifiable infinities provide answers to "why" questions about the stability of protectorates that more fundamental theories devoid of such divergences are incapable of answering. Emergent protectorates are thus considered to be regulated by higher organizing principles that are independent of microscopics, which Batterman considers "the universal behaviour of lower energy protectorates" (1041). What renormalizable group theory does is to relate what Batterman calls an "intractable problem" concerning an infinite correlation length to a tractable problem characterized by a correlation length reduced by some factor. In this way we may reduce coupled degrees of freedom but in the process demonstrate "scaling or self-similar behaviour" (1042).

Concerning unity in physics, Morrison (2013) has argued that *effective field theory* shows "phenomena at critical points have an underlying order. Indeed, what makes the behaviour of critical point phenomena predictable, even in a limited way is the existence of certain scaling properties that exhibit 'universal' behaviour" (413). In other words, when various systems (e.g. liquids, solids, paramagnets, ferromagnets, and superconductors) approach a phase transition they bear a common "cooperative behaviour" irrespective of their microscopic constituents or spatial scale, and so belong to the same *universality class*; for example, liquid-vapour systems' correlation length diverges in precisely the same way as of ferromagnets. Morrison points out that this kind of unity has no reliance upon analysis of "microstructure" and, is in this way, quite different from reductive unification. To my mind, effective field theory thereby provides one way of exemplifying the kind of *fractalesque unity-in-diversity* characterized by phase transitions that Harris attempted to articulate.[19]

Considering Harris's contention that the respective Levels of Nature should reflect one another and some higher-level unity, Gleiser's (2013) proposal of the oscillon is particularly fitting. Gleiser claims that progress regarding the generation of spatiotemporal complexity can be made by discovering the "fundamental properties of non-linear field models which are common to several areas of physics" (113). Towards this end he proposes the soliton *wave*, whose partial differential equations appear in many areas of physics. For example, "symmetry-breaking during phase transitions can give rise to solitons" (114). He argues that if we "relax" our criteria to include time-independent but spatially localized systems; we find "oscillating soliton like structures" or *oscillons* in a wide class of fundamental models including "symmetry breaking in elementary particle physics, cosmology, and in condensed matter systems" (115). He defines the oscillon as "any long-lived, oscillating coherent field configuration found in non-linear field models in arbitrary spatial dimensions. These can involve a single real scalar field [...] or several interacting fields" (116).

In line with Harris's suspicion that scales will involve self-organized attractors, Coffman has argued that the strange attractor has indeed enriched conceptions of causation, "an attractor is a final cause accessed by the regulatory network, which is in turn a formal cause established by organization that developed via the selective agency of autocatalytic cycles" (2011, 300). Gleiser shows that mathematically, oscillon systems "evolve into an attractor configuration" and "emerge spontaneously as a system attempts to regain thermal equilibrium after a sudden change in the nature of its interactions" (117). He goes on to say that "[i]t is during the early stages of this evolution to a new equilibrium state that self-organized oscillon-like patterns emerge" (122). It is because oscillons tend to emerge in phase Gleiser argues, that they act as "bottlenecks to equipartition", meaning that oscillons "lock within them long-wavelength modes, impeding them from interacting with shorter wavelength modes and thus from reaching equilibrium with the rest of the system" (123).

Considered together, Morrison's appeal to "universality classes", Batterman's "protectorates", and Gleiser's "oscillons" all agree with Harris's conception of "ϕ" and AE emergence insofar as each reflects a common capacity for instantiating self-directed constraints (laws) at particular spatiotemporal scales. Hence, a way of further evaluating the applicability of ϕ is to consider the extent to which oscillons and phase transitions may provide a means of modelling and unifying the dynamics of a range of phenomena including gravitational fields, cells, ecosystems, brains, and societies. It must be noted that at this point it remains an open question as to whether respective ϕ are ontological or merely designate the epistemic joints of our theoretical and cognitive capacities, an issue to which I return in Chap. 8.[20]

5.3 Conclusion

From the above discussions, I conclude that Harris's depiction of a necessary scale of forms is alive and well, albeit spread throughout various scientific and philosophic camps. As discussed in Sect. 5.2, Harris's appeal to process metaphysics, dialectical relations, self-organization, and monism to describe cosmogenesis has anticipated a number of philosophical and scientific theories of complexity involving symmetry breaking, teleology, and evolving laws. In Sect. 5.3, I have proposed that Harris's conception of ɸ can be identified in a number of interrelated theses concerning (i) the interdependence of scales, (ii) second- and third-order emergence, (iii) the oscillon, (iv) protectorates, and (v) universality classes in physics. Taken together these theses provide the first sketch of a metaphysics founded in GST. Consequently, I have also proposed that the resulting synthesis can be considered the logical extension of AE into cosmology.

Combining the lessons of these two sections it should be clear that Harris's conception of cosmic evolution ($\mathfrak{E}$) is one of symmetry breaking, in which each bifurcation is characterized by unique modes of governance (ɸ) that are partial *reflections* of the whole (Concrete Universal). According to dialectical holism, it should be possible to trace this proposed scale of forms from the abiotic to the biological and the cognitive, revealing an uninterrupted chain that, by Harris's reasoning, should render life and human consciousness continuous with the organizing principles of matter. In the following chapter, I examine the extension of ɸ and $\mathfrak{E}$ to theories of life and evolution respectively, thus laying a foundation for addressing Harris's response to the hard problem of consciousness.

Notes

1. Interestingly, Harris's point regarding homogeneity may be restated as symmetry via the phenomenon of ferromagnetism. As Rosen (2008) describes, "If a magnet is heated, the many small magnets of which it is composed become agitated and begin jiggling about erratically. When this happens, the mini-magnets lose their common orientation and their axes become randomly distributed. With no preferred orientation, the many different orientations can be described as 'cancelling each other out.' It is this overall absence of difference that is said to constitute symmetry" (182).
2. This thesis further agrees with Gunn's (2011) contention that fundamentally matter is "pneumatical", which he defines as self-moving, holistic, and self-creating. While much of his argument lends support for Harris's thesis, at this point Gunn's extensive analysis of the history of philosophy of physics is too technical for anything but a passing recognition of the evident parallel between the respective theses.
3. To make his point Brender exemplifies 1, a hexagram; 2, a circle; and 3, a uniform field (homogenous space). He maintains that 1–3 increase in symmetry but decrease in rotations and reflections: 1, six reflections; 2, infinite reflections; 3, unbounded reflections (266).

4. Chaisson's account of ϕ_m is not unlike Ingram's (2007) conception of "entropy management" (103) as a framework for understanding a range of phenomena. A former graduate student of the University of Canterbury, Ingram also demonstrated particular sympathy for GST and enactivism more specifically.
5. Though unnoticed by Kurakin, Smolin (1997) has made a very similar argument: "if there is no time at which the universe will come to equilibrium, then it might be useful to view it permanently as a self-organized nonequilibrium system" (159). Like Harris, this line of reasoning leads Smolin to contend that life is expected to proliferate throughout the history of the cosmos and that the theoretical underpinnings of this self-organization justifies the Gaia theory (discussed in Chap. 6). Although cyclic cosmologies are usually mutually exclusive, many appear compatible with the basic premise of the Universe as an open thermodynamic system, for example, Rosen's *self-evolving cosmos* (2008, 206) and Penrose' *conformal cyclic* cosmology (2010, 147).
6. I thank Stephen Pike for helpful discussions on this issue.
7. Ellis (2006) argues that top down action can be seen in the relationship between the Universe as a whole and variables like the arrow of time. He argues that the arrow of time is not determined by fundamental physical laws but by the boundary conditions at the beginning and end of the universe that "select which solutions of the fundamental physical laws are accepted as physically allowed solutions" (90). He emphasizes that the reasons underlying the choice between respective contingent possibilities for the universe cannot be explored scientifically however, but can only be examined through metaphysics (Ellis 2007, 1236).
8. Again, Unger and Smolin (2015) recent work apparently agrees with these points: "Laws evolve. Our most important point is that taking laws to be mutable and subject to evolution rather than timeless and immutable brings questions as to the choices nature has had to make about which laws govern the single universe within the domain of scientific explanation" (2015, 358). In this way Smolin, like Davies, emphasizes the presupposition of order or "choice" at the level of the universe as a whole, which leads to a model of evolving laws of nature.
9. Coffman defines information as a *decrease in* symmetry: "that which brings about a reduction of uncertainty or indeterminacy. Information is a manifestation of asymmetry (i.e., reduction in degrees of freedom, mutuality between events, or expectation of one thing given another, and is the basis of predictability, the holy grail of the scientific enterprise" (292, emphasis omitted). Formally he defines it as I = H – Hr, where H is *a priori* uncertainty and Hr is the residual uncertainty after measurement or the "development of a system". It follows that the max information is also the maximum a priori uncertainty, what he terms Imax = H.

10. Barrow (2008) has argued that in a chaotic system small perturbations can result in exponential amplification of trajectory uncertainty after each collision, which prima facie seems to support Gould's conclusion that the evolutionary process should have a different outcome if history were to be replayed with a slightly different starting point. However, if the motions of molecules in a system are chaotically unpredictable, we still have Boyle's law (PV/T = constant) in the large scale. "Thus, although the individual molecular motions are chaotically unpredictable, their average behaviour is entirely predictable and satisfies a particular probability distribution (the Maxwell–Boltzmann distribution). [...] If we reran the 'tape' of the history of our gas, we would find essentially the same average behaviour, in accord with Boyle's law, even though the individual trajectories of the molecules would be quite different" (146). Barrow's appeal to Boyle's law has essentially the same structure as Harris's argument for the primacy of $\mathfrak{C}$.
11. Indeed, Thompson (2007, ff. 430) appears to be well aware of this mistake and like Harris (2006), devotes the better part of his own appendix to criticizing Kim (1999) for similar reasons.
12. Writing on the relation between classical and quantum domains from an orthodox perspective, Joos (2006) agrees that macroscopic properties "emerge from, or are created by, irreversible interactions with the environment [...] the local classical properties with which we are so familiar have their origin in the nonlocality of (entangled) quantum states" (71). In line with Bohm and Wilson, Joos goes on to argue that "micro- and macro-objects are so strongly dynamically coupled that we do not even know where the boundary between the two supposed realms could possibly be found" (74-75). McGivern and Rueger (2010) likewise maintains the occurrence of macro classical behaviour arising from a quantum mechanical micro level is a candidate for emergence. Here "classical behaviour" is considered "novel with respect to the quantum mechanical base. In our classification, this would be a case of synchronic emergence, and, indeed, the mathematical treatment of the relation between the levels involves taking the 'classical limit' h→ 0 of the Schrödinger operator h2/2m (∂2/∂x2)" (229).
13. Empirical examples include the surface tension of liquid, laminar flow, viscosity, the generic properties of gases, liquids and solids, chemical bonds such as H_2O, conductivity, and heat capacity. Boyle's law, for instance, states that the absolute pressure of the mass of an ideal gas is inversely proportional to its volume if the temperature and amount of gas remains constant in a closed system, which provides knowledge of a global level of description. While this law (PV = K) follows from the underlying constituents, it cannot be known *a priori*, but only from considering the micro constituents.

14. Bishop (2009) argues that a primary quality of chaotic systems is that they are "holistic". He posits three primary distinctions between chaotic and classical explanations: (i) Due to SDIC chaotic explanations are predictions of qualitative behaviour rather than quantitative detail. (ii) This is a description of geometric mechanisms rather than causal processes. Since Chaotic explanations are resistant to reductive analysis, models and systems that exhibit similar patterns of behaviour are grouped together without regard for their underlying causal differences. (iii) These include patterns (e.g. bifurcation points, attractors, and fractals) rather than law-like necessity. "Chaos explanations do not rely on nomic considerations at all; rather, they rely on patterns of behaviour and various properties characterizing this behaviour. In brief, chaos studies search for patterns rather than laws" (Bishop 2009). Chaos does not propose any revision to laws in the same manner as quantum mechanics and general relativity. Debate continues as to whether chaos is a product of the world, or only our models of the world.
15. Further examples include sand dune formations, magnetism, cellular automata, autocatalytic cycles, gravitational structure formation, and laser physics. In these cases, statistical physics becomes inadequate (Ellis 2006, 99). Davies (2006) cites Mach's principle as an example of whole-part influence: the force of inertia captured by a local particle derives from gravitational interactions with all the matter within the universe. Likewise, the second law of thermodynamics requires that we appeal to the global concept of *closed system*: "implicit in the second law is some sort of global constraint (or compulsion) on what happens locally" (41). Davies also includes the phenomena of condensed matter physics; global restrictions in the instance of a black hole, whose event horizon prevents a naked singularity; closed time-like lines that prevent time travel into the past; and *Pauli's exclusion principle*. Wallace (2013) adds vibrations of quantum crystals (quasi-particles).
16. A great deal of theoretical work has been devoted to accounting for this geometric *elasticity* universal to sufficiently soft solids. These characteristics have apparently been utilized in biological evolution and development to give rise to innumerable forms and functions. For example, the surface of the human brain may have formed because the cerebral cortex grows faster than the interior and results in the more efficient gyri folding patterns as a result of soft matter dynamics. http://www.softmatterworld.org/; http://www.durhamemergenceproject.co.uk/understand/3.
17. Indeed, the PEP states: "No two electrons can ever be in the same quantum state; therefore, no two electrons in the same atom can have the same set of quantum numbers" (Serway, et al., 2013, 1127). Both Davies (2006) and Clayton exhibit this principle for the same reason as Harris. For example: "PEP specifies that no two electrons of an atom can have the same set

of four quantum numbers. Thus, a maximum of two electrons can occupy an atomic orbital. This requirement on the way electrons fill up orbitals is basic for understanding modern chemistry [...] A rather simple principle thus has as its outcome the complex chemical distribution of the elements" (Clayton 2004, 581).

18. The present consideration of scales satisfies the typical criteria of *natural kinds*: (1) all members have some (natural) properties in common, (2) each class permit inductive inferences, (3) each participates in laws of nature, (4) each member contributes to a kind, (5) each form a hierarchy, and (6) each kind is categorically distinct (Bird and Tobin 2015, 3-4). However, the formal meeting of scales and natural kinds will have to wait until a later work.
19. For a critical assessment of both Batterman's and Morrison's arguments for emergence see Crowther (2015). She presents a positive account of emergence, but argues that the physics do not reveal that emergence obtains in opposition with reductionism. Rather, the issue concerns "dependence" and "independence" between high- and low-energy physics.
20. Though it is to my awareness not mentioned by any of the above authors, *autocatakinetics* is perhaps the oldest and most general theory of self-organization. I do not mention this theory any further however because those who appeal to it (e.g. Swenson 1991, ff. 215; Davis and Turveyy 2016, 331-32) tend to misunderstand and inappropriately dispute autopoiesis. As will be argued below, autopoiesis does not result in relativism and is consistent with many such theories of self-organization.

Works Cited

Alexander, Samuel. *Space, Time and Deity*. Macmillan, London, 1929.

Barrow, John D. "Chemistry and Sensitivity." In Barrow, John D., Simon Conway Morris., Stephen J. Freeland., Charles L. Harper Jr, eds. *Fitness of the Cosmos for Life: Biochemistry and Fine-Tuning*. 132-150. New York: Cambridge University Press, 2008.

Batterman, Robert W. "Emergence, Singularities, and Symmetry Breaking." In *Foundations of Physics*. 1031-1050. 41 (6), 2011.

Batterman, Robert W. "The Tyranny of Scales." In Batterman, Robert, ed. *The Oxford Handbook of Philosophy of Physics*. 255-286. New York NY: Oxford University Press, 2013.

Bergson, Henry. *Creative Evolution*. London, Macmillan Press, 1911.

Bickhard, Mark H. (2011). "Systems and Process Metaphysics." In Cliff Hooker, ed. *Philosophy of Complex Systems*. 91- 104. The Netherlands, Amsterdam: Elsevier Pub.

Bird, Alexander, and Tobin, Emma. "Natural Kinds." In Edward N. Zalta ed, *The Stanford Encyclopedia of Philosophy*, 2015.
Bishop, Robert. "Chaos." In Edward N. Zalta, ed. *The Stanford Encyclopedia of Philosophy*. 2009.
Bishop, Robert. "Metaphysical and Epistemological Issues in Complex Systems." In Cliff Hooker, ed. *Philosophy of Complex Systems*. 105-136. The Netherlands, Amsterdam: Elsevier Pub, 2011.
Bohm, David., and Hiley, Basil J. *The Undivided Universe: An Ontological Interpretation of Quantum Theory*. New York NY: Routledge, 1993.
Bouligand, Yves. "Liquid Crystals and Morphogenesis." In Bourgine, Paul., and Lesne, Annick, eds. *Morphogenesis: Origins of Patterns and Shapes*. 49-86. New York, NY: Springer Pub, 2011.
Brender, Noah. Moss. "Sense-Making and Symmetry-Breaking: Merleau-Ponty, Cognitive Science, and Dynamic Systems Theory." In *The Canadian Journal of Continental Philosophy*. 246-270. 17 (2) (Fall/Autumn), 2013.
Capra, Fritjof., and Luisi, Pier L. *The Systems View of Life: A Unified Vision*. Cambridge University Press, 2014.
Chaisson, Eric J. "Using Complexity Science to Search for Unity in The Natural Sciences." 68-79. In Charles H. Lineweaver, Paul C. W. Davies, and Michael Ruse, 2013.
Clayton, Philip. *Mind and Emergence: From Quantum to Consciousness*. New York, NY: Oxford University Press, 2004.
Coffman, James A. "On Causality in Nonlinear Complex Systems: The Developmentalist Perspective." In Cliff Hooker, ed. *Philosophy of Complex Systems*. 287-309. The Netherlands, Amsterdam: Elsevier Pub, 2011.
Crowther, Karen. *Decoupling Emergence and Reduction in Physics*. In *European Journal for Philosophy of Science*. 419-445. 5 (3). Springer, 2015.
Davies, Paul C. W. "Algorithmic Compressibility, Fundamental and Phenomenological Laws." In Weinert, Friedel, ed. *Laws of Nature: Essays on the Philosophical, Scientific and Historical Dimension*. 248-267. Berlin Germany: De Gruyter, 1995.
Davies, Paul C. W. *The Cosmic Blueprint: New Discoveries in Nature's Creative Ability to Order the Universe*. Templeton Foundation Press, 2004a.
Davies, Paul C. W. "John Archibald Wheeler and the Clash of Ideas" In John D. Barrow., Paul C.W. Davies., and Charles L. Harper, Jr., eds. *Science and Ultimate Reality: Quantum Theory, Cosmology and Complexity*. 3-24. New York, NY: Cambridge University Press, 2004b.
Davies, Paul C. W. "The Physics of Downward Causation." In Philip Clayton., and Paul, C.W. Davies, eds. *The Re-Emergence of Emergence: The Emergentist Hypothesis from Science to Religion*. 35-52. New York. Oxford University Press, 2006.

Davies, Paul C. W. "Universes Galore: Where will it all End?" In Bernard Carr, ed. *Universe or Multiverse?* 487-505. New York: Cambridge University Press, 2007.
Davies, Paul C. W. "Directionality Principles from Cancer to Cosmology." In Charles H. Lineweaver, Paul C. W. Davies, and Michael Ruse, eds. *Complexity and the Arrow of Time.* 19-41. New York, NY: Cambridge University Press, 2013.
Davis, Tahran J., and Turveyy, Michael T. "One World, Multiple Organisms: Specificity/Autocatakinetics versus Enactivism/Autopoiesis." In Tom Froese, Vörös, Sebastjan, and Alexander Riegler, eds. *Exploring the Diversity within Enactivism and Neurophenomenology.* 330-332. *Constructivist Foundations*: Vol 11, 2, 2016.
Deacon, Terrance. W. "Emergence: The Hole at the Wheel's Hub." In Philip Clayton., and Paul C.W. Davies, eds. *The Re-Emergence of Emergence: The Emergentist Hypothesis from Science to Religion.* 111- 150. New York. Oxford University Press, 2006.
Ellis, George. "On the Nature of Emergent Reality." In Philip Clayton., and Paul C.W. Davies, eds. *The Re-Emergence of Emergence: The Emergentist Hypothesis from Science to Religion.* 79-107. New York, NY: Oxford University Press, 2006.
Ellis, George. "Issues in the Philosophy of Cosmology." In Jeremy Butterfield., and John Earman, eds. *Handbook for the Philosophy of Physics.* 1183-1282. The Netherlands, Amsterdam: Elsevier Pub, 2007.
Ellis, George. *How Can Physics Underlie the Mind? Top Down Causation in the Human Context.* Berlin Heidelberg: Springer Pub, 2015.
Gillett, Carl. "The Hidden Battles over Emergence." 801-818 In *The Oxford Handbook of Religion and Science.* New York, NY: Oxford University Press, 2006.
Gleiser, Marcelo. "The Three Origins: Cosmos, Life, and Mind." In John D. Barrow., Paul C.W. Davies., and Charles L. Harper, Jr., eds. *Science and Ultimate Reality: Quantum Theory, Cosmology and Complexity.* 637-653. Cambridge University Press, 2004.
Gleiser, Marcelo. *Emergent Spatiotemporal Complexity in Field Theory.* In Charles H. Lineweaver, Paul C. W. Davies, and Michael Ruse, eds. *Complexity and the Arrow of Time.* 113-131. New York: Cambridge University Press, 2013.
Gunn, David. *Pneumatology of Matter: A Philosophical Inquiry into the Origins and Meaning of Modern Physical Theory.* Winchester UK. Iff Books, 2011.
Harris, Errol E. *The foundations of Metaphysics in Science.* London: Allen and Unwin Pub, 1965.
Harris, Errol E. *Hypothesis and Perception: The Roots of Scientific Method.* London: Allen & Unwin Pub, 1970.
Harris, Errol E. *Formal, Transcendental and Dialectical Thinking: Logic and Reality.* State University of New York Press, 1987.
Harris, Errol E. *The Reality of Time.* Albany NY: State University of New York Press, 1988.

Harris, Errol E. *Cosmos and Anthropos: A philosophical Interpretation of the Anthropic Cosmological Principle*. Atlantic Highlands, NJ: Humanities Press International Inc, 1991.

Harris, Errol E. *The Spirit of Hegel*. Atlantic Highlands, NJ: Humanities Press International, Inc, 1993.

Harris, Errol E. *The Restitution of Metaphysics*. Amherst New York: Humanity Books, an imprint of Prometheus Books, 2000.

Harris, Errol E. *Reflections on the Problem of Consciousness*. Dordrecht, Netherlands: Springer Pub, 2006.

Ingram, David. *Using Systems Theory to do Philosophy: One Approach, and Some Suggested Terminology*. Christchurch, NZ. University of Canterbury, 2007.

Jenny, Hans. *Cymatics*. NH: Newmarket. MACROmedia Publishing, 2001.

Joos, Erich. "The Emergence of Classicality from Quantum Theory." In Philip Clayton., and Paul C.W. Davies, eds. *The Re-Emergence of Emergence: The Emergentist Hypothesis from Science to Religion*. 53-78. New York. Oxford University Press, 2006.

Krauss, Lawrence M. *A Universe from Nothing: Why There Is Something Rather Than Nothing*. 3rd ed. New York, NY: Free Press, 2012.

Kurakin, Alexei. "The Self-organizing Fractal Theory as a Universal Discovery Method: The Phenomenon of Life." In *Theoretical Biology and Medical Modelling*. 8/4, 2011.

Lange, Marc. *Laws & Lawmakers: Science, Metaphysics, and the Laws of Nature*. New York: Oxford University Press, 2009.

McGivern, Patrick, and Rueger, Alexander. "Emergence in Physics." In Antonella Corradini, Timothy O'Connor, eds. *Emergence in Science and Philosophy*. 213-232. New York, NY: Routledge, 2010.

McKenzie, Kerry. *On the Fundamentality of Symmetries*. In *Philosophy of Science*. 1090-110281 (5), 2014.

Meyer-Ortmanns, Hildegard. "On the Success and Limitations of Reductionism in Physics." In Brigitte Falkenburg, and Margaret Morrison, eds. *Why More is Different: Philosophical Issues in Condensed Matter Physics and Complex Systems*. 13-39. New York, NY: Springer Pub, 2015.

Morrison, Margret. "Unification in Physics." In Batterman, Robert. ed. *The Oxford Handbook of Philosophy of Physics*. 381-414. New York NY: Oxford University Press, 2013.

Peacocke, Arthur. "Emergence, Mind, and Divine Action: The Hierarchy of the Sciences in Relation to the Human Mind-Brain-Body." In Philip Clayton., and Paul C.W. Davies, eds. *The Re-Emergence of Emergence: The Emergentist Hypothesis from Science to Religion*. 257-278. New York. Oxford University Press, 2006.

Penrose, Roger. *Cycles of Time: An Extraordinary View of the Universe*. London: Random House, 2010.

Rosen, Steven M. *The Self-Evolving Cosmos: A Phenomenological Approach to Nature's Unity-in Diversity*. Singapore: World Scientific Publishing Co, 2008.
Serway, Raymond A., Jewett, John W., Wilson, Kate, and Wilson, Anna. *Physics Vol. 2. Asia-Pacific Edition*. Melbourne: Cengage Learning Australia, 2013.
Silberstein, Michael. "Emergence, Theology, and the Manifest Image." In Philip Clayton, ed. *The Oxford Handbook of Religion and Science*. 784-800. New York NY: Oxford University Press, 2006.
Silberstein, Michael. *Why Neutral Monism is Superior to Panpsychism*. In Mind and Matter. 239-248. Vol 7, (2), 2009.
Smolin, Lee. *The life of the Cosmos*. New York, NY: Oxford University Press, 1997.
Snyder, Carolyn W., Mastrandrea, Michael D and Schneider, Stephen H. "The Complex Dynamics of the Climate System: Constraints on our Knowledge, Policy Implications and The Necessity of Systems Thinking." 467-505 In Hooker, Cliff, ed. *Philosophy of Complex Systems*. The Netherlands, Amsterdam: Elsevier Pub, 2011.
Steurer, Walter, Deloudi, Sofia. *Crystallography of Quasicrystals: Concepts, Methods and Structures*. Springer Pub, 2009.
Swenson, Rod. "Autocatakinetics, Yes—Autopoiesis, No: Steps toward a Unified Theory of Evolutionary Ordering." In *General Systems*. Vol 21. 207-228. Gordon and Breach Science Pub, 1991.
Thompson, Evan. *Mind in life: Biology, Phenomenology, and the Sciences of Mind*. Cambridge, MA: Belknap Press of Harvard University Press, 2007.
Unger, Roberto M. and Smolin, Lee. *The Singular Universe and the Reality of Time: A Proposal in Natural Philosophy*. Cambridge UK: Cambridge University Press, 2015.
Vilenkin, Alex. *Many Worlds in One: The Search for Other Universes*. New York: Hill and Wang, 2006.
Wilson, Mark. "What is Classical Mechanics Anyway?" In Batterman, Robert. ed. *The Oxford Handbook of Philosophy of Physics*. 43-106. New York NY: Oxford University Press, 2013.
Wolkenhauer, Olaf and Muir, Allan. "The Complexity of Cell-Biological Systems." In Cliff Hooker. ed. *Philosophy of Complex Systems*. 355-385. The Netherlands, Amsterdam: Elsevier Pub, 2011.
Wu, Richard. "Transformation Emergence, Enactive Co-emergence, and the Causal Exclusion Problem." In *Philosophy Studies*. Springer Pub, 2016.
Yin, Gao, and Herfel, William. "Constructing Post-Classical Ecosystems Ecology." In Cliff Hooker, ed. *Philosophy of Complex Systems*. 389-420. The Netherlands, Amsterdam: Elsevier Pub, 2011.

CHAPTER 6

Towards a Teleonomic Philosophy of Biology

6.1 Harris's Scale of Life

In the previous chapters, I have outlined Harris's contention that the Universe must be conceived in terms of the dialectical whole. In Chap. 4, I have discussed Harris's TAP concerning his argument for the *a priori* necessity of consciousness. In these sections I have revealed significant empirical support for the unifying principle (or meta-law), but I have also shown that despite his ontological conclusions, Harris's reasoning has thus far only permitted epistemic restrictions in how consciousness and the Universe can be discussed in certain philosophical camps. In Chap. 5, I have argued that Harris's original conception of cosmogenesis anticipates numerous theories in the philosophy of physics and reveals wider implications of AE metaphysics. In this chapter, I assess Harris's success in extending his conception of unifying principles (ϕ) and explicative evolution ($\mathfrak{E}$) to biological phenomena.

It is widely accepted by biologists today that self-organization is explicable in terms of physical laws and chemistry, whereas self-preservation remains resistant to this kind of analysis. This difficulty has led to arguments that living systems cannot be understood mechanistically, and that they require a theoretical treatment akin to third-order emergence, which recognizes purposes. In the present section I will show that a wide range of research in contemporary biology now supports Harris's push for a

J. Schofield, *A Phenomenological Revision of E. E. Harris's Dialectical Holism*,
https://doi.org/10.1007/978-3-030-65029-2_6

paradigm shift concerning the nature of minimal life not as irreducibly complex but as ***naturally teleonomic*** and ***holistic***. Specifically, I argue that Harris anticipates recent arguments for what can be called *formal governance* endorsed by AE and systems biology more broadly.

For Harris, biological self-preservation is an invaluable topic that any philosophy of Nature is required to address because if neglected, we are left with a glaring chasm between accounts of cosmos and mind. For Harris, self-preservation both is both the mark of minimal life and closes the gap between *is* and *ought*. In other words, life is considered yet another phase in the scale of forms that is characterized by ***norms***. Anyone who follows this line of thought is charged with providing arguments for the following: (1) the necessary and sufficient conditions of life, (2) an explanation as to why satisfying these conditions should introduce norms, and (3) a systematic account of life as an instance of ϕ.

In his attempt to identify the primary qualities that distinguish living from non-living systems, Harris notes four possible criteria but then adds that each can also occur in the case of non-living systems: (i) specific bodily form and structure (e.g. crystals); (ii) growth and development (e.g. ice); (iii) metabolic change, or continuous exchange of energy and chemical transformation within a stable form (e.g. fire); and (iv) reproduction (e.g. viruses) (1965, 163). Living systems, he contends, are both "self-regulatory" and "adaptive" (1965, 164). Following Schrodinger, Harris holds that in the scale of forms the growth and replication (i.e. self-organization) of "aperiodic crystals" such as chromosomes provide the transition step from the inorganic to those that distinguish living systems (1965, 171). These structures establish "new principles of combination" that are "selective and directive in special ways" (ibid.). However, the line between living and non-living phenomena is marked by "process rather than simply of spatial pattern" (172) and denotes an "open system" involving continuous matter and energy flow (173). For example, systems as simple as a *Paramecium* "actively pursues and accumulates energy supplies" (178). Harris recognizes that the system maintains lower entropy within its boundary than that of its external context; the key questions are how and why does the system do so?

Harris claims the "principle of order" (ϕ) for living systems is one of "self-maintenance" that at this stage is "termed functional", such that "structure becomes secondary and instrumental" (1965, 180). His original proposal runs as follows:

> I propose to use the word "auturgy," now rarely seen, to indicate the capacity of the living system to "work upon itself" and spontaneously to adjust its own internal processes to external conditions so as to maintain its systematic coherence. I shall call an organization of chemical processes auturgic the interrelations of which are regulated according to an overriding principle of order in such a way as to maintain the system in being as an integral whole despite variations in external conditions. The activity of an auturgic system is regulated so as to maintain a steady state, and the system is nothing more nor less than the active process, mutually adjusted within it, that constitute this steady state. The various phases and aspects of these constitutive processes are in consequence mutually ends and means. Anything less than a system of this kind fails to qualify as an example of life. (1965, 180-181)[1]

In other words, the function is considered one of ***adaptation*** via "negative feedback" in the face of environmental forces that would otherwise destroy the system (1991, 64-65).

Harris claims that the living system is a "polyphasic unity" that achieves "self-maintenance [...] not by eliminating or abolishing physico-chemical and thermodynamic laws, but by using them" (1965, 182). Consequently, he holds we will not attain an adequate understanding of minimal life through analysis of a system's parts, but must attend to its function. To do so he begins with the conception of a whole and its relations with an environment. Here the "whole is *ab initio* prior to the parts, and the nature and interrelation of the parts are explicable only by reference to the systematic structure of the whole" (1991, 68). This of course is an epistemic point about part and whole. In ontological terms Harris later makes two further claims: (i) there is "a constant interchange of energy and material between the living organism and its surroundings, so that strictly it is impossible to draw a sharp distinction between them; they are in organic interrelation" (2000, 121); and (ii) the "dominance of the whole over its constituent parts is more obvious and marked than in any of the inorganic forms. The organism is a dynamic, open system in which none of the metabolic processes can maintain itself, or even exist, apart from the totality" (ibid.).

This is to say that directing our attention to the system's boundary (fuzzy as it may be), affords a means of distinguishing and possibly characterizing the living via its functions not found in the inorganic. Claiming the whole is "more marked" in the living system than lower forms is to further emphasize that the living whole is its own scale in virtue of this function. In accordance with the scale of forms, life is thus "a new type of organization more fully and successfully exerting the general tendency of things in nature [...] to form and elaborate wholes of ever-increasing stability and integration" (1965, 184). This capacity of auturgic systems he considered to be "a higher degree of self-reference and individuality" (1991, 64). By "self-reference" Harris means that metabolism and "self-determination" are linked in the organism and "foreshadow consciousness":

> Within the organism, structure and functions are merely two aspects of a single process, constructive of the organ in the course of its functioning. The function, meanwhile, is always and only contributory to the maintenance and integrity of the system. Consequently, it is the system or whole that determines the nature and activity of the parts, and not *vice versa*, and it is this that constitutes the self-determination (or freedom) of the organism and distinguishes it, as a successor and further degree of wholeness to any simply physico-chemical entity. (1991, 66)

In considering these claims, it is important to remember that for Harris, wholeness at any scale is a process of relation. Hence, we may understand the function of the organism to be the system's mode of relation, one that is to Harris's mind, for the sake of the organism's maintenance. This means the organic "activity" is an ongoing distinction of the system from its background *for sake of itself*, thus introducing *the first form of freedom*, or self-determination. Hence, it is in virtue of the adaptive self-governance that the activities of the organism are believed to "transcend" those "processes occurring below the organismic level" and to provide an explanation of the organism's parts (1965, 192).[2]

In line with discussions from the previous chapter, what Harris here considers the simultaneous complementarity and opposition of organism and environment may now more elegantly be considered a further symmetry break in the scale of forms. He later adds, "the existence of feedback is always a sure sign of auturgic wholeness and integral, polyphasic organization" (1965, 224). Harris is claiming that the living system, no matter the degree of complexity, must maintain its homeostasis by way of negative feedback, whereby disparate states can be identified and counteracted with respect to a "norm" or steady state specified by the whole:

> Organism and environment are at one and the same time mutually continuous and mutually complementary, while also in opposition and conflict [...] Life relates to the non-living as opposite, and equally as complement, and also as a new and higher degree of specification of the universal organizing principle of reality, now manifesting its unity in its own active and auturgic self-maintenance at a higher level of self-sufficiency and individual self-determination. (1991, 75)

For example, "living growth is never the mere accumulation of material, but is always nicely regulated to fit the requirements of the total system" (1991, 69). Life then is considered a complexification and enfoldment of a pre-existing self-organizing nisus of Nature as a whole, that is, adaptive freedom marks the ϕ of life.

To Harris's mind, it follows that auturgic wholeness introduces its own "nisus", one directed "towards self-elaboration" (1965, ff. 208). He claims examples of this kind of *adjustment* include reconstructive processes in response to organic damage and removal of systemic components. Processes such as morphogenesis and embryogenesis he holds, not only begin but (perhaps more importantly) also stop in such a manner that mere efficient causation is inadequate for their explanations (1965, 210-13). By Harris's lights, these are example of form that gives primacy to the field over the constituent effects of DNA, providing support for formal and final causation in individual organisms:

> The higher-level morphogenetic field restricts the probabilities of the lower levels. Accordingly, it would seem, not only are the structures hierarchical, but also the formative causation. Thus, in the case of developing structures, such as organisms, the higher-level morphic units (the system in its more complex forms) will exercise a directive influence upon the simpler and more germinal forms. In short, the causation is *final*, the efficient causation being provided, presumably, by physico-chemical activity. (1991, 73)

Additionally, he suggests, "the strange attractor, with the forms it generates (of which the Mandelbrot set is the compendium), is the same thing as the morphogenetic field. Biological forms were, in fact, just what these scientists succeeded in generating in their computer graphics" (1991, 74). This is to claim the forms of life, just as an unfolding fractal, differentiate in in accord with an overarching principle of order.[3] Ultimately, he reasons, "we may regard self-maintaining systems, and explanation in terms of organizing relations, as teleological without ceasing to be scientific"

(1965, 263). Thus, Harris extends his *teleological explanation* discussed in the context of cosmology to living systems.

As is discussed in the following sections, Harris's broadening of causation in biology finds sympathy in current philosophical and scientific theories. Nevertheless, his theory of the living system as a scale of nature has certainly left a number of gaps that although partially addressed by AE, still require a great deal more argumentation to arrive at Harris's conclusion.

6.1.1 *The Nature of Living Being*

Towards an understanding of the nature of life, it is instructive to begin by examining some of their necessary and sufficient conditions. Beginning just below the threshold of life, Dehmelt and Bastiaens (2011) have argued that "organization based on information stored in templates, blueprints, or recipes, such as the base sequence in DNA [are] not related to self-organization. In contrast, self-organized processes display global pattern formation, based on dynamic interplay and teamwork of lower-order system components" (220). They go on to say that in such a self-organizing system the "organization is not imposed form outside", rather what we may consider a second-order emergent pattern at the global level results from the rules specifying dynamic interactions among the system's lower-level components using only local information (ibid.). For this reason, they contend such a global pattern (e.g. calcium waves, sand ripples, termite colonies, or Benard convection cells) can be understood as increasing the complexity or information of the system just in case it is open and far from equilibrium, because energy is needed to sustain the global patterns that are created.

However, Dehmelt et al. continue, external regulators, receptors, and growth factors can be understood to execute pre-determined or template-based processes: "The information stored in the base sequence of genomic DNA is used to generate a complementary sequence of messenger RNA, which is then decoded to generate a protein. In that way, the genomic DNA can be understood as a recipe for generating a protein" (221). Self-organization of the DNA and its context of transcription is thus a prerequisite for living systems and their composite functions. Nevertheless, if self-organization of the organism can be analysed as wholly bottom-up processes, as many authors maintain, the complex functions of living systems may eventually be reduced to the relations among what has been described above as first- and second-order emergent phenomena. However, what appears to be missing in analyses of this type is an account of *periodic feedback* that arises between the emergent form and the underlying constituents.

Gleiser contends that life may be identified when "a certain level of morphological complexity is reached that allows for the spontaneous unfolding of functional complexity" while the function itself arises from form and again, "optimization may indeed play a key role" (2004, 642). He goes on to say that over time "[l]iving organisms must also keep their coherent structures, their 'identities'" (ibid.). He apparently agrees with both Thompson (2007, 212) and Harris (1991, 65-66) that replication is clearly required for a species to survive but not for an individual organism to be alive: "Preservation of biochemical identity and interaction with the environment seem to suffice" (Gleiser 2004, 643).[4] As a precursor to replication and genetic influence, Gleiser proposes, "some of the key physical mechanisms of living systems may be modelled with oscillon-like structures" (2013, 127). For example, oscillons demonstrate longevity and coherence of spatiotemporal structure in the face of environmental perturbation. Additionally, he points out, in the theory of phase transitions a perturbation can lead to nucleation, which permits the formation of crystals and bubbles, which can grow, fission, and fuse. This he believes would provide a "metabolism first" framework for protocells, within which "more complex chemical reactions can occur" (2013, 127).[5]

Following his SOFT theory (see Sect. 5.2.2) Kurakin (2011) proposes that living and non-living systems can be understood as respective *phases of matter* characterized by the system's electron transfer chains. Under far-from-equilibrium conditions, Kurakin says there is competition between alternative electron-conducting pathways and respective conformations within the intervening media. The pathways and conformations that are actually "selected and stabilized" within the animate media further depend upon the environment and internal state of a particular conducting medium. He argues that in a stable environment, the pathways and conformations that are optimized via electron transfer (ET) efficiency will persist longer than those that have proven to be less efficient:

> [I]n order to persist and grow in size and complexity, any living organization must be coupled to a source of and a sink for energy/matter. Moreover, to survive and prosper in conditions of continuous competition with other living organizations and life forms for energy/matter flow, any living

organism/organization will strive to couple to its environment, both physical and social, in such a way as to maximize the rate of energy/matter flow through its internal organization, while minimizing the cost of maintaining and managing this flow. (26)

Accordingly, living matter originates from a plasma-like *organizational phase* and preserves essential characteristics of this phase as it develops and expands over ever larger "spatiotemporal scales in the form of an evolving multiscale hierarchical organization of energy/matter flow/circulation" (56). Kurakin finds that both living systems and electrons move between ionized centres that are situated close enough to influence their nearby neighbours, which in both cases leads to collective dynamics. Based upon his principle of scale-symmetry, electrons passing between ionized centres in far-from-equilibrium conditions are believed to be responsible for the central characteristics of living systems. To illustrate this scale-free process shared by living and non-living alike, he notes that the plasma lamp is a common example of this very phenomenon:

Plasmas are ionized gases, whereas living matter can be conceptualized as an ionized fluid-like condensed phase. Plasmas are electrically conductive, whereas living matter appears to exhibit both insulating and conductive properties and, thus, can be conceptualized as a disordered conductor. (57)

In clear support of Harris's aim of modelling life as continuous with the cosmological 𝕮, Kurakin argues further that if the difference between living and non-living matter is that of organization, then "life is a natural consequence of the evolution of non-living matter. In other words, life is likely to be a rule rather than an exception in the Universe at large" (26).[6]

Similarly, Perunov et al.'s highly acclaimed (2016) paper advanced the argument that it is possible to identify a "general tendency in driven many-particle systems towards self-organization into states formed through exceptionally reliable absorption and dissipation of work energy from the surrounding environment" (1). They find that while this tendency is independent of self-replication and natural selection (14), physical systems with sufficient energy input should be expected to develop towards adaptation and life. They contend that the structures that are most likely to form are the result "of competing pressures to fall to lower internal energy, to further increase heat dissipation through work absorption, and to dissipate that heat as reliably as possible" (11). Thus, their model purportedly shows

> why driven systems should tend preferentially to traverse organized states that have exceptional properties in terms of the reliable absorption and dissipation of energy for the particular driving environment they experience [...] the driven system progresses to states that are better adapted to the drive over time, and it is clear that this emergent adaptation should be accompanied by organization into particular special shapes whose physical properties would appear to us to be tailored to the features of the particular environment. (14)[7]

Concerning self-organisation from the standpoint of complex systems theory, Hooker (2011) finds that in many cases, there is no self that is doing the organizing, for example, the emergence of crystal lattices in iron or convection rolls. He finds that while all living systems are "ipso facto non-linear, irreversible, open, self-organizing, globally constrained", non-living systems may also manifest one or more of these qualities (35). As Hooker agrees however, the macro constraint of biological systems results in intrinsic global coherence, meaning functions of the system as a whole—such as the metabolic and chemotaxing processes of a cell—are irreducible via an analysis of the system's components (i.e. supporting third-order emergence):

> Though the detail, especially the dynamical boundaries, vary in graded ways across living organisms, this autonomy requirement picks out all and only living individuals—from cells, to multicellular organisms to various multi-organism communities, including many suitable organized business firms, cities and nations. But because only autonomous systems have their functions serve their own physical regeneration, in turn supporting their functioning (they are 'recursive self-maintenant' [...]), they represent a distinctively new category of complex system organization. (Ibid.)

Walker (2014) has elaborated upon this premise, providing an argument that biological systems are a unique class of physical system characterized by their manner of creating and managing information. This conceptualization of living systems provides a novel way of framing the emergence of living from non-living, which she believes can be understood as a transition in *causal* information *flow*. Accordingly, living systems do not passively store information but they also process and use it to influence the physical world and themselves. Moreover, Walker argues that insofar as living systems are capable of "universal construction" that is, recreating all of their own components, the living being is "self-referencing"

(427). Taking these points together, she contends the influential aspect of information flow provides a top-down causal structure for the living system.

Walker notes that "while information is abstract, in the sense that it involves one entity symbolically representing another, it is nonetheless physical and only exists when physically instantiated (it therefore holds similar ontological status to energy…)" (430). So top-down causation by information control is purportedly achieved via the flow of information among and between the systems' primary functions and has the goal of conserving these very functions. "Top-down causation operates through functional equivalence classes. Functional equivalence occurs when a given 'higher-level' state leads to the same high-level outcome, independent of which 'lower-level' state instantiates it" (430). This means that there is a "decoupling of 'software' (information controlling the dynamics of the chemical system) and 'hardware' (the chemical substrate)" (435). She claims this decoupling signifies the transitions in the complexity of chemical systems that "enabled the emergence of new layers of control architectures (feedback from state to dynamics or top-down causation from 'software' to 'hardware')" (436). Walker concludes, living systems are considered "a unique state of matter" characterized by self-creation, top-down causation, adaptive selection, and information control.[8]

The research cited in this section has shown a number of mutually reinforcing ways that the processes characteristic of living being may be understood as an instance of third-order emergence. More specifically, these processes evidently make use of second-order emergent structures and as a result, life has been widely considered a (sometimes inevitable) "phase of matter" resulting from processes of energy dissipation. These findings clearly agree with many of the founding premises of Harris's theory of auturgy as a scale and are consistent with the arguments of the previous chapter. Specifically, Walker's proposal for living process to be depicted as a "phase of matter" that introduces functionality (multiple/universal realizability) fits with the above line of reasoning in that the "decoupling" of higher and lower levels via functional equivalence may be considered a further symmetry break, continuous with previous levels of complexity. Though much of her account is agreeable, it is not yet clear how exactly Walker's posit of a decoupling between information and material constituents relates with Harris's approach. Additionally, there remains a deeply puzzling question about the *agent* of causation: in what way can the organism as a whole be understood to *use* information?

6.1.2 *Agency and Formal Governance*

In equating living systems with a scale, what Harris has claimed is that we must recognize a new level of (formal) governance characterized by freedom, *adaptation*, and the instantiation of *norms*. To more precisely evaluate these claims, I now turn to a number of contemporary works that either explicitly or implicitly posit a theory of living systems in these terms.

In a later work Harris (2006, 118-19) follows Kauffman (2000) arguing that the transition from autocatalytic (or self-organizing) to living process is to be distinguished by "actions made on one's own behalf", what both writers are happy to call "agency". For example, Kauffman (2004) considers it reasonable to say a bacterium swimming upstream in a glucose gradient, "is acting on its own behalf in an environment", it is what he calls an "autonomous agent" (654). Positing agency, he maintains, demands something beyond the mere analysis of efficient relata. He claims that even in understanding matter, energy, entropy, and information we still cannot adequately account for this "organization of process" (2004, 656). Apparently unaware of AE, Kauffman holds that in discussing cellular construction, propagation, and division, "we have, as yet, no developed language in physics, chemistry, or biology" (661).

For his argument, Kauffman has since appealed to Kant's view of the organism, for which parts exist in virtue of their whole and vice versa. He holds that "the function of a part is its causal role in sustaining the existence of the Kantian whole" (2013, 168). However, the simplest kind of Kantian whole, *collective* autocatalytic *sets*, Kauffman maintains, cannot be captured by mechanistic Newtonian and efficient laws. Rather, he concludes, "the theory of the spontaneous emergence of collectively autocatalytic sets is a new kind of law: a formal cause law" (2013, 171). Kauffman's sustained thesis has been that positing agency on the basis of formal causation has significant implications for scientific explanation. For Kauffman, the autonomous agent introduces "a new state of matter", in which the "function of a part of an organism is a subset of its causal consequences" (2004, 664). The very postulate of "function" thereby depends upon the context of a self-creating, non-equilibrium system, which bring us for the first time (over the course of cosmic evolution) to consider "doings" rather than mere "happenings". To understand the function of an organism's agency, he concludes, "we must study the whole organism in its environment", which he recognizes implies "an unavoidable holism to biology" (ibid.).

Articulating a position that is to my mind more precise than any of the above in relation to Harris, following Varela, Thompson (2007) defines an autonomous system as one in which "the constituent processes (i) recursively depends on each other for their generation and their realization as a network, (ii) constitute the system as a unity in whatever domain they exist, and (iii) determine a domain of possible interactions with the environment" (44).[9] According to AE, the simplest systems to count as living such as prokaryotic bacteria cells, are *more* specifically considered *autopoietic*: the system is *thermodynamically far-from-equilibrium* (constantly exchanging matter and energy with its surroundings), *organizationally closed* (the system is self-referential circular and recursive), *operationally closed* (the products of the system are re-entrant) (45), and *coupled to an environment* (the system's motor activity loops through its environment to affect and enable its sensory capacities) (49).

In alignment with Harris's understanding of wholeness as a process of completion, Thompson follows Varela and Goguen (1978), for whom the wholeness of a system is "embodied in its organizational closure". In this vein, the whole is not the sum of its parts but rather, the "organizational closure of its parts" (Thompson 2007, 449). For example, accounting for how the system orients itself with respect to sugar concentrations, "the local molecular effects happen as they do because of the global organizational context in which they are embedded. And it is this global level that defines the bacterium as a biological individual and sucrose as food" (75). So, he continues, self-preservation instantiates a relation "between organism and environment that is meaningful and normative" (ibid.). This also means the system is *adaptive in maintaining autopoietic coupling to its environment* and consequently, Thompson endorses teleology: (i) because an organism is a self-organized being (being both cause and effect of itself), it is "intrinsically teleological"; (ii) teleology, he claims cannot be subsumed under efficient causation or mechanism of non-living systems, since the two are "incommensurable in principle"; and (iii) any extension of efficient causes in biology would still have to be understood within the context of the teleological organism (2007, 130).

Hence, *environmental coupling* implies an instantiation of meaning and norms in all living systems for both AE and Harris. In agreement with Harris and AEs (though citing neither), Walsh (2008) maintains "[u]nreduced teleology is a legitimate, wholly natural explanatory mode; moreover, it is not only available to evolutionary biology, it is indispensable [...] It is naturalistically acceptable, metaphysically unimpeachable, and explanatorily autonomous" (113). In the case of living systems, he argues, because the organism is the cause and consequent of its own self-organization, self-reproduction, and self-nourishment, such *goal-directed processes* "cannot be accounted for by pure mechanism alone. Teleology is needed to account for the distinctive characteristics of organisms" (115).

Concerning the relationship between parts and whole of the organism, Walsh endorses dialectical relations and scales without recognizing it. The parts of an organism, he holds, are created by the organism itself and have the properties they do "because of the goals of the organism that it subserves [...] each part is explained by the overall functioning of the organism, and the function of the organism is realized by and explained by the parts" (ibid.). Walsh proposes formal causation as a means of retaining teleological explanation in biology: "The matter of which an organism is constituted is organized by the form. At the same time, the form of an organism is realized in, and constrained by, its matter. In biology, as in any other natural science, one must know both the form and the matter" (119). By implication, individual functions cannot be used to prescribe teleology to a system, rather he contends, "adaptiveness" affords the irreducible teleological description (130). While statements of this kind agree with Thompson, Kauffman, and Harris's notion of teleology, the physical locus of organic causality remains unclear.

In apparent agreement with the above, Hooker (2011) holds that the functions of a living system are both cause and effect of itself, which requires appealing to "holistic, organisational features" including "autonomy" that are "central to being alive" and "cannot be captured by analysis into machine mechanisms" (45). Nevertheless, he voices reservations with respect to autopoiesis on account of its strict requirement of *operational closure*. These concerns are however with reference to much earlier articulations of the theory (e.g. Maturana 1981, Sect. 2.2) that did not emphasize *adaptive* meaning *making* or *environmental coupling*, which implies an additional *openness of dynamic states*. In clear agreement with current AE conceptions however, Hooker holds "[a]utonomous self-regeneration constitutes the fundamental mental basis for normative evaluation because it is the *sine qua non* and reference point for all else" (46-47). Moreover, he claims, meaning depends upon adaptivity, which is the system's "capacity to alter its specific traits in mutually coordinated ways so as to [...] satisfy autonomy in, a wider range of life-environments than its current one" (47).

Hooker's latter point is indeed quite loaded and will require further consideration in the following section, but at this point it is useful to consider whether the present account is descriptive or if it provides explanatory value as well. Recent works concerning adaptation and learning in relatively simple organisms may provide an example of explanatory import, or rule it out. For example, Mitchell et al. (2009) have argued that anticipation analogous to "*Pavlovian conditioning*" has been identified in single celled organisms and acts as an adaptive trait. Their research purportedly demonstrates that over the span of a few generations, E. coli can be conditioned to anticipate changes of sugar types within their environment by activating the agentic pathways appropriate for metabolism based upon a learned response to otherwise unrelated stimuli:

> [P]re-exposure to the stimulus that typically appears early in the ecology improves the organism's fitness when encountered with a second stimulus. Additionally, we observe loss of the conditioned response in E. Coli strains that were repeatedly exposed in a laboratory evolution experiment only to the first stimulus [...] Our work indicates that environmental anticipation is an adaptive trait that was repeatedly selected for during evolution and thus may be ubiquitous in biology. (220)

To my mind, this is precisely what should have been hypothesized on either Harris or AE's accounts. Insofar as further evidence can be provided in support of such adaptive anticipation or "learning", the organism may be understood to exert a dynamic downward constraint on its internal functions on the basis of recognized stimuli. If sound, this would vindicate a kind of causal efficacy in the guise of adaptive constraints that occur on broader spatiotemporal scales than moment to moment responses to stimuli.

Getting to the heart of the issue, considering the organization of living systems Tabaczek (2013) calls for pluralistic causality. By Tabaczek's lights (and in agreement with Thompson's remarks, Sect. 5.2.1) the *downward* causation of living systems cannot be understood in terms of efficient causation without conflicting with the efficient causation already apparent at lower levels. The way beyond this impasse is to hold that the causal influence of an emergent system on its components is not efficient but formal, in that the form of a system *constrains* the interactions of its constituents. However, "the higher level is not a concrete substance, but an organizational level [...] understood by analogy to the concept of 'attractor' in a dynamical system that is a steady state toward which the system may evolve" (396).

Taking Hooker, Tabaczek, and Mitchell et al. together appears to provide a possible dialectical holist response to Walker's above comments on top-down causation and information flow. The steady state may be established not just through organic response to recognizable information, but rather, the information and maintenance of the steady state would have to reciprocally define and create one another. This is to say that the organism *makes and uses* information in "meaningful" or "normative" ways precisely because it is *sensitive* to and may make adjustments *for the sake of* its steady state. This steady state can further be considered the *form* or *organisational closure* of the system that specifies a viable space of environmental coupling, beyond which the system dies. Again, this means the organism does not supervene upon its lower-level constituents or efficient causal chains because as a whole, the *living system only exists in and through time* as an open-ended (4D) attractor trajectory specified by its steady state (see also Grytsay and Musatenko 2013).[10] In possible contrast with Walker's potential appeal to abstracta however, the information here exists for the organism as a viable trajectory *sensed* through time and space.

It should be clear at this point that Kauffman, Thompson, Walsh, Hooker, and Tabczek's arguments are really clarifications of third-order emergence, as introduced in Sect. 5.3. Positing unique forms of governance for the living system, each thesis offers support for Harris's contention that the living system can be considered a scale characterized by minimal *freedom* and *norms*. With these perspectives to consider, the living process may be more intimately related to non-living systems by instantiating a higher order of self-organization and downward constraint, that is, what Harris considered another instance of ϕ. Although much more discussion is needed to assess this possibility, it can be concluded that Harris finds significant support in contemporary philosophers of biology concerning his conceptions of agency, self-organization, and self-preservation. What must now be considered in greater detail are conceptions of *sentience* and *goal directed* behaviour in simple living systems.

Conceptualizing Minimal Cognition and Intentionality

In this section I aim to reconcile what appears to be an un-intentional *contradiction* committed by Harris in discussing the relationship between life, intentionality, and minimal cognition. According to AE, an autopoietic (or auturgic) system demonstrates minimal intentionality and cognition in virtue of adaptation and discrimination (as previously discussed in Sect. 2.3). I claim that although he initially disagrees with this conclusion, Harris's own reasoning leads him to the same conclusions. To reveal this contradiction, I examine discrepancies across Harris's earlier (1965, 1991) and later (2006) works and compare his reasoning with Thompson's (2007) arguments.

In some of his earliest remarks on the issue, Harris claims that the existence of a simple type of sensation can be established on the basis of behavioural analysis:

> In its most elementary form, conscious behaviour is that which depends upon sensitivity. It is response to a stimulus and the behaviouristic formula S-R, is [...] a tacit admission of the presence at least of sensation, for what is not felt cannot stimulate behaviour. In a mere chemical reaction there is neither stimulus nor response but only interchange of matter and energy. (1965, 296)

For Harris, "feeling" broadly "covers sentience of every kind prior to discrimination of modalities or the distinctions between sensation, emotion and pleasure-pain tonality" (310). From the above we obtain the argument, *there is* behaviour *iff there is some* kind *of feeling*. Additionally, Harris appears content to extend behaviour to any living system that acts on its own behalf, for example, a *Paramecium* swimming towards oxygen (1991, 107). Thus, for Harris, if a system is living and acting on its own behalf, it is endowed with feeling.

In later writings. Harris reserves the term *intentionality* for referring to higher levels of conscious awareness and deliberate choice (e.g. 2000, 85; 2006, 40). Harris reasons that because "in primitive sentience there is no discrimination of feeling, self and other (subject and object) [...] there is no intentionality" (1991, 107). Harris argues that "[i]f sentience is a higher degree of unity than physiological organicism, it will transform physiology into behaviour, which is a single activity of discriminative response to a presented situation" (108). He contends that with primitive sentience, "the organism has internalized the diversity of the physical and biotic world as a single and confused whole of feeling [...] what may be called the psychical field" (109). Importantly, this "*primitive psychical field*" is believed to be "non-discriminative" and "homogenous" until attention is introduced (110). He reasons that a primitive sentient system "being indiscriminate, is necessarily pre-perceptual, which means that it is precognitive" (111). Such systems endowed with primitive sentience are non-intentional and non-cognitive. With these remarks, Harris finds that if sentience introduces a further degree of wholeness to the scale of forms, it should be identifiable by an increase in behaviour.

In the following passage, Harris establishes a more in-depth definition of attention as a requirement for intention and cognition:

> Because sentience has a diverse and variegated content that (as merely sentient) is indiscriminate, it is in contradiction with itself. On the one hand, its unity demands the amalgamation of its constituent qualities into a whole, while, on the other, its wholeness implies and requires differentiation. The contradiction needs to be resolved, and the sentient organism fulfils this need by turning its sentient activity back upon itself. The felt impulse of the organism to maintain its systematic coherence, differentiated as it is into specific urges, directs the sentient activity upon special features that answer to its present demands, so that it concentrates itself in the selective operation of attention. (1991, 111-12)

Here Harris is arguing that attention, while differentiating a field of sensation, serves to unify the sensing organism into a self in contrast to the outside world. By Harris's view, such attention is the hallmark of some degree of consciousness. He further qualifies this point: "Only after a minimal figure-and-ground complex has been singled out by attention from the sentient field can the experience be described as cognitive" (1991, 113).

Harris's claim that consciousness requires cognition of one's self (the subject-object distinction) appears unobjectionable. However, his separation of figure-ground distinction from the primitive sentience of living systems simply does not follow from his previous remarks. For if behaviour is linked with a "discriminative response" to a sensed situation and simple living beings are understood to be capable of *goal-directed* behaviour on the basis of *self-preservation*, they already meet Harris's conditions for intentionality and cognition. The "pull in the direction of self-maintenance" constitutes what Harris calls a felt interest for the system to direct "attention to the relevant element in the psychical field" (1965, 326). This is to say that attention acts as the means of organizing the field and provides what he calls an "auturgic nisus that directs the organization of the living system as a whole" (ibid.). He concludes that "[a]ttention, felt want or appetite, leading to desire, are the expressions of auturgy raised to the psychical phase [...] They are, as Spinoza averred, the conatus *in suo esse perseverare*, expressing itself on the level of life and mind" (1965, 328).

In a later text, Harris appears to realize that he had previously created an unnecessary gap between phases of life and mind and grants this very conception to all living systems. He proposes that "attention" is "agency operating on the psychical level. If this is correct, the innate ability to direct attention to arresting sensations is neither more nor less than being alive. It is the innate urge to self-preservation" (2006, 119). Hence, in all cases of living systems he maintains, there must be some level of attention to a diversity of feelings in response to internal and external stimuli:

> On the hypothesis I am putting forward, it would be legitimate to presume that even a single celled animal, being already a highly complex, integrated organic system, might be in some degree sentient. If so, a presumed feeling of privation and discomfort in *Paramoecium* might be assuaged by chance contact with a bubble and the protozoan, being alive, might be capable of "learning" by association of this feeling of relief with bubbles. (Ibid.)[11]

Ultimately, the *contradiction* Harris identifies at the biological level is here not the same as that of his dialectical scales, but only in his use of terms. Indeed, discriminative sensation and action are exactly what the *Paramecium* above is claimed to achieve and consequently, it meets Harris conditions for cognition. In addition, the system may be understood to be *about* the sensed differences of its environment on the basis of its self-preservation, thus instantiating minimal intentionality.

According to AE, all living systems are said to exchange matter and energy with their environment and orient themselves according to what they sense. Moreover, what they sense is said to depend upon how they are able to orient themselves within their environment. This sensorimotor feedback loop establishes the self-world distinction:

> Whereas autopoietic closure brings forth a minimal "bodily self" at the level of cellular metabolism, sensorimotor closure produces a "sensorimotor self" at the level of perception and action. In the one case the passage from network closure to selfhood (and correlative otherness) happens at the level of an active semipermeable boundary or membrane, which regulates interaction with the outside environment. In the other case it happens at the level of behaviour and intentional action. In both cases we see the co-emergence of [...] selfhood and a correlative world or environment of otherness. (Thompson 2007, 49)

By this reasoning, adaptive living systems satisfy the necessary and sufficient conditions of autopoiesis, thus fulfilling the criteria for mind: "First, the instantiation of the autopoietic organization in an actual, concrete system entails a cognitive relation between that system and its environment. Second, this cognitive relation reflects and is subordinated to the maintenance

of autopoiesis" (124).[12] Certain aspects of the organism's environment hereby count as *relevant* cues for the alteration of its behaviour; that is, *the* organism *as a whole is informed.*[13]

This means cognition "is behaviour or conduct in relation to meaning and norms that the system itself enacts or brings forth on the basis of its autonomy" (126). Thompson later summarizes:

(i) Autopoiesis entails the emergence of a bodily self.
(ii) The emergence of a self entails the emergence of a world.
(iii) The emergence of a world is equated with sense making/enaction.
(iv) This, "[i]ntentionality arises from the operational closer and interactive dynamics of autopoiesis" (159).

Hence, for AE meaning-making (cognition) and intentionality are believed to be indispensable in describing the functionality of the living system. It can thus be seen that both Harris and AE agree that if minimal sentience is the regular response of a physical self-organized system to elements of its external environment, then minimal intentionality is no more or less than such a system's capacity for differentiated response to elements of its environment on the basis of self-preservation.

Rather than an intrinsic property of a system, cognition is constituted by a necessary and sufficient *dynamical* (and *dialectical*) *relation* between a living system and the wider system to which it belongs. Importantly, as AEs maintain, there exists an "interactional asymmetry" between organism and environment because the former can adaptively "modulate" its coupling to the environment (Thompson 2011, 121). Illuminated by Harris's metaphysics, the formal governance and intentionality of living systems (i.e. the co-emergence of self and *Other*) can be understood to introduce a further instance of ϕ.

To my mind, the most sympathetic interpretation of Harris's line of thought regarding the natures of cognition and intentionality is twofold: First, one should exclude his earlier conclusion that non-discriminating, non-intentional sentience may be found in auturgic systems because this claim is inconsistent with the rest of his system; and second, permit Harris's later claims regarding differentiation of "attention" and "feeling" to extend to any behaving, living system as per his (2006) assertions. For Harris, at a certain level of functional complexity we find the emergence of purposeful discrimination on the basis of self-preservation, whereas for AE linking this discrimination to self-preservation in its simplest physiological instantiation entails that minimal cognitive intentionality is attributed to all living systems. Hence, the two accounts remain consistent and mutually supportive to this point.

6.2 Dialectical Evolution

Extensive debates surround the nature, parameters, mechanisms, and possible trajectory of biological evolution. With this in mind, in clarifying dialectical evolution (as with previous sections) I confine myself to explaining Harris's ideas and elucidating connections these ideas bear to contemporary research. More specifically, I aim to show that various conclusions regarding the unit of selection and teleological evolution (i.e. formal governance) follow from Harris's depiction of the emergence of living systems (Sect. 6.2) and are consistent with recent works in systems biology. For Harris, the goal of this stage of his thesis is to reconsider 𝕰 in light of biological processes in order to establish a seamless bridge from processes of cosmology to life, and ultimately to consciousness. Harris's strategy to address the increase of biological complexity takes a number of steps that ultimately conclude with his proposal for directed evolution.

(1) *Mechanisms of evolution*—from the outset, Harris argues against Neo-Darwinism and the notion that random variation and natural selection are sufficient to create the life forms we observe today. As with the rest of his philosophical discussions, Harris maintains we must begin by positing a dialectical whole, which in this instance is the self-organizing auturgic system. He claims the "presumption of its existence must be made before we can speak of mutation or selection or even of accidental change" (1965, 230). It is by careful examination of this presupposed system that Harris believes we might arrive at a more accurate understanding of evolution. Harris holds, "selection by itself cannot produce specific changes. Only mutation can do that" (1965, 227). Though mutations do occur randomly, he maintains that not all known mutations are random, and we have yet to establish all the causes of mutation. In this vein, Harris argues that if natural selection presupposes reproducing organisms capable

of "inherent adaptability [...] however much may be effected by random mutation and natural selection, they cannot be the only determining principles in the evolutionary process" (1965, 232). Additionally, he claims that if mutations were to occur independently, they would likely be deleterious to the organism at each step and selected against. Moreover, he considers it improbable that mutations "should occur one by one *in the right order*, even if interspersed with disadvantageous modifications" (1965, 235). Such stepwise random development he claims, would have taken far longer than the time available over evolutionary history, which implies some other mechanism is needed.

(2) Information—for Harris, considering the relationship between information and the organism provides insight into the nature of evolving forms. "Information is equivalent to form. It is technically defined as negative entropy (or order) and corresponds, in general, to structure or organization. That the organism is a system with a specific form means that it contains information" (1991, 66). This information, Harris maintains, is what natural selection acts upon and importantly "without specific form there would be nothing to select" (1991, 66). Harris understands Neo-Darwinian selection as a negative force, whereby the unfit is eliminated but "makes no positive contribution" to the organism that is "fit" (1991, 80). The negative selection effect of Neo-Darwinism, he claims, "applies only to the occurrence of disadvantageous variations [...]", but selection must also provide "a positive effect in making successful varieties predominant in the population, so that fresh mutations will occur, for the most part, in forms already best adapted to the environment" (1991, 80-81). Towards this end he contends that the genome influences traits as a whole system, which precludes piecemeal random mutations being responsible for the increase of complexity. "Geneticists have established that single genes do not control or determine single characteristics, but that the chromosome functions as a whole, as does, in fact, the entire genome" (1991, 84. See also 1965, 243; 2000, 123). In this way he proposes that the information or form of an organism provides a directing constraint or positive/physical selection effect (PSE) on future mutations.

(3) *Neo-Lamarckianism*—concerning the "interplay of structure and function" he claims, a "mutation affecting any aspect of either must *relevantly* affect some fact in the other for the system to work" (1965, 237, emphasis added). He maintains that heredity of adaptive traits is not limited to genetic material but crucially involves behaviour, for example, *Venusia Veniculata* tucking its antennae to aid in camouflage (238).

According to Harris, what is utterly missing from the Neo-Darwinist picture of "random mutations and selection", is a process of integration and diversification, a "principle" that is present for the "organism as a whole" in relation "with its environment" (ibid.). Here, Harris intends to unify his conceptions of self-preservation and development with evolution under the guiding notion that the same principles apply, albeit on different scales: "the process of evolution, like metabolism, growth and development [...] may be described as a process giving rise, for the most part, to progressively more highly organized systems, more diversified yet more closely unified, more efficient and auturgically self-maintaining" (242). Harris admits that by this reasoning "Lamarckian evolution is simulated" (246). This means that an organism's acquisition of traits through adaptive environmental interaction (intentionality) plays a central role in evolution.

(4) *Collective* Organization—by this view, "reproduction" is considered another "process of self-maintenance", through which selection (of another form) can operate (1965, 247). Harris argues that what is selected are the systems that "decrease entropy and increase information. As this goes counter thermodynamic trends, it requires for its explanation some *nisus* towards greater order and complexity, some organismic principle" (1991, 84). In a later work, he finds that the arrival of multicellular organisms is an example of just such an increase in wholeness and information. He argues that as evolution proceeded,

> to maintain themselves more effectively, single cells collected together in colonies and adjusted their functioning one to another. The colonial creatures subsequently become more organically integrated as multicellular individuals with organs functioning mutually as means and ends one to another and each to the whole. (1988, 67)

Harris says mutations are "integrated" into their wider ecosystem and then "what is selected is the more efficiently self-maintaining and self-determining whole" (1991, 84). In the case of multicellular systems this means that we are witnessing the emergence of a higher-order constraint, that is, a new unifying principle (ϕ). As with individual cells, self-determination of a collective system is characterized by an increase in "versatility and freedom" in virtue of the organism's "behaviour" (ibid.). Behaviour is again considered "informed" and "purposive" activity in the service of increasing coherent wholeness (1991, 90).

(5) *Gaia*—Harris extends this reasoning concerning the wholeness of an individual cell to the biosphere, originally advocating for "*biocoenosis*" (1965, ff. 251). He claims that the ecological niche within which organisms live is itself to be considered a biologic whole (1988, 76). In addition, the "world of life as a whole is thus a single system of interlocking life-processes [...]" and he maintains that the biosphere "supervenes in time", rather than space, upon the concentric inorganic levels of the earth (1965, 252). Organisms thus "belong to larger and more comprehensive biological unities in the sphere of life, just as in that of physics we find an over-arching polyphasic unity composed of members which are themselves polyphasic unities, and these are so united as to form a hierarchy" (253). Evolution then, is an "on-going directional process" involving the "continuous self-organization of an entire world" (ibid.). Harris concludes that "organism and environment are inseparable, and in the last resort we are dealing with a single totality—the biosphere—the morphology of which is generating and incorporating focal organisms in a single process" (277). Ultimately Harris recognizes that just as each living system is in a "symbiotic" relationship with its environment, the same must be said of individual ecosystems, thus he endorses the "Gaia theory" (1991, ff. 94).

Harris summarizes his conclusion stating that at the scale of biological evolution we again find the same unifying principle considered heretofore: "the concrete universal which differentiates itself into its own particulars. The process of life, of metabolism, reproduction and evolution, all mutually continuous, constitute a single indivisible process of self-differentiation of one whole, in effect, a cosmic organism" (1988, 78). Hence, for Harris, evolution is understood as an instance of the same self-organizing cosmic process, now manifest across individual and collective levels of living systems. To vindicate his contentions then requires identifying common self-directing principles of individual organic behaviour and ecological constraint that supports the sort of Neo-Lamarckian feedback loop Harris has in mind.

6.2.1 *Evolution According to Systems Biology*

The goal of this section is to consider the impact of systems biology upon the traditional Neo-Darwinian view, and to assess what bearing the paradigm has in relation to Harris's line of reasoning as outlined in the previous section. Towards this end it is instructive to begin with a summary of the prevailing account of *Neo-Darwinism* (e.g. Lennox 2007, 80; Schuster 2011, 45):

1. There are several mechanisms of evolution including natural selection, random genetic mutation, and genetic drift.
2. Characteristics are inherited as discrete entities called genes and population variation is due to different alleles of a gene.
3. Changes in genes are due to mutations that occur *randomly*; that is, they are unpredictable.
4. The traits of an organism (phenotype) are the direct expression of difference in genetic information (genotype), that is, adenine, cytosine, thymine, and guanine.
5. Natural selection acts on the information of genes composed of base pairs that make up chromosomes by excluding those genetic variants with lower reproductive fitness from succeeding generations.
6. Speciation is due to the (typically gradual) accumulation of individual genetic changes in accordance with natural selection.

The changes that occur are believed to have no inherent directionality and are constrained only by the requirement that they produce viable organisms. The upshot, Newman (2011) claims, is that the "possible evolutionary steps that can be taken in any lineage, therefore, are only those phenotypic modifications that occur as a consequence of genetic variation" (235-36).

Arguments have begun to accumulate however that the Neo-Darwinian picture of evolution by these mechanisms is antiquated. Shapiro (2011) finds that this view has been retained by contemporary writers not for scientific, but for philosophical reasons. In place of this view he maintains, we must make room for the paradigm of systems biology as supported by the most recent empirical research. Accordingly, genetic variation is not the only factor influencing phenotype modification and genetic change along with other mechanisms can occur after the organism's primary form and functions have developed. This implies the relationship between the whole system and its environment plays a crucial role in any sufficient explanation of biological evolution. In this vein, theoretical chemist Peter Schuster (2011) holds that while reduction is a necessary *methodological* tool for understanding complex biological phenomena via molecular mechanisms, ontological irreducibility reigns in contemporary biology (46).

Appealing to systems biology, AE has gone some way to embracing this paradigm ("enactive evolution") and in doing so resonates with Harris's original thesis. Thompson maintains, rather than seeing the gene as information, "information is what counts as information for some process at some time" (2007, 191). Likewise, Hutto et al. (2013) note that instead of being "directed by information" in genes, "the relevant structures become manifest only through a developmental process in which factors belonging to the organism and factors belonging to its environment play equally important roles" (33-34). Thompson goes further to argue, "information is context-dependent and agent-relative; it belongs to the coupling of a system and its environment. What counts as information is determined by the history, structure, and needs of the system acting in its environment" (2007, 51-53). In this vein, AE agrees with and elaborate upon Harris's view of information.

By extension, Thompson recognizes the chicken and egg state of affairs we are left with concerning DNA and proteins: "proteins can arise only from a DNA/RNA "reading" process, but this process cannot happen without proteins". He goes on to say that "the DNA "writing" and "reading" processes must be properly situated within the intracellular environment, but this environment is a result of those very processes" (55-56). As a result, he contends that the information of DNA is *embodied* in autopoietic unity:

> it is unacceptable to say that DNA contains the information for phenotypic design, because this statement attributes an intrinsic semantic-informational status to one particular type of component and thereby divests this component of its necessary embedding in the dynamics of the autopoietic network. It is this network in its entirety that specifies the phenotypic characteristics of a cell, not one of its components, and it is this network as a whole that serves as the precondition and causal basis of DNA replication [...] information exists only as dynamically embodied. (57)

Central to an AE critique of Neo-Darwinian, or "genocentric evolution", is the claim that evolution will continue just in case there are heritable traits, irrespective of the mechanisms (176). For example, phenotypes can be transmitted via mechanisms other than DNA replication, called "epigenetic inheritance systems".[14] Indeed, as he points out, many organisms including humans, pass on their symbiotic bacterial partners directly to their offspring.[15]

Concerning the software/hardware metaphor for information and phenotype, Thompson finds that while the two can be considered independent in a computer, in the case of living systems the hardware must pre-exist the software. In addition, according to autopoiesis, the hardware and software must reproduce one another reciprocally. "DNA replication and gene activation are entirely dependent on the autopoiesis of the cell. They contribute enormously to this process, but they also owe their existence to it" (2007, 180). Hence, AE stresses that the network as a whole "specifies the phenotype characteristic of a cell", functioning as the "precondition and causal basis of DNA replication and protein synthesis" (182). In this way, appeal to information *of* or *in* DNA results in an ***unconfessed*** dualism (187). This dualism results from the act of ***abstracting*** the supposed information (e.g. DNA) from the rest of the system (i.e. from the informed) and then applying an intrinsic value to it. Information is thus believed to pre-exist the system and as Thompson says, pass through bodies, affecting them while remaining unaffected itself. This he calls a "reification that has no explanatory value. It is information idolatry and superstition not science" (187).

Similarly, Moreno et al. (2011) hold that the reductive geneocentric approach to biology has hit a dead end, requiring that biologists change their focus from "a program inscribed in DNA analysis to a new distributed (namely, more holistic) program in which DNA, RNA and protein components operate alternatively as instructions and data" (315).[16] In direct contrast to the traditional reductive view that claims a one-to-one mapping of genotype → phenotype, this approach requires positing multiple levels of self-organization, each with their own evolutionary influences:

> evolution does not operate upon an abstract functional phenotype space but through a highly constrained (occasionally discontinuous) space of possible morphologies […] whose formation requires acknowledging the environmental, material, self-organized and often random processes that appear networked at different scales. (320)

Newman (2011) points out an interesting implication that there can be "more than one phenotype for organisms of a given genotype, or the same phenotype for different genotypes, i.e., one-many and many-one relationships among levels of organization" (236). Schuster (2011) elaborates upon this point claiming that current molecular biology reveals epigenetic mechanisms are what make possible the inheritance of characteristics "that are not encoded by the DNA of the individual […] epigenetics and environmental influences provide additional effects that are indispensable for understanding and modelling the relations between genotypes and phenotypes" (56-57). Far from the geneocentric view, the origin of complexity in evolution, he claims "results primarily from genotype-phenotype relations and from the influences of the environment" (76-77). Hence (reiterating third-order emergence), the systems biological paradigm requires that we de-couple the *one-way* efficient causal chains of genotype and phenotype:

> The solution of long-standing evolutionary puzzles concerning the rapid appearance of body-plans and organ forms with conservation of a molecular toolkit and the apparently abrupt recruitment to unprecedented functions of novel tissue motifs and protein complexes, can be advanced by incorporating evidence for a causal disconnection between phenotype and genotype. This includes recognition that developmental systems and proteins (and their complexes) each have inherent dynamical properties that are only loosely related to the genes associated with them, and that each type of system is capable of undergoing changes in their form and function, often in a discontinuous fashion, through means other than genetic variation. (Newman 2011, 348)

Supporting this view, recent research has shown that the genome is no longer to be understood as *read-only* memory, but a read-write information-processing network (Shapiro 2011, 25-26). This distinction means that individual organisms preform what Shapiro calls "natural genetic engineering", which involves all the biochemical mechanisms (i.e. retrotransposons) cells use to cut, splice, copy, repair, polymerize or otherwise manipulate DNA:

> RNA + protein + chromatin → new DNA structure and sequence (retrotransposition, retroduction, retrohoming, diversity-generating retroelements)
>
> Protein + ncRNA + chromatin + signals + other molecules + structures ↔ phenotype and genotype and epigenotype. (26)

Hence, natural genetic engineering purportedly occurs via complex chemical feedback loops across levels of DNA, RNA, proteins, and phenotypic function, in direct conflict with classical Neo-Darwinism. This can involve cells manipulating their own DNA molecules, transporting DNA from one cell to another, or acquiring DNA from the environment. Though for some reason unnoticed by AE or Shapiro, this line of reasoning supports a kind of Neo-Lamarckianism, for which evidence has been steadily accumulating.[17]

Thompson (2007) finds that a further implication of systems thinking is that the dichotomous view of "nature and nurture", "innate versus acquired", and so on are replaced by an ontology of process and dialectical relations. He writes,

> [L]ife cycles propagate from one generation to the next by constructing and reconstructing themselves like a path laid down in walking), instead of unfolding according to any transmitted, genetic blueprint or program. The process of reconstruction involves numerous interdependent causal elements, which relate to each other reciprocally as process and product, rather than belonging to the dichotomous categories of genetic nature versus environmental nurture. (188)

Information then is not passed from one generation to the next, but is reconstructed in and through development. From this vantage point, we are to understand organism and environment as "product and process" that reciprocally select each other, as opposed to a dichotomous conception; hence natural selection *is not an external force acting on organisms.* Similarly, Moreno et al. (2011) have noted, "it is not the environment that poses problems for genetically instructed phenotypes to solve, but a complex organism-environment interaction process that defines the adaptation and survival of an organism" (320).

In sum, according to the systems view of evolution the organism's self-organization and functionality at numerous spatiotemporal levels provides a necessary contribution to not only the process, but also the trajectory of evolution. Thus, there exists a measure of verification for Harris's argument that auturgic behaviour directs evolution beyond the mechanisms of random mutation, natural selection, drift, and so on.

The Telos of Mind and Life

In this section I elaborate upon the above results with respect to conceptions of teleology and value in living systems. I argue that Harris's line of reasoning (Sect. 6.3, contention 3) is in agreement with the AE contention that there exists an inherent relationship between autopoiesis, teleology, self-transcendence, and value. In addition, I consider one means by which the above conception of cognition may be integrated into a systems theory of evolution.

In evolution Harris maintains, modifications that provide "improvements" to the system (i.e. increasing its "fitness") introduce "value" to the system (1965, 249). Value here consists in the modification contributing to the organism's "external adaptiveness and (*ipso facto*) to its internal wholeness" (1965, 250). Harris defines adaptation as a "harmonious co-ordination and integration of functions within the organism with reference to its environment. It is at once internal coherence and adjustment to external conditions" (ibid.). Hence, he claims the coherent relation between organism and environment is what makes an organism "fit" and it is valuable for both. Natural selection is thus considered a consequence of adaptation and value, not their cause. The cause is rather, "the auturgy of the living system which exerts a constant nisus towards integration and order" (1965, 251). Harris thus contends that the cause of selection, adaptation, and value is the auturgic *nisus* of all living systems.

In a later work Harris asserts that "recognition of values is consequent upon purposive activity, which is characteristic of life" (1987, 245). However, he goes on to say that these

> values are the goals of purposive action, and purposes derive from desire, which is the conscious phase of natural appetite and instinctive impulse. But these again are no more (nor less) than the expression of the urge intrinsic to all life to maintain itself [...] all purpose is the endeavour to realize a design or structure of activity—the fulfilment of a whole. (Ibid.)

On the one hand, purposeful behaviour is considered basic to all life because it is the very maintenance of organic being. On the other hand, Harris has it that purposes derive from desire, which is characteristic of consciousness. This creates a contradiction. To reconcile this conflict he could either (i) maintain that value obtains only in later stages of consciousness or (ii) permit that all living systems act on value. In other works, he elaborates that to be aware of value changes appetite into desire, for example: "Appetite is thus transformed in the light of self-consciousness, for to desire an object, we must be conscious of ourselves as seeking satisfaction in its attainment" (2000, 136). I maintain Harris has again made the mistake of relegating qualities (in this case value) to later stages of consciousness than befits his own system, thus the reconciliation requires the latter strategy in accord with Sect. 6.2.2.

As discussed above, the inherent adaptiveness of organic being is for Harris the very same telos, or "universal pressure toward increasing wholeness" discussed heretofore (1965, 273). Harris claims evolution likewise appears progressive "from simpler organisms to more complex" ("orthogenetic"), only insofar as there is a global tendency "towards integration and co-ordination" of form and function (274-75).[18] In a later work, he writes:

> Sentience then becomes the means of assimilating and in some measure domesticating the environment [...] as living beings evolve they are able to penetrate further afield and they develop distal sense modalities [...] so progressively the organism assimilates wider and wider regions of its surrounding world. The environment is thus progressively inwardised (*wird erinnert*). (2006, 147)

Taking these points together, the simplest form of mind is instantiated by purposive action towards *fulfilment of appetite*, which renders the living system cognitive, intentional, and now imbues its activity with value. The "assimilation" or "inwardization" to which Harris refers is constituted by both environmental (domestication) and biochemical alteration in response to *recognizable* stimuli. Hence, behaviour on the basis of value results in what Harris considers a progressive broadening of the organism's sensitivity to stimuli and capacity for environmental alteration. What must now be understood in greater detail is how this process of *inwardization* can be conceived in both phenomenological and scientific terms.

Taking a phenomenological approach to biology, Thompson (2007) maintains "[m]ental life is animated by an intentional striving that aims toward and finds satisfaction in discloser of the intentional object. In this way, intentionality is teleological" (24, emphasis omitted). This same principle is later extended to all living systems: "living beings embody an immanent purposiveness and this purposiveness manifests itself in the two complementary modes of autopoiesis (the intrinsic teleology of self-production) and sense-making (the projective teleology of adaptivity and cognition)" (148-49). Accordingly, the difference between extrinsic and intrinsic functions is that the former presupposes and supervenes upon the larger context of a system (e.g. the purpose of a frying pan relies upon social context), whereas intrinsic purposiveness is a feature of the organization of the system itself. It is this "twofold purposiveness" of identity through self-production, and sense-making through adaptivity and cognition, which makes and reshapes the sensed environing world.

In phenomenological terms Thompson argues, "[i]f the organism must change its matter in order to maintain its identity, then the organism must aim beyond itself, beyond its present condition or point-identity in the here and now" (155). This he recognizes, is precisely what Spinoza called "conatus, the effort and power of life to preserve itself; to stay in existence" (ibid.). The consequence has been that, "the horizons of biological space and time have been greatly expanded over the course of evolution" (ibid.). Following Jonas (1966), this expansion, he claims, has been due to the organism opening "outward into space because its metabolism propels it forward in time, and this forward trajectory is fuelled by want, concern, and need" (156). Thus, Thompson extends existential phenomenology as an "incessant self-transcendence" to the very root of mind and life (157) and concludes, "[l]iving is a process of sense-making, of bringing forth significance and value" (158).

For AE, adaptive self-preservation is *the response to what is sensed in the world on the basis of what is sensed within the system's bounds*, which is considered inherently teleological and self-transcendent. In other words, the symmetry breaking of self and world is not merely structural and functional, but an ongoing phenomenological process of self-transcendence. Since it is this nature of life that is responsible for the broadening of phenomenological horizons, it appears reasonable to conclude that this conception is congruent with Harris's "*environmental internalization*"—what may now be considered the *sentient explication of the world* for dialectical holism. Of course, projecting phenomenological descriptions to the living process may result in a greater philosophical cost than payoff since phenomenology itself remains to be naturalized. Even if only a heuristic at present, I claim establishing continuity across scales is key for Harris and more specifically, linking cognition and evolution is a consequence of AE. Thus, I consider one means by which processes of cognition and evolution may be unified.

Taking a strictly scientific approach to evolution, Krakauer's (2013) proposal goes some way towards accommodating the kind of teleology and cognition supported by Harris and AE. Krakauer contends that there is something "fundamentally lacking" in a definition of evolution based solely upon the mechanistic change in frequencies of genotypes over time as a result of fitness (229). Krakauer considers organic complexity, "a measure of inferential or predictive capacity: the ability for a population to store adaptive responses to a variety of states of the environment" (231). He reasons that new information can only be acquired through mutation and fixation, "but the greater the rate of mutation *p*, the lower the total storage capacity of the genome *L*. This is an inescapable constraint operating on a genome, and requires mechanisms of selection or learning other than natural selection in order to be overcome" (236).

For Krakauer (in line with Kurakin 2011; Mitchell, et al. 2009; and Shapiro, 2011), evolution includes "the changing frequency of alternative environment-predictors", a process of "overcoming fundamental constraints" through the development of "basic cognitive capacities: mechanisms that augment the predictive capacity of genomes alone" (ibid.). Evolution he claims, is the same as cognitive learning to the extent that we find "inferential dynamics that build representations in order to predict their environments" (243). Agreeably, writing on the AE paradigm, Stewart (2010) has recognized that it follows "from the equation 'cognition = life' that the historical evolution in forms of life, ever since its origin, is an evolution in forms of cognition" (4-5).

In order to overcome genetic constraints and acquire and store more information to survive, Krakauer proposes three functions utilized in systems as widespread as "single cells, and in large tissue systems such as brains":

i. "self-similar structures" generated by genetic regulatory networks, that is, fractal compression of hierarchical source data;
ii. "modification of these structures within a single generation" in response to "environmental cues" (237); and
iii. the same structures must be able to be "written" and "read" in many different ways (240).[19]

Cognitive mechanisms on Krakauer's account lend themselves to the same definition as evolution because behaviours that receive rewards will be reinforced whereas others will be extinguished. In addition, "learning will tend to maximize the amount of information that an individual extracts from the environment. The best behaviour will fix in the population excluding all others until the environment changes" (241).

By this reasoning, biological complexity arises as "a series of cognitive innovations, all aimed at supplementing or overcoming the adaptive limitations of genomes, allowing for more efficient representation and predictions of environmental regularities inducing fitness consequences" (242). Krakauer further argues however that there are necessary conditions for an increase in cognitive capacity:

> These are when regularities in the environment present themselves at a rate that exceeds the ability of natural selection to record an adaptive response in the genome. When environments change slowly, cognition offers little advantage and genomes should be sufficient to encode adaptive responses. Complexity therefore increases on a "need to know" basis, and does not follow an inexorable trend but a wavering pathway governed by the rate of information production by the environment, where the environment includes organisms and their products. (243-44)

Taking Thompson and Krakauer's contentions together, this section has demonstrated that Harris's conception of "internalizing the environment" finds support from contemporary phenomenology and evolutionary theory. For Harris and Thompson, the living process entails value and teleology akin with Spinoza's conatus. With this in mind, the dialectical holist conception of cognition can be phrased as *the* sense-making *of a living system that drives a* transcendence *of its (internal and external) physiological constraints based upon an inherent value of* self-maintenance. To this philosophical conception, Krakauer adds a much-needed scientific framework that provides one way by which the very process of *cognition can be analysed as the same process as evolution but occurring upon a different spatiotemporal* scale. In the following section, I consider in detail what implications result for the theory of evolution if autopoiesis (ϕ) is extended to *collective biological systems.*

6.2.2 *Ecopoiesis and Evolutionary* Selection

In this section, I argue that a consequence of both maintaining the range of evolutionary mechanisms noted above and linking cognition with evolution is that the same formal governance posited at the level of individual organisms is extended to *ecosystems.* I claim this entails a revision of evolutionary selection by endorsing multi-level selection, convergent evolution, and symbiogenesis. While these implications have been only partially recognized by Harris and proponents of AE, Harris's conception of the environment as a scale (Sect. 6.3, contention 4), I maintain, provides further clarification concerning why these theories should be accepted in concert.

In the AE account, there is an emphasis on structural coupling, a co-construction between organism and environment in both ontogeny and phylogeny. In agreement with Krakauer and Kurakin above, the organism is not merely subject to selection because part of the inheritance of each successive generation is the organism's environment, one "structured into viable niches by the organisms themselves" (Thompson 2007, 203-4).[20] The result is a "co-determination" in which "organism and environment enact each other through their structural coupling" (204). The AE approach to evolution is based on the notion that the collective behaviour of a species changes the environment through the production of a niche that creates a reflexive loop between organisms and their environment (i.e. third-order emergence). Thompson summarizes his conception of *developmentalist evolution* as follows:

1. The unit of evolution at whatever level (genomic, cell lineage, individual, social group, and so on) is a developmental system (in the widest sense of a milieu-embedded, propagative unit).
2. Developmental systems are composed of ecologically embedded autonomous networks, which exhibit rich repertoires of self-organizing configurations.
3. Such networks are analysable not in terms of *optimality* of design or fittedness to the environment, but rather in terms of *viability* in the face of an unpredictable or unspecified environment.
4. Through reproductive structural coupling with their environments, these networks generate selection in the sense of the differential retention of inherited variation (206, emphasis in original).[21]

This conception of evolution provides support for Harris's thesis because it posits that the units of selection are simultaneously the *agents* or forces of selection at some relevant scale. Thompson recognizes the following three implications:

> (1) that the replicator/interactor distinction is no longer useful for conceptualizing evolutionary processes: (2) that there is no intelligible distinction between inherited (genetic) and acquired (environmental) characteristics; and (3) that there is no intelligible distinction between nature and culture. (192)[22]

In this way, natural selection results from autonomous systems enacting viable trajectories effected by the autonomous networks, which result from the individuals' dynamical coupling with their environments. Thompson holds natural selection is an "emergent consequence of autopoiesis", so while natural selection requires reproduction, "reproduction presupposes autopoietic unities that reproduce" (212).

Reminiscent of Kurakin's electron transfer chains, rendering natural selection in the above light means the notions of co-emergence and co-evolution are again particularly apt. As Thurner (2011) explains in a broader context:

> Any element, a biological species or an industrial good, whose (re)production rate falls below a critical value, will vanish over time. This again can lead to a re-adjustment of (re)production rates of existing elements. The key aspect of evolutionary systems is that the set of existing elements and their corresponding reproduction rates (fitness landscapes) *co-evolve*. (120, emphasis in original)

Formalizing dynamical systems in which elements co-evolve with their (re)production rates is however considered especially challenging. Even if the dynamics of elements could be exactly described with differential equations Thurner explains, "the boundary conditions of evolutionary systems constantly change as they evolve; they are coupled to the dynamics of the system itself" (ibid.). It is for this reason that such systems are inaccessible with traditional mathematical methods.

Thurner points out that the parameters of these evolving systems are typically more numerous than can be identified, let alone accounted for. This results in a high degree of unpredictability, meaning that it is unfruitful and problematic "to predict future fitness of species from their present fitness. Instead, one has to understand how species and their fitness landscapes co-construct each other, how they co-evolve" (121). By extension, he claims that the notion of a niche in ecology (and economics) is actually an "*a posteriori* concept" because the niche itself cannot be predicted. "Consequently, a *physical*, fully causal and useful description of evolution has to consider abandoning the concept of fitness as the central concept and to look for alternatives that operate with the information actually available at a given time" (ibid.). Hence, *fitness* (of either a system or a trait) and *niche* are related such that the meaning of one depends upon the nature of the other; to have an understanding of either one, we must first identify a specific *context* where both niche and fitness provide a means of *mutual explication*.

Kauffman (2013) has gone considerably further than Thurner to propose a number of implications for evolution that may follow from both Harris's and Thompson's positions. "If we are right, entailing law, the centrepiece of physics since Newton, ends at the watershed of evolving life" (173). Thus, for Kauffman, evolution cannot be understood mechanistically and so he proposes an alternative explanation. He argues that much like the boundless functions of a screwdriver, functionality between a cell and its environment is likewise boundless because both are continually altering each other: "we cannot prestate the actual niche of an evolving cell by which it achieves task closure in part via that niche" (175). Moreover, "the actual niche can be considered as the boundary conditions on selection" (ibid.). This implies that adaptations that occur through evolution, or "the very phase space of evolution" changes in "unprestatable" ways (ibid.). Hence, we cannot know what will or can happen because we "do not know the sample space" (179). So, he finds that if organisms, as Kantian wholes, "co-create their worlds with one another", mechanistic laws of evolution are precluded (183). Niches then are systems of living and non-living structures-processes that are not themselves explained by selection but permit a range of unprestatable possibilities that collectively constrain selection and are equally applicable to social systems (186). Here Kauffman reiterates the AE thesis of co-evolution and advocates for the incompleteness of physics reflected in third-order emergent systems.[23]

Closely associated with the developmentalist paradigm, many authors have begun to highlight the importance of macroevolutionary theories including symbiogenesis and multi-level selection.[24] Sepkoski (2008) argues evolution "takes place at or above the level of species" (211). He claims that taxonomic groups can function as "individuals" with their own selection pressures and "evolutionary trajectories" (212). Thus, this view is hierarchical, but does not rely upon qualitatively "different causal mechanisms":

> [S]election is seen to operate at both the organismic and species (or higher) levels according to what are effectively the same principles. In this formulation, however, the evolution of higher taxa is not strictly reducible to microevolution, since higher taxa have organizational principles, which determine the outcome of selection, that are not simply the aggregate of their constituent organisms. (217)

By this view, species as well as "higher taxa", can be understood to operate as individual evolutionary units that have their own adaptive traits (221).[25] As was previously discussed concerning the causal de-coupling of phenotype and genotype, macro-evolution is thus considered "decoupled from microevolution" because the mechanisms of selection operating on the level of individual organisms "cannot be extrapolated to explain evolution at the higher taxonomic levels" (226). This is all to say that the "processes that explain macroevolution—whether species selection, sorting, extinction, or something else—are fundamentally irreducible to neo-Darwinian microevolutionary mechanisms" (ibid.).

Following Lynn Margulis's (1993, 1998, 2004) pioneering works, Peacock (2011) shows that symbiogenesis provides a deeper insight into multi-level selection. In symbiogenesis networks of cooperative behaviours can be successful enough that "they are amplified by natural selection into a coherent, reproducing whole" (235). If such a network is sufficiently organized that it reproduces as a whole, then the entire genetic code is considered a replicator. As a result, "what is 'seen' by natural selection, is not simple; it is not just the gene (whatever that is) unless by 'gene' one means simply any replicator. A sufficiently well coordinated symbiotic association can itself become a unit of selection" (235-36). Peacock elaborates that even those authors most sympathetic for the theories do not tend to notice the connection between symbiogenesis and multi-level selection:

> Apparently-altruistic behaviour need not be explained merely as "kin selection"; an organism need not be in a mutualism merely with its cousins. It could be in mutualism with any other conceivable form of life at all, so long as the net effect is to provide a modality of survival for the organism. Here, by the way is the basis for so-called group selection, which is nothing more than selection in favour of mutualistic symbiosis. (238-39)[26]

Considering the ecosystem as emergent in its own right, Yin and Herfel (2011) describe such a system as complex and dissipative, a flux network that permits quantitative and qualitative analysis but is nevertheless relatively independent of its constituents. "1. the structure is an emergent property that cannot be reduced to the sum of its component species; 2. The structure requires constant energy and material throughput to maintain its existence" (399). Taking what they call a *post-classical dynamic perspective*, ecological models they argue, should possess a number of necessary parameters: (i) lack of steady state, that is, dynamic equilibrium; (ii) novelty and unpredictability; (iii) nonoptimality; (iv) fundamentally cyclic variables; (v) emergent constraints such as selection effects; (vi) historicity or path-dependent dynamics; (vii) diversity; and (viii) locality or site-specific dynamics (404). The environment—including other organisms, weather patterns, and nutrient availability—establish self-organizing and cyclic dynamics (e.g. seasons) that constrain the form and function of individual organisms (e.g. the second-order emergence of circadian rhythms and migratory patterns).

Taken together, the above evidence suggests that the "classical" view of ecology is antiquated and a richer view is required. Such a framework must account for the capacity of ecological systems to self-organize and establish many of their own boundary conditions "such as the intrinsic rates of growth, carrying capacities and interaction coefficients" (Odenbaugh 2011, 427). Towards this end, Beschta and Ripple (2009) have defended *trophic cascades* as a theory of ecological self-organization. They provide significant evidence that top-down influence from predators (carnivores) to herbivores (consumers) and plants (producers) play essential roles in determining ecosystem structure and dynamics via feedback relations. Estes et al. (2011) have more recently argued that the "topology of ecosystem dynamics is now understood to be nonlinear and convoluted, resulting in distinct basins of attraction" (301).[27] They maintain this demands a "paradigm shift in ecology"; rather than being seen as "ecological passengers", predators "exert strong cascade effects" via what they call "top-down forcing" (306). In this way, trophic cascades may denote feedback loops of second and third-order emergence at the level of ecosystems. Here, predators act as *order parameters*, determining the *phase transition of the ecological system* to which they belong.[28] With these arguments in mind, the ecosystem satisfies the necessary criteria for a ϕ.

A still further means of identifying formal governance in natural selection can be seen in instances of convergent evolution (e.g. echolocation in dolphins and bats involving over 200 genes).[29] Coffman (2011) argues that according to *developmentalism*, "morphological homoplasy and convergent evolution would be far more common than is generally assumed owing to the existence of (perhaps cryptic) developmental or structural attractors generated by ecological organization" (305). Walker (2014) has likewise argued that top-down causation is an important mechanism for adaptive evolution and is particularly apparent in cases of convergent evolution. "Convergent evolution thus provides a clear example of top-down causation via adaptive selection, where causal influences run from macroscopic environmental context to microscopic biochemical structure" (430). The point is that the ecological system as a whole specifies viable forms and functions (or modes of coupling) that act as attractors for those creatures that remain alive. At this point, a fair consideration is that the recurrent forms and functions of convergent evolution are likewise indicative of scales.

To summarize this discussion, it is instructive to consider Davies's (2013) view of complexity. Davies holds that evolution can be understood as "a random walk through the possibility space which is bounded by a 'wall' of minimal complexity. The wall exists for the elementary reason that there is a lower limit, but no obvious upper limit, to the complexity of a living organism" (32). As many have noted above however, this boundary is created by the total organization of the ecological system itself, is unpredictable, and establishes *a viable trajectory of form and function* (or fittedness). Accordingly, evolution follows a series of paths of entropy management created by feedback relations between ecosystem organization on the one hand and organism adaptation on the other. If evolutionary selection can be considered *about* the forms and functions that *more efficiently use energy and manage entropy*, then it would be expected that symbiogenesis is necessary for speciation and acts as a form of natural selection. This would warrant the notion of *ecopoiesis* and sufficiently qualify the ecosystem as an indication of ϕ.

The Gaia Theory

The concept of Gaia is by no means new, having been articulated in various forms by many philosophers and scientists since the Ancient Greeks. In its modern version, the claim is that biosphere, hydrosphere, atmosphere, and lithosphere constitute a semi-independent level of nature, an emergent function that is auturgic/autopoietic, demonstrating sentience and self-directed behaviour through self-regulation (see Crist and Rinker 2010). Here I outline what support can be gathered for Harris's argument (Sect. 6.3, contention 5) for the culmination and telos of biological phenomena at the macro scale.

According to AE, a further means of characterizing a living entity is by *order* of autopoietic systems: first-order is proper autopoiesis demonstrated by the paradigmatic cell, whereas second-order is a *network* of first-order systems such as insect colonies and human societies. A crucial question here is whether these second-order systems can meet the criteria of first-order autopoiesis. In this case, Thompson recognizes that the idea of boundary takes on a wider definition. He contends that indeed such second-order systems can count as first order as long as the system produces and regulates "its own internal topology and functional boundary"

(Thompson 2007, 107). By this definition, the totality of living systems on earth is considered a self-organizing whole, "with the self-regulation of the planetary climate and atmosphere occurring as emergent phenomena" (119).

Thompson anticipates the most common means of refuting this theory, namely the lack of reproduction and selection involved in the operation of such a system. He contends that with regard for a definition of "minimal life" such systems do not require a hereditary lineage:

> Reproduction requires an individual to be reproduced, and hence is logically and empirically secondary to the process whereby an autonomous system constitutes itself as an individual. Therefore, a self-producing but non-reproducing planetary system could count as a genuine biological individual. Furthermore, its self-producing and self-maintaining mechanisms, though not the result of natural selection acting on a population of similar systems, would nevertheless hardly be "accidental." Indeed they would presumably reflect lawful principles of self-organization and emergence in complex systems. (120)

Thompson contends that considering the criteria of autopoiesis listed above, Gaia can clearly be understood to produce its own internal topology and functional boundary, circularly producing the components that compose the system as a whole. Interestingly, if the same symmetry breaking considered in cosmological evolution is utilized here, a natural consequence is that ecological systems are the consequent, not the cause of a planetary Gaia. In this vein, rather than reproduction being necessary, we find that *self-differentiation* is necessary for the continuity and congruence of not just the biosphere, but of all life.

Extending his conception of coupled self-organization and electron transfer flow to its further conclusions, Kurakin articulates a view of ecosystem form and function that renders Harris's $\mathfrak{E}$ continuous with the Gaia theory:

> As a result of molecular self-organization, driven and sustained by electron flow, various living structures, cells, organisms, and ecosystems continuously emerge, metamorphose, and transform one into another in an eternal process of transformation of forms, which unfolds simultaneously on multiple scales of space and time within the multiscale whole of the planetary life, held together and integrated by the invisible threads of moving electrons. As a consequence of such an arrangement, molecules, cells, organisms, and ecosystems function as scale-specific constituents of one multiscale whole of

> energy/matter flow/circulation, where they represent both means and ends, at one and the same time. (2011, 26, emphasis omitted)[30]

This very process of organization and exchange of energy/matter has the result of creating ever more complex and efficient paths of energy consumption, "with the ultimate purpose being the perpetuation and expansion of living matter as a whole" (42).

Among others, Peacock (2011) has set out what is perhaps the most realistic argument for the Gaia theory in terms of organic interrelation, that is, *mutualistic* symbiosis.[31] Whereas commensalism or parasitism at a global level leads to a reduction of free energy and ultimately, extinction, he contends that reinforcing feedback relations may provide selective pressure for cooperative behaviour sufficient for the global level of organization required (238-39). He finds sufficient evidence to posit a "distributed genome" in symbiotic systems of numerous scales up to and including a planetary Gaia. This is to say that if the interactions of symbiotic, mutualistic organisms are capable of supporting the reproduction of each separately, and the self-differentiation of the system as a whole, then Gaia is a viable theory.

Kauffman (2013) articulates a view that is in many ways indistinguishable from Harris and AE by extending his conception of the *Kantian whole* (including autocatalytic systems and cells) to the biosphere. Interestingly, he argues that although such wholes are themselves unpredictable, they can "become collectively autocatalytic", and the process by which this occurs is that of "the phase transition"—the same as for all chemical reactions (187). So, while such wholes are not *irreducibly* complex (due to our limited predictability), they are not captured by efficient causal laws. Hence, as we have seen, he proposes the "Formal Causal Law". Evidently his extension of such a law to ecosystems and to the biosphere, is all that is required to lend credit to Harris and bring Kauffman into accordance with dialectical holism, but he goes still further.

In his conclusion, Kauffman contends that laws of this kind may not only "help explain the emergence of complexity in the diversifying biosphere", but also reveal that a whole range of "non-equilibrium systems such as Benard cells, whirlpools, and perhaps even stars and galaxies, are linked sets of cross-coupled processes that jointly cause one another's continued co-creative, non-equilibrium existence in the universe" (188). Kauffman's theory that systems as diverse as DNA, the biosphere, economic systems, and galaxies, are linked into a mutually reinforcing whole, clearly resembles Harris's theory of 𝕰, in which each system is a partial reflection of the self-specifying Concrete Universal. Hence, Kauffman's "Formal Causal Law" is not new, but is rather a repackaged version of the presently examined theory of ϕ.

Four conclusions can be reached from the above: (i) AE provides arguments for considering Gaia in the same terms as the autopoietic system irrespective of its lack of reproduction and difference in boundaries. (ii) Kurakin shows how Gaia may follow from the above conception of evolution (𝕰 or electron transfer) that is continuous across living and non-living systems alike. (iii) Irrespective of its soundness, the same formal governance posited at the scale of the self-organizing cell has been depicted as the selection effect of biological evolution, and the biosphere as a whole. (iv) To the extent that this principle is held responsible for an increase in complexity (via symmetry breaking and symbiogenesis) and broadens the Neo-Darwinian paradigm, Harris's teleonomic philosophy of biology receives significant support from contemporary systems biology.

6.3 Conclusion

In this chapter I have argued that there is a clear parallel between the conceptions of life, cognition, and evolution articulated by Harris and contemporary systems biology. Both involve formal governance operating at numerous levels (e.g. genome, the organism as a whole, the ecosystem, and biosphere). Additionally, in agreement with Harris's conception of unifying principles (ϕ), each of these levels have in recent years been modelled as a respective phases of matter.

In Sect. 6.2.2, I have argued that the most coherent (and sympathetic) interpretation of Harris's views regarding the essential parameters of mind and life result in the attribution of both sentience and intentionality to any (auturgic) living system, in line with AE. In Sect. 6.3.1, I have argued that there is significant support for a broadening of evolutionary mechanisms that involve individual sensorimotor engagements and genomic modification. The argument rehashed in favour of Harris's theory of ***environmental internalization*** is that if self-maintenance (entropy management) of living systems entails a basic sensation of need and continuous re-evaluation of values (adaptive meaning-making) on the basis of self-preservation, then organic forms and functions are expected to generate new modes of sensation, behaviour and increasingly efficient social organizations (symmetry breaking).

Thus, it appears to be a logical extension of both Harris's and AE's evolution that the cognitive *process* may play a role in driving and constraining evolutionary development via *learning* at shorter time scales than natural selection. As discussed in Sect. 6.3.2, though not emphasized by Harris, organism-environment coupling, multi-level selection theory, and ecological emergence respectively, may be understood to introduce evolutionary selection effects (PSEs) irreducible to any other level of analysis. Hence, regardless of whether there is a teleological drive to evolution as Harris and others have claimed, were this kind of argument to receive ongoing empirical corroboration it would provide pragmatic evidence of third-order emergence (ϕ) in ecosystems and the biosphere alike.

The above research has revealed that Neo-Darwinism is not incorrect per se, but merely incomplete in light of systems biology, some philosophical implications of which Harris has clearly anticipated. The resulting teleonomic philosophy of biology reveals a continuity between living and non-living systems, remains consistent with AE, and, I maintain, captures what one should expect from extending Bohm's implicate *order* to biological phenomena. In the final two chapters I examine Harris's arguments for extending his metaphysics into a theory of consciousness.

Notes

1. Harris purposes that initial emergence of living from the non-living may have been due to a "non-local faster than light influences" akin to the EPR experiment, which at "some threshold in the concrescence of convoluted polymers may have occurred in the primeval seas, which spontaneously initiated the generation and fecundation of living bacteria" (1991, 68). Harris made this appeal to defend his conclusion that living unity is implicit within and necessitated by prior processes of quantum mechanics. Though this line of consideration is beyond the scope of the present thesis, Harris appears to have anticipated what is now a flourishing area of research concerned with the Quantum Aspects of Life (Abbott, et al. 2008).
2. With this assertion, Harris (1991, 64) and later Thompson (2007, 152) follow Hans Jonas (1966), who articulated this very conception of freedom.
3. To this end, Harris follows Rupert Sheldrake's speculative postulate of the *morphogenetic field* to be supported by a marriage between chaos and biology (1991, 74). Since the connection of fractal geometry to biology has

already demonstrated itself to be a fruitful (e.g. Scott, 2007; Hancock, 2014), Sheldrake's postulation of a (metaphysical) morphogenetic field appears redundant to the extent that it succeeds and excessively speculative to the extent that it posits as yet unsubstantiated forces in nature.

4. Wolpert (2013) contends embryogenesis is a clear example of an increase in biological complexity over time, but one that reveals why evolution cannot be held responsible. "At best, one might argue that the embryo's increase in complexity arose via competition occurring in the past, between its ancestors and their antagonists. This is a rather torturous connection between a current rise in complexity of a system and the process of natural selection. It also doesn't address the possibility of hand-crafting an artificial embryo so that its complexity will rise in an artificial womb, without that womb having any ancestors" (247). In other words, biological self-organization whether of an individual cell or development through ontogenesis is presupposed by natural selection rather than its effect.
5. Towards a metabolism-first model, Davies (2008) considers the "citric-acid cycle" to be one example of an autocatalytic system serving as a self-organizing precursor of living beings, "which forms the basis of intermediary metabolism" (109). Zeravcica and Brennerc (2017) have made great strides towards such a model with their research on soft-matter self-organization involving colloidal spheres.
6. In this vein Ball (2009) has argued extensively that among other phenomena, lightning, convective roll cells, and river networks are structures "that arise because they offer 'channels' for relieving energy stress and producing entropy efficiently". Following a number of predecessors, he goes on to write, "life itself is an example of non-equilibrium regularity and structure", what he considers one of a number of "inevitable ordered forms, waiting to burst forth as soon as the universe gets the chance" (208). Additionally, see de Duve (2008), in which he considers life *as a cosmic imperative*.
7. Perunov et al. contend that "an endless variety of far-from-equilibrium many-particle Newtonian systems are capable of exhibiting self-organization phenomena in which strikingly patterned structures emerge in the presence of dissipative external drives" (2016, 23). The types of phenomena they have in mind include sand dunes, snowflakes, hurricanes, spiral bundles of protein filaments, and motors, a list that appears redundant after the above discussion concerning emergence. Indeed, the "nonequilibrium world offers many test cases for the general hypothesis that organized, kinetically stable structures emerge and persist because their formation is reliably accompanied by extra work absorption and dissi-pation" (ibid.).
8. Walker's (2014) theory is in line with Tononi's *integrated information theory* of consciousness. In either case, she holds "Φ quantifies how much

information is generated by a physical system when it enters a particular state through the causal interactions among its elements, above and beyond the information generated independently by its parts" (433). I return to this theory in Chap. 8.

9. According to Thompson, whereas an ant colony may be autonomous, it is not autopoietic because the components do not bear a circular causal relationship with its boundary. Similarly, *allopoietic* systems such as ribosomes or vesicles are not autonomous since they lack this circularity, their end product being different from themselves. Viruses are exemplified as systems that are also non-autonomous because outside of a host, they "do not exchange matter or energy with the environment" and are "completely inert [...] subject to the vicissitudes of the environment" (123).
10. Though space prevents my further discussion of the issue, Bohm's extension of the implicate order to the nature of living systems is largely in agreement with the contentions of this section and the previous. For further details see Bohm (1969).
11. In all his works, we find only one instance in which Harris mentions autopoiesis. He makes reference to Maturana and Varela's (1980) Autopoiesis *and* Cognition, to which he remarks that it is by no means new, originating from Aristotle, but that it is certainly in line with his understanding of the irreducibility of an organism. Nevertheless, he finds fault with the theory on account of its being potentially *relativistic* concerning how such a system makes meaning (2006, 171). For Harris's nearest anticipation of this theory see (1965, 327-28).
12. *Minimal* autopoiesis consists only of this maintenance of identity in the circular process of material exchange but it lacks a metabolic network. Autopoiesis plus adaptivity (i.e. active monitoring of internal homeostatic or homeodynamic mechanisms) are required for meaning-making according to Thompson (2007, 148-49). As he further admits, a living material system requires the proactive quality of the latter (e.g. searching for food). To my mind, the former may just as well be considered autocatalytic.
13. *Minimal* autopoiesis consists only of this maintenance of identity in the circular process of material exchange but it lacks a metabolic network. Autopoiesis plus adaptivity (i.e. active monitoring of internal homeostatic or homeodynamic mechanisms) are required for meaning making according to Thompson (2007, 148-49). As he further admits, a living material system requires the proactive quality of the latter (e.g. searching for food). To my mind, the former may just as well be considered autocatalytic.
14. These include a steady-state inheritance system (consisting of self-regulating and gene-regulating elements), structural inheritance (e.g. cytoskeleton characteristics being passed on through mitosis/meiosis),

chromatin marking (e.g. methylation patterns on DNA), and social structures including symbiotic functionality (Thompson, 2007 176-77).

15. Bacteria outnumber human cells 10 to 1 but compose only a few percentages of our body mass, while their number of genes outnumber human genes by 360 times. (http://www.nih.gov/news/health/jun2012/nhgri-13.htm). Hence, the human being is an ecosystem by definition!
16. Lineweaver and Davies (2013) likewise emphasize that information *in* DNA should be understood as being *about* its corresponding niches (i.e. like the organism, it is context dependent) and that it has even come from those same niches via natural selection. Davies contends that a gene is "a coded set of instructions for a ribosome to make a protein. Instructional information is clearly more than a mere bit string. There has to be a molecular milieu that can interpret and act on those instructions. In other words, biological information is contextual [...] Context is manifestly a global property. You cannot tell by looking at the local level whether this or that base pair in DNA is instructional information or just junk" (Davies, 2013, 33-34). Dehmelt and Bastiaens (2011) provide further sympathetic accounts to this end.
17. For further example, Koonin and Wolf (2009) have found that both Lamarckian and Darwinian views are important and capture *different* modalities of evolution: "in principle, the backward flow of specific information from the phenotype – or the environment viewed as extended phenotype – to the genome is not impossible owing to the wide spread of reverse transcription and DNA transposition. Highly sophisticated mechanisms are required for this bona fide Lamarckian scenario to work, and in two remarkable cases, the CASS and the piRNA system, such mechanisms have been discovered" (8). Additionally, Sarkies and Miska (2014) have proposed that mobile RNAi may provide short-term adaptation sufficient for the heritability of acquired characteristics (532).
18. Though not emphasized by Harris, it appears to follow from his articulation of "freedom" that as the species or "swarm" increases its specific capacity to alter an environment; over time it establishes a positive selection effect for specific functions in future generations. In this vein, Harris's position seems to assert that organic systems, both individual and collective, work toward transcending their own environmental and biological constraints by establishing their own. This issue will be further addressed in the following section.
19. Recognizing that his account is counter the central dogma of Neo-Darwinism, Krakauer argues from what appears to be just the kind of "genetic engineering" mentioned above: "DNA is always associated with histone proteins forming the chromatin complex. Histone proteins possess long tails that can be biochemically modified. Changing the chemical envi-

ronment of the cell leads to changes in the chemical composition of chromatin. Thereby DNA can be secondarily tagged along a repeating structure that is potentially as information-rich as the genome proper, and possesses the benefit of being directly modifiable within a single cell or organismal generation" (239).

20. Recent discoveries by Chen et al. (2017) reveal evidence that the self-organization of viable niches in bacteria, for example, is achieved by "large-scale collective oscillation" that (in agreement with Gleiser above remarks) once again takes the form of "travelling waves" via "symmetry breaking".
21. Rothschild (2006) concludes her analysis of emergence in biology precisely in line with an interpretation of "natural selection" as formal governance: the "proper locus of natural selection", or "evolutionary individuals could very well include the individual gene, organism, deme, and species [...] Indeed, one could argue for even more levels, such as the level of a tissue or subcellular organelle" (160).
22. Krakauer (2013) draws a similar conclusion concerning his own conception of evolution to be considered below, finding that the "nature" versus "nurture" dichotomy rests upon the false premise that "some traits are fixed whereas others are variable". He claims that this premise is false "because nothing about a species remains fixed over suitably long periods of time" (242).
23. In complement to this argument of organism-environment coupling, Krakauer has claimed that "there can be no information in the genome that is not already present in the environment", which implies that evolutionary complexity "is not a property of an individual but a population" (2013, 233).
24. For further discussions on *levels of selection* see Wilson (2002); Wilson and Wilson (2007); Lloyd (2007); and Serrelli and Gontier (2015).
25. Here Sepkoski claims macroevolution supports Gould's theory of punctuated equilibrium: exemplified through the case of a mass-extinction, much like the case of a forest fire, space is cleared for new life to emerge.
26. For a further treatment of the role symbiogenesis plays in evolution see Gontier (2015).
27. Estes et al. (2011) argue that perhaps the greatest human influence on ecology has been the widespread elimination of predators across the globe, which has been largely responsible for the mass extinction currently taking place. It is ironic, they point out, that due to the complex web of species interdependencies, "we often cannot unequivocally see the effects of large apex consumers until after they have been lost from an ecosystem, at which point the capacity to restore top-down control has been lost" (302).

28. Li's (2002) paper provides significant detail concerning how environmental variables are to be modelled such that "*a theoretical framework of ecological phase transitions*" is scientifically viable.
29. Further examples include amino acids, photosynthesis, wings, antifreeze proteins in the northern sea cod *Boreogadus saida* and Antarctic *Dissotichus nawsoni* (consisting of repeating threonine, alanine, and proline), sensitivity to long-wavelength green and red visual pigments, and serine protease molecules in bacteria (subtilisin) and vertebrates (trypsin), which have different structures but the same active site (Chela-Flores 2008, 158-59). Interestingly, Schloss (2008) has argued that convergent adaptations such as lungs, gills, guts, kidneys, organelles, and whole-organism branching morphologies create an internal, "three-dimensional" surface area that effectively endow living systems with an additional, "fourth dimension" (ff. 330).
30. Though he does not cite Kurakin or the Gaia theory, Smith (2013) has argued that "phase transitions" provide the appropriate paradigm for understanding the biosphere. He claims some universal patterns of life should be understood as the order parameters of these transitions. As the phase transitions of the biosphere are dynamical rather than equilibrium, he argues (consistent with Bohm and Harris) the "individuality" of living systems that compose it is "derived" and "emergent" from the "core metabolism" of the biosphere (210).
31. For further discussion concerning the connection between symbiosis and Gaia, see Hird's (2010). In this work she argues that the dependence of environmental activity and regulation upon bacteria through symbiosis and symbiogenesis can be seen as Gaia's fundamental actants, connecting life and non-living matter in biophysical and "biosocial entanglements". Hird also provides a lucid historical summary of how the Gaia theory has subsumed the theory of autopoiesis.

Works Cited

Abbott, Derek, Davies, Paul C. W., and Pati, Arun K, eds. *Quantum Aspects of Life*. Singapore: World Scientific Pub, 2008.

Ball, Philip. *Branches. Nature's Patterns: A Tapestry in Three Parts.* New York, NY: Oxford University Press, 2009.

Beschta, Robert L, and Ripple, William J. *Large Predators and Trophic Cascades in Terrestrial Ecosystems of the Western United States.* In Biological Conservation. Elsevier, 2009.

Bohm, David. *Some Remarks on the Notion of Order.* In Conrad H. Waddington, ed. *Towards a Theoretical Biology, 2 Sketches.* (Aldine, Chicago), 1969.

Chela-Flores, Julian. "Fitness of the Cosmos for the Origin and Evolution of Life: from Biochemical Fine-Tuning to the Anthropic Principle." In Barrow, John

D., Simon Conway Morris., Stephen J. Freeland., Charles L. Harper Jr, eds. 151-166. *Fitness of the Cosmos for Life: Biochemistry and Fine-Tuning*. New York: Cambridge University Press, 2008.

Chen, C., et al. "Weak Synchronization and Large-Scale Collective Oscillation in Dense Bacterial Suspensions." In *Nature*. Springer Pub, 2017.

Coffman, James A. "On Causality in Nonlinear Complex Systems: The Developmentalist Perspective." In Cliff Hooker, ed. *Philosophy of Complex Systems*. 287-309. The Netherlands, Amsterdam: Elsevier Pub, 2011.

Crist, Eileen and Rinker, Bruce H. *Gaia in Turmoil: Climate Change, Biodepletion, and Earth Ethics in an Age of Crisis*. Cambridge, MA: MIT Press, 2010.

Davies, Paul C.W. "Fitness and the Cosmic Environment." In Barrow, John D., Simon Conway Morris., Stephen J. Freeland., Charles L. Harper Jr, eds. *Fitness of the Cosmos for Life: Biochemistry and FineTuning*. 97-113. New York, NY: Cambridge University Press, 2008.

Davies, Paul C.W. "Directionality Principles from Cancer to Cosmology." In Charles H. Lineweaver, Paul C. W. Davies, and Michael Ruse, eds. *Complexity and the Arrow of Time*. 19-41. New York, NY: Cambridge University Press, 2013.

Dehmelt, Leif and Bastiaens, Philippe "Self-Organization in Cells." In Hildegard Meyer-Ortmanns., and Stefan Thurner, eds. *Principles of Evolution: From the Plank Epoch to Complex Multicellular Life*. 219-38. (The Frontiers Collection). Heidelberg, Germany: Springer Pub, 2011.

Estes, James A., et al. "Trophic Downgrading of Planet Earth." In *Science*. 301-306. vol 333., 2011.

Gleiser, Marcelo. "The Three Origins: Cosmos, Life, and Mind." In John D. Barrow., Paul C.W. Davies., and Charles L. Harper, Jr., eds. *Science and Ultimate Reality: Quantum Theory, Cosmology and Complexity*. 637-653. Cambridge University Press, 2004.

Gleiser, Marcelo. "Emergent Spatiotemporal Complexity in Field Theory." In Charles H. Lineweaver, Paul C. W. Davies, and Michael Ruse, eds. *Complexity and the Arrow of Time*. 113-131New York: Cambridge University Press, 2013

Gontier, Nathalie. *Reticulate Evolution: Symbiogenesis, Lateral Gene Transfer, Hybridization and Infectious Heredity. Interdisciplinary Evolutionary Research. Vol 3*. Switzerland: Springer Pub, 2015.

Grytsay, V.I., and Musatenko, I.V. "The Structure of a Chaos of Strange Attractors Within a Mathematical Model of the Metabolism of a Cell." In Phys. 677-86. Vol. 58, No, 2013.

Hancock, Ronald, and Jeon, Kwang W. eds. *New Models of the Cell Nucleus: Crowding, Entropic Forces, Phase Separation, and Fractals*. Academic Press, 2014.

Harris, Errol E. *The foundations of Metaphysics in Science*. London: Allen and Unwin Pub, 1965.

Harris, Errol E. *Formal, Transcendental and Dialectical Thinking: Logic and Reality*. State University of New York Press, 1987.
Harris, Errol E. *The Reality of Time*. Albany NY: State University of New York Press, 1988.
Harris, Errol E. *Cosmos and Anthropos: A philosophical Interpretation of the Anthropic Cosmological Principle*. Atlantic Highlands, NJ: Humanities Press International Inc, 1991.
Harris, Errol E. *The Restitution of Metaphysics*. Amherst New York: Humanity Books, an imprint of Prometheus Books, 2000.
Harris, Errol E. *Reflections on the Problem of Consciousness*. Dordrecht, Netherlands: Springer Pub, 2006.
Hird, Myra J "Indifferent Globality: Gaia, Symbiosis and 'Other Worldliness.'" In *Theory, Culture & Society*. Vol. 27 (2–3). 54-72. (SAGE, Los Angeles, London, New Delhi, and Singapore), 2010.
Hooker, Cliff. "Introduction to Philosophy of Complex Systems Part A: Towards a Framework for Complex Systems." In Cliff Hooker, ed *Philosophy of Complex Systems*. 3-90. The Netherlands, Amsterdam: Elsevier Pub, 2011.
Kauffman, Stuart A. *Investigations*. New York: Oxford University Press, 2000.
Kauffman, Stuart A. *Autonomous Agents*. In John D. Barrow., Paul C.W. Davies., and Charles L. Harper, Jr., eds. *Science and Ultimate Reality: Quantum Theory, Cosmology and Complexity*. 654-666. Cambridge University Press, 2004.
Kauffman, Stuart A. "Evolution Beyond Newton, Darwin, and Entailing Law: The Origin of Complexity in the Evolving Biosphere." In Charles H. Lineweaver, Paul C. W. Davies, and Michael Ruse, eds. *Complexity and the Arrow of Time*. 162-190. New York: Cambridge University Press, 2013.
Koonin Eugene V., and Wolf, YI. "Is evolution Darwinian or/and Lamarckian?" In *Biology Direct*. 4:42, 2009.
Krakauer, David. "The Inferential Evolution of Biological Complexity: Forgetting Nature by Learning to Nurture." In Charles H. Lineweaver, Paul C. W. Davies, and Michael Ruse, eds. 224-245. New York: Cambridge University Press, 2013.
Kurakin, Alexei. "The Self-organizing Fractal Theory as a Universal Discovery Method: The Phenomenon of Life." In *Theoretical Biology and Medical Modelling*. 8, 4, 2011.
Lennox, James G. "Darwinism and Neo-Darwinism." In Sahotra Sarkar., and Anya Plutynski, eds. *Cambridge Companion to the Philosophy of Biology*. 77-98. New York, NY: Cambridge University Press, 2007.
Li, Bai-Lian. "A Theoretical Framework of Ecological Phase Transitions for Characterizing TreeGrass Dynamics." In *Acta Biotheoretica*. 50, 3. Springer Pub, 2002.
Lloyd, Elizabeth. *Units and Levels of Selection*. (pp. 44-65) In Sahotra Sarkar., and Anya Plutynski, eds. *Cambridge Companion to the Philosophy of Biology*. New York, NY: Cambridge University, Press 2007.

Margulis, Lynn. *Symbiosis in Cell Evolution*. W.H. Freeman, New York, Second, 1993.

Margulis, Lynn. *Symbiotic Planet: A New Look at Evolution*. Basic Books, New York, 1998.

Margulis, Lynn. "Serial Endosymbiotic Theory (SET) and Composite Individuality: Transition from Bacterial to Eukaryotic Genomes." In *Microbiology Today*. 172–174. 31, 2004.

Maturana, Humberto R., and Varela, Francisco. *Autopoiesis and Cognition*. Boston, MA: Reidel Pub, 1980.

Maturana, Humberto R. "Autopoiesis." In Milan Zeleny ed. *Autopoiesis: A theory of the living organization*. Westview Press, Boulder, 1981.

Mitchell, Amir. et al. "Adaptive Prediction of Environmental Changes by Microorganisms." In *Nature*. Jul 9; 460 (7252): 220-4, 2009.

Moreno, Alvaro, Ruiz-Mirazo, Kepa, and Barandiaran, Xabier. "The Impact of the Paradigm of Complexity on the Foundational Frameworks of Biology and Cognitive Science." In Cliff Hooker. ed. *Philosophy of Complex Systems*. 311-333. The Netherlands, Amsterdam: Elsevier Pub, 2011.

Newman, Stuart A. "Complexity in Organismal Evolution." In Cliff Hooker, ed. *Philosophy of Complex Systems*. 335-354. The Netherlands, Amsterdam: Elsevier Pub, 2011.

Odenbaugh, Jay. "Complex Ecological Systems." In Cliff Hooker, ed. *Philosophy of Complex Systems*. 421-439. The Netherlands, Amsterdam: Elsevier Pub, 2011.

Peacock, Kent A. "Symbiosis in Ecology and Evolution." In Kevin DeLaplante,, Bryson Brown., and Kent A. Peacock, eds. *Philosophy of Ecology*. 219-250. Amsterdam, The Netherlands. Elsevier, 2011.

Perunov, Nikolai, Marsland Robert, and England Jeremy. "Statistical Physics of Adaptation." In Phys. Rev. X. 6, 2, 2016.

Rothschild, Lynn J. "The Role of Emergence in Biology." In Philip Clayton., and Davies, Paul C. W., eds. *The Re-Emergence of Emergence: The Emergentist Hypothesis from Science to Religion*. 151-165. New York. Oxford University Press, 2006.

Sarkies, Peter, and Miska, Eric. "A. Small RNAs Break Out: The Molecular Cell Biology of Small RNAs." In *Nature Reviews Molecular Cell Biology*, Vol 15. 525-535. Macmillan Publishers Limited, 2014.

Schloss, Jeffery P. "Would Venus Evolve on Mars? Bioenergetic Constraints, Allometric Trends, and the Evolution of Life-History Invariants." In Barrow, John D., Simon Conway Morris., Stephen J. Freeland., Charles L. Harper Jr, eds. *Fitness of the Cosmos for Life: Biochemistry and Fine-Tuning*. 318-464. New York: Cambridge University Press, 2008.

Schuster, Peter. "Physical Principles of Evolution." Hildegard Meyer-Ortmanns., and Stefan Thurner, eds. *Principles of Evolution: From the Plank Epoch to Complex Multicellular Life*. 45-79. (The Frontiers Collection). Heidelberg, Germany: Springer Pub, 2011.

Scott, Alwyn C. *The Nonlinear Universe: Chaos, Emergence, Life* (The Frontiers Collection). Heidelberg Germany: Springer Pub, 2007.

Sepkoski, David. "Macroevolution." In Ruse. Michael, ed. *The Oxford Handbook of Philosophy of Biology*. 211-237. New York: Oxford University Press, 2008.

Serrelli, Emanuele, and Gontier, Nathalie, eds. *Macroevolution: Explanation, Interpretation and Evidence. Interdisciplinary Evolution Research. Volume 2.* Switzerland: Springer Pub, 2015.

Shapiro, James A. "Bacteria are Small but not Stupid: Cognition, Natural Genetic Engineering and Sociobacteriology." In Stud Hist Philos Biol Biomed Sci. 807-19. 38, 2007.

Shapiro, James A. *Evolution: A View from the 21st Century*. Upper Sattle River, NJ: FT Press Science, 2011.

Smith, Eric. "Emergent Order in Processes: The Interplay of Complexity, Robustness, Correlation, and Hierarchy in the Biosphere." In Charles H. Lineweaver, Paul C. W. Davies, and Michael Ruse, eds. *Complexity and the Arrow of Time*. 191-223. New York: Cambridge University Press, 2013.

Stewart, John. "Foundational Issues in Enaction as a Paradigm for Cognitive Science: From the Origin of Life to Consciousness and Writing." In John Stewart., Oliver Gapenne., and Ezequiel A. Di Paolo, eds. *Enaction: Toward a New Paradigm for Cognitive Science*. 1-31. MIT Press, 2010.

Tabaczek, Mariusz. "The Metaphysics of Downward Causation: Rediscovering the Formal Cause." In *Zygon*. 380-404. Vol 48, Issue 2, 28 May, 2013.

Thompson, Evan. *Mind in life: Biology, Phenomenology, and the Sciences of Mind*. Cambridge, MA: Belknap Press of Harvard University Press, 2007.

Varela, Francisco J., and Goguen, Joseph A. "The Arithmetic of Closure." In *Progress in Cybernetics and Systems Research*. 48-63. vol. 3. New York: Wiley Hemisphere, 1978.

Walker, Sara I. "Top-Down Causation and the Rise of Information in the Emergence of Life." In *Information*. 424-39. Issue 5, 2014.

Walsh, Denis M. "Teleology." In Michael Ruse ed. *The Oxford Handbook of Philosophy of Biology*. 113-137. New York, NY: Oxford University Press, 2008.

Wilson, David S. *Darwin's Cathedral: Evolution, Religion, and the Nature of Society*. Chicago: University of Chicago Press, 2002.

Wilson, David S. and Wilson, Edward O. "Rethinking the theoretical foundation of sociobiology." In Quarterly Review of Biology, 2007.

Wolpert, D. H. "Information Width: a Way for the Second Law to Increase Complexity." In Charles H. Lineweaver, Paul C. W. Davies, and Michael Ruse, eds. *Complexity and the Arrow of Time*. 246-275. New York, NY: Cambridge University Press, 2013.

Yin, Gao, and Herfel, William. "Constructing Post-Classical Ecosystems Ecology." In Cliff Hooker, ed. *Philosophy of Complex Systems*. 389-420 The Netherlands, Amsterdam: Elsevier Pub, 2011.

PART IV

Dialectical Holism and the Metaphysics of Mind

CHAPTER 7

Harris's Reformation of the "Hard Problem"

7.1 Harris's Anticipation of an Enactivist Strategy

In the remaining chapters I evaluate Harris's appeal to the dialectical whole as a means of establishing a theory of consciousness. For Harris there are two "critical" points of transition in the scale of forms: "The first is from the physicochemical to the metabolic, marking the emergence of life. The second is from the sentient and perceptive to the fully self-conscious and self-reflective" (1991, 139). Neither of these transition points is abrupt, however. Each is considered an enfoldment of the preceding scales. In this section my aim is to elucidate Harris's conception of this second transition to mind in the context of human cognition.

As discussed in Sect. 6.2, the first critical transition from non-living to living systems is characterized by a constant adaptive response to elements of the system's environment based upon the sensation of meaningful stimuli. For Harris, the capacity of living systems to create information thus demonstrates more explicitly the principle of order and progression inherent in our universe. However, as with the previous scales, life demonstrates this principle "deficiently". He claims this is because in its more primitive forms living systems behave automatically and instinctively, and lack a *holistic cognition*, or awareness of itself-as-a-whole. This lack of holistic cognition means that the organism's "wholeness is still implicit, and its centreity, while recognizable by us as observers is not for itself a subject"

J. Schofield, *A Phenomenological Revision of E. E. Harris's Dialectical Holism*,
https://doi.org/10.1007/978-3-030-65029-2_7

(1991, 103). The simple organism's parts adapt to one another (establishing a self) and the organism can adapt to its environment (establishing awareness), but it nevertheless lacks *self-awareness*. Harris claims that the "mere organism, as such, is self-contradictory" (1991, 104).

In his use of the term "contradiction" Harris aims to mirror earlier phases of development involving the reconciliation of opposition leading to *sublation*, or synthesis of scales. At this stage, the organism is inevitably struggling against the system to which it belongs in pursuit of its own survival. Hence, Harris finds that the organism strives to reconcile this contradiction through an evolutionary process (Sect. 6.3.1). This process involves "an inherent urge" of the organism "to appropriate and internalize its other, to unite and integrate the outer world into its own self" (1991, 104). This he claims to be achieved in two ways: (i) "by absorption and ingestion of material substances required for the maintenance of its bodily integrity", and (ii) "in the quest for this sustenance, through sentience, the feeling and registration of external influences and conditions" (ibid.).

Taken together, these processes establish the "next phase transition in the self-differentiation of the universal principle of order, the next step in the scale of forms—what constitutes the matrix of succeeding phase, mentality" (ibid.). Harris has argued that the adaptive character of life arises from the requirements of self-maintenance of the organic unity so that living activity itself results in a progressive development, that is, increasing the agent's capacity to respond to environmental elements. His starting premise then is that the progressive enfoldment of complex systems (an aspect of $\mathfrak{E}$) leads to the second transition of mind. For the remainder of this section I outline six facets of Harris's theory of consciousness that will guide the discussions to follow:

(1) *The crucial problem*—Harris proposes our initial question thus: "How is what the brain does converted into what the mind does?"; and "how is what the mind does related to its body and the external world?" (2006, 21). These questions compose what Harris denotes the "crucial problem".[1] In his earlier work he considers whether observable behaviour "is a sufficient index of the presence of consciousness in the creature so behaving", but concludes that it "does not seem to be wholly sufficient" (1965, 293). One page earlier Harris supports this conclusion by reasoning that for any behaviour one considers indicative of mind, a *mechanism* can be

conceived that most would agree to be devoid of mind. In this way he finds the problem of mind is a discrepancy between material behaviour and mental awareness, such that behaviour will never be sufficient to derive conscious experiences.

(2) *Judgement*—building upon the *cogito*, Harris contends that judgement may be a non-mechanical, non-behavioural function of mind, sufficient for guiding further investigation:

> To doubt or deny is not to pronounce a form of words, to write them or in some other way to make them communicable (operations that might be performed by a machine). It is to judge, without necessarily communicating anything. It is, in short, to be in a state of mind or awareness with respect to some object. And if the subject of the judgment is one's own awareness, all doubt of its occurrence must be bogus and any denial false. (1965, 293)

Accordingly, judgement in the form of *valuation* provides a non-mechanical and pre-linguistic function of mind. Towards a clarification of this process, Harris compares it with the innumerable cases of previously learned intellectual and physiological skills that can be performed without immediate awareness. He maintains that no matter the marker of consciousness, it is "not *sui generis* but emerges from processes, which are of the same general kind and may with equal right be called mental, but which occur on a level of activity which is not conscious" (1965, 295). Judgement is here considered scalar, arising from numerous unconscious dynamics, it provides a *necessary* contextualization (or "*synthesis*"), and performs an *evaluation* upon some sensation(s) (374). It is crucial to bear in mind that such valuation purportedly occurs as an organizational feature of experience, that it is transcendental, and not to be confused with the judgement of analytic thought.

(3) *Scale of forms*—Harris hopes to dissolve the problem by investigating consciousness as a phase of evolving material systems on the one hand and of non-conscious *mental* processes on the other (1965, 296). Harris assumes that "this is not a relationship between two different substances or separate entities. It is the relationship between different phases of a continuous organic process" (ibid.). To support this line of reasoning, Harris partially follows the

gestalt neural-identity theory, which he uses to depict both mental and physical domains as respective *fields*:

> [N]either behaviour nor experienced phenomena can be adequately explained in terms of the conjunction and conglomeration of single units, whether simple reflexes or single sensations [...] both the physiological aspect and the phenomenal must be treated as Gestalten, wholes of parts which exist inseparably and only in the whole. (302)

Harris hereby dismisses any *corpuscular* conception for the units of mind or matter maintaining there is an "identification of the psychological with the neural 'field' [...] an identity of structural order—an isomorphism" (302–3).[2] In further agreement with gestalt psychology he maintains the sensum is created by a selective activity of attention and it is never simple, rather it is "constituted by interrelations" of neural and psychical fields, each exerting internal (albeit unconscious) "'forces' of organization" (334).

(4) *Dynamic constraint*—Harris reasons that neural and sensory processes are subject to thresholds of impulse intensity resulting from what he calls "a fusion—a summation of subliminal pulses" (1965, 305). Although it is not yet know how this process of summation takes place, he suggests it may be "rooted in the metabolism of the cell body" (ibid.).[3] Likewise, he claims that the nervous system acts as an organic whole that is governed by the wider whole of the total organism, which is in turn "governed by an auturgic principle which unifies it in all its multiform variety" (ibid.). He goes on to say that "at a definite level of integration this unity can only be expressed as, and become effective in, behaviour which is 'informed,' in both the senses 'given form' and 'guided by knowledge'" (ibid.).

Harris later elaborates that although form and structure are generally synonymous, as complexity increases in the scale of forms the difference between these terms becomes significant and "mere spatial pattern is transcended [...] The proposition now being advanced is that the integration of physiological processes at a high degree of complexity and intensity assumes a new form, the experience of feeling" (1991, 105). This claim may be more fruitfully cashed out to say that as a kind of ***organization***, conscious behaviour is not so much a matter of ***capability***—since, as noted above, all ***capacities of a particular level*** might be mechanically instantiated—but a matter of ***dynamical constraint*** across numerous ***levels of activity***:

> There are not two agencies influencing each other, but two or more levels of activity, the lower ancillary to, sublated in, and integrated with, the higher, which at a certain stage of development is conscious (or mental) activity. Each level, of course, has effects upon every other for they all (in the higher phase) belong to a single, though complex, dynamic totality. (1965, 309)

This is to say that "feeling" (phenomenality) and "knowledge" result from and eventually guides, the ***sublation*** of disparate phases or wholes (both biological and mental). The "***form*** of organized physiological functioning", is what Harris considers ***conscious behaviour*** (2000, 180). As with the emergence of preceding scales, Harris claims that such "integration" is analogous to the unity of disparate lines into a new form, which suggests that analysis in terms of ***dimensional transformation*** may be relevant (2000, ff. 178).

(5) *Anticipation of Neurophenomenology*—Harris (1965) argues that while insights from phenomenology have significantly influenced psychologists, their insights have been overlooked by philosophers. Towards a unity of phenomenal experience and the analysis of behaviour, he again appeals to the gestalt conception of a psychical field. Harris argues that the perceived "units" of sensation must not only be considered dialectically related to its gestalt, but that this field is likewise inseparable from the spatiotemporal world:

> For the alleged datum, be it colour-patch, sound or tweak of pain would not be the whole of what was presented but would necessarily occur against a contrasting background and in succession to contrasting predecessors. In awareness there are no mere blanks, the absence of stimulation is itself a stimulus. Consequently, the so-called sense-datum cannot be taken as separable from its spatio-temporal context together with which it makes up an organized field. (322)

Harris here contends that the *unit* of mind must be associated with the spatiotemporally extended field of awareness as a whole. Thus, he recognizes that the epistemic aim to "build up our knowledge of the external world from atomic data" is one that has been long since antiquated (324).

Harris argues that mental organization is "continuous with biological and biological with physical". Each he maintains reveals "some active *nisus* to wholeness and integration recognizable even in the spatio-temporal matrix of the physical world. The principles sought are those of polyphasic unity, no less immanent in the operation of the physical laws than in the exercise of conscious mind" (341). In consequence, he finds that the activity of mind involves both practical behaviour and cognitive processes, each characterized by its own scientific methods, and each inseparable from the other within a sufficient theoretical account of consciousness:

> There is no reason to regard either of these methods as superior to the other in scientific rigour [...] neither the behavioural nor the phenomenological approach is sufficient by itself. Each demands and illuminates the other, as is to be expected if the object of study is at once both a pattern of movements and a form of experience. (341–42)[4]

(6) *Antithetical Ontologies*—Harris recognizes the relations between his own position and a number of alternative theses of mind. First and foremost, Harris holds that the worldview of twentieth-century science is radically different from the mechanistic picture that informed Descartes's *dualism* and so the respective views are incompatible (1965, 290). Harris disavows the traditional *identity theories* that mind is *nothing but* a neurological process or bodily behaviour because the agreeable *correlation* is insufficient for *causal explanation* (1965, 300–1). If (as discussed above) nature manifests in degrees of holism, then at lesser degrees of integrity qualities "prefiguring" mind should be identifiable (1991, 107). Harris argues his thesis is also incompatible with Alexander's *emergentism* because positing a disparate supervening quality "breaks the continuity of evolutionary process" (1965, 309). He admits that his monistic account bears resemblance to *panpsychism* in "its insistence on continuity between the graded phases of self-specification of the organizing principle, which is universal to the entire scale" (1991, 107). Panpsychism is dismissed however, because on Harris's account the Concrete Universal is considered non-mental,

> since mind requires "a special and specific degree of wholeness before it emerges" (ibid.). *Epiphenomenalism* is likewise rejected since by Harris's lights, what mind contributes is influential (though non-causal) for the system that has it (1991, 108).

In sum, the emergence of mind is believed to have resulted from the explicative process of self-differentiation "that has been operative throughout, active now with a higher degree of unified complexity and self-determination" (1991, 109). Harris holds that "a monistic position can be set out, to which the neural-identity theory is a sort of halfway house, which is better able to account for our common experience and which can still accept the dualist's reason for wishing to separate the mind from the body" (2000, 168). What Harris means is that while there is nothing beyond the physical, mind is considered irreducible to *bodily behaviour*. Nevertheless, at a particular point in the system's development, "self-maintenance of the system is possible only if the organism's activity is informed by the character of the total situation within which it is reacting. At this stage, its organic responses must be so closely integrated that they become fused into a single unity" (2000, 176). In other words, self-maintenance is possible if the system is *informed* by the relations necessary for the maintenance of its respective complexity. Here Harris aims to describe bio-physical self-maintenance in such a way that bridges the Cartesian chasm. Hence, he must provide a valid line of reasoning demonstrating how self-maintenance (as a bio-physical process) gives rise to phenomenal awareness (i.e. mental *states*).

7.2 *Ideatum*: The Idea of the Body

In this section I compare Damasio's (2010, 2003) works with Harris's original (1973, 1995, 2006) arguments that contemporary neuroscience supports a metaphysics of mind based on key principles of Spinoza's system; a thesis Harris maintains follows from the wider metaphysics considered heretofore.[5] In addition, I compare these results with more recent enactivist theories. My aim is not to defend a particular interpretation of Spinoza, nor to establish soundness of the resulting theory of mind, but rather to demonstrate a common and coherent thread that runs through the works considered, that is, a *Spinozian embodiment*.

Harris finds Damasio's (1994, 1999, 2003) works particularly illuminating (unlike Dennett) because Damasio "makes no attempt to deny or explain away consciousness" (2006, 63). For Spinoza, Harris explains, there is a distinction between *attributes* as apparent aspects of nature, and *Substance*, which is the actual monist reality. "Body and mind are one thing (*res*), one entity, not two. They are substantially one and the same; but the essence of the one substance is expressed differently under different attributes, and what is, under the Attribute of Extension, a body, under the Attribute of Thought, is an idea or mind" (1995, 53). Damasio later writes that by "keeping separate levels of description" he is not endorsing substance dualism but instead appealing to Spinoza's "*aspect* dualism" (2010, 44). Both Damasio and Harris vehemently refute the homunculus, direct realism, and Cartesian dualism. Towards this end they depict the mind in terms of what Spinoza calls the *ideatum*.

Positing an interpretation of perception and learning in line with embodied cognition, Harris claims that "perceptual discrimination of objects is not an immediate revelation but a skill which we learn in early infancy by a gradual process [...] an activity of organization, of selection, of distinction, correlation and synthesis in which objects are constructed out of the deliverances of feeling" (1965, 336). As noted above, he here claims that feelings and their synthesis are fundamental to mind. Following Spinoza, he later argues mind is the idea of the body, "not the conception of the body, nor its mental image, not a complex model of its physiological functioning constructed by abstract scientific thinking, but the feeling or self-sentience of the body—sensory-awareness" (1995, 63). Bodily feeling is thus (coincident with Peirce's *firstness*) "the basic level of consciousness" (2006, 111).

Importantly, in one of his earliest texts on Spinoza, (drawing from the Ethics II, xiii) Harris noted that despite our perceptions to the contrary, an implication of the ideatum is that, "when one speaks or thinks of oneself as 'feeling' other bodies (e.g. tactually), it is only when and because one experiences a specific sensation in some part of one's own body" (1973, 81). Harris hereby recognizes that our sense organs are transparent to the subject. Positing that the body is primary in awareness means questioning our very *conception* of bodies beyond our own in all cases—not merely those in which we suspect that our body is *malfunctioning*.

> The mind is the idea of the body as a whole, not only of the nervous system; and, no doubt, different physiological activities are reflected in it by differences in idea, but the awareness of the body that constitutes the mind is not

> a clear and distinct idea of the body and its internal processes, but (at any rate, in the first instance) a confused idea. (82–83)[6]

If awareness is always of whatever is being processed through the body, then the *idea* of self and world is mistaken because we do not readily have awareness of its nature in Substance. Harris proposes that Damasio's *protoself* is a neuro-cognitive analogue of the *ideatum* (2006, 64).

Damasio (2010) agrees that "the body is a foundation of the conscious mind. We know that the most stable aspects of body function are represented in the brain, in the form of maps [...] located below the level of the cerebral cortex, in [...] the upper brain stem" (14). What Damasio calls "maps" are "not a mere copy, a passive transfer from outside the brain toward the inside" (44). Rather, maps consist of interoceptive, proprioceptive, and exteroceptive *primordial feelings that* together compose the first stage of his model of consciousness: the "*protoself*" (51–52). Damasio claims the operation of upper-brain-stem nuclei are responsible for bringing the body to mind, which "is the ultimate expression of the brain's intrinsic *aboutness*, its intentional attitude regarding the body" (63).

Importantly, Damasio maintains that over time mappings of the body permanently influence the body itself. This he claims, is because "neurons are about life and about managing life in other cells of the body, and that that *aboutness* requires two-way signalling" (2010, 62). Such feedback systems Damasio believes to be based upon a "biological value" ever working towards *homeostasis*, whether in the case of a cell, an individual or a social system. This *value*, he maintains, is a principle fundamental to mind and evolution:

> Biological value has influenced the evolution of brain structures, and in any brain it influences almost every step of brain operations. It is expressed as simply as in the release of chemical molecules related to reward and punishment, or as elaborately as in our social emotions and in sophisticated reasoning. Biological value naturally guides and colours, so to speak, almost everything that happens inside our very minded, very conscious brains. Biological value has the status of a principle. (Damasio 2010, 17)

Following his arguments presented in Chap. 6, on Harris's (1965) account, biology reveals a "value judgement" under the guise of "survival value", which has become a criterion of success in evolution. This, he claims, has shown that value is "consequent upon auturgic wholeness".

He goes on to argue that if we identify the intelligence of "social interaction and theoretical formulation" with the same self-organizing *nisus* heretofore considered, we are justified in holding intelligence to be a higher phase of *evolution* (425–26).

Although Damasio denies *mind* to individual cells, in addition to *value*, he grants that *intentionality* and *will* are evident in neurons (25). Consequently, he recognizes that by granting intentionality to neurons, the "body-brain separation" appears "somewhat exaggerated since the neurons that make up the brain are body cells," a point, he recognizes, that has significant "bearing on the body-mind problem" (203). In this vein, self-maintenance may be a means to bridge the mind-body chasm if, as Damasio contends, the process by which cells of the body maintain life is one of self-sentience, a process that gives rise to our felt conception of self and world.[7]

Both Damasio and Harris follow Spinoza's (*Ethics* II) conception of "*idea ideae*", or the idea of ideas. Damasio considers this to be crucial for "representing relationships and creating symbols" and he holds that it "opens a way for creating an idea of self" (2003, 215). He goes on to claim that the most basic kind of self (in humans) is a "second-order idea" because

> it is based on two first-order ideas—one being the idea of the object that we are perceiving; the other, the idea of our body as it is modified by the perception of the object. The second-order idea of self is the idea of the relationship between the two other ideas—object perceived and body modified by perception. This second-order idea I call self is inserted in the flow of ideas in the mind, and it offers the mind a fragment of newly created knowledge: the knowledge that our body is engaged in interacting with an object. (Ibid.)

In neuro-cognitive terms, this capacity depends upon *reentrance and recursion*, which refers to signals that in addition to "going forward along a single chain, also returns to the origin, looping back to the ensemble of neurons where each element of the chain begins" (2010, 60). This is part of what he calls the "massive interconnectivity" of "mind-making regions" in which, "a high complexity of cross-signalling is achieved, a feature that in the case of the cortex is amplified by cortico thalamic interlocking" (ibid.).

Harris appears sympathetic to this approach, but he is philosophically more fine-grained in his discussion. He claims the "active" idea of the human body

> is conscious of itself, is self-reflective and so self-transcendent, because being idea it is (*ipso facto*) *idea ideae*. Thus it can grasp in idea the relation between ideas and the structural principles of configurations, which are registered in, but not apprehended by, bodies as such. This reflectiveness (as we are told in TdIE) enables the human mind to clarify itself of confusion and to make its knowledge adequate. (1995, 86)

While most phenomenologists agree that all consciousness is self-conscious, contrary Harris, they would add that it is conceptually "pre-reflective" (Gallagher and Zahavi 2012, 70). For Harris, the *idea ideae* manifests as a *judgement* of one's relation to their *object*, but does not involve an infinite regress of ideas. Such a regress, Harris asserts, is unnecessary for consciousness and impossible in practice (1973, 88). While this "judgement" is initially non-conscious, it involves a cognitive *decoupling of object from background that is coincident with negation*—for example, what if not x relation between y and z—thus permitting "intelligent" clarification of the self-world relation (coincident with Peirce's *secondness*). Hence, Harris appears consistent with the contemporary phenomenological account given that his meaning of "reflectivity" only concerns an inherent recursive dynamic within consciousness.

What follows from this recurrent picture of cognition Damasio calls the "not even *unpopular* view" in neuro-psychology that mind begins in the brain-stem (further supported by cases of hydranencephaly). The philosophical implication of the proto-self is thus a monistic view of consciousness described as a continually regenerated neurobiological process identified with the *homeostasis*, which generates feeling-centred maps of the body-in-relation. The body is thus conceived "as the rock on which the protoself is built, while the protoself is the pivot around which the conscious mind turns" (Damasio 2010, 15). Consciousness, for both Harris and Damasio, is achieved when the idea of the body becomes *self-referential*, thus achieving a greater level of constraint on the dynamics of the body. In what follows I consider how this line of thought conceptualizes emotion in comparison with more recent accounts from AE.

7.2.1 Sensorimotor-Enactivism and Affectivity

On Harris's (1965) account, cognition is not "all there is to consciousness for every cognition is, at the same time and by the same token, emotionally toned and conatively propulsive" (367). In clear anticipation of sensorimotor contingencies, "perception" Harris maintains, "is conditioned and limited by the organic neuro-muscular conditions of sensibility and the spatio-temporal circumstances of the percipient" (429). He avers that muscular tensions and bodily movements "do (perhaps always) accompany thinking and that internal problem-solving and reasoning are sometimes assisted, though at other times hindered, by them" (440). However, he recognizes that "some thinking occurs even in the absence of specific muscular tensions", which brings him to conclude that "there is a sense in which we think with our whole bodies, but it is because thinking is the organization of total behaviour, not the functioning of special organs" (ibid.).

What is meant by "organization"? In his later (1988) work, Harris elaborates this point by arguing that there is a "close association of cognition with impulse and emotive tone [...] even in the coldest and driest theoretical speculations of developed consciousness"; and consciousness, or perception, cannot "emerge independently of instinctive urges and the biological needs" (81). Developing upon his interpretation of the *ideatum*, Harris holds that cognition is dependent upon emotions that are further developments of underlying biological needs. Hence on his view, emotion serves as an organizational constraint that facilitates both biological survival and cognitive awareness.

With regard for Spinoza, Harris elaborates that just as the body is an "active metabolic and physiological system" (by maintaining dynamic equilibrium), for Spinoza the idea is not a passive replica of its object, but an "activity of thinking" (1995, 53–54). He goes on to say "self-sentience of the body, what for Spinoza is the idea of the body, includes everything of which I am sensuously aware; and it is from this, and on this as foundation, that all my 'ideas' of whatever kind are elaborated" (1995, 55). Following Spinoza (*Ethics* III), the physical *affects*, such as joy and desire, are primary, with the *passions* (e.g. lust, fear, shame) considered secondary and constructed out of these *ideas*. Again, it is because our ideas are fragmentary that our practices result in error and suffering (1973, ff. 113). That is, "they are only partial and leave out significant aspects or factors belonging to the objects they present" (1995, 59).[8] To ground his view in neuropsychology, Harris appeals to Damasio's conceptions of the *somatic marker*, *convergence-divergence zones* (CDZ), and *core-self* respectively (2006, 72–79).

On Damasio's account, the body (and structures with which it relates) acts as a distributer of cognitive functions, speeding up reaction time by using physiological cues corresponding to cognitively constructed *dominions*. Following William James, Damasio claims the mind distinguishes such dominions of body, mind, past, and present by attributing emotions and feelings to their perception, feelings that serve to separate self from not-self:

> The somatic marker does not need to be a fully formed emotion, overtly experienced as a feeling. (That is what a "gut feeling" is.) It can be a covert, emotion-related signal of which the subject is not aware, in which case we refer to it as a bias. The notion of somatic markers is applicable not just to high levels of cognition but to those earlier stages of evolution. The somatic marker hypothesis offers a mechanism for how brains would execute a value-based selection of images and how that selection would translate in edited continuities of images. In other words, the principle for the selection of images was connected to life management needs. (2010, 123)

Similarly, Thompson (2007) has argued these "emotion-based signals" or "affects" serve as an "allure" operating prior to conceptual or conscious information processing in order to differentiate our field of awareness into a "dynamic gestalt or figure-ground structure" (374). Following Thompson, Maiese (2014) contends this produces a "caring-contoured mapping of one's surroundings, so that one can immediately focus one's cognitive attention. This serves to directly bias the competition for processing resources in favour of information one feels is important" (237). By implication, Maiese concludes in apparent agreement with Harris that to the extent AI systems

> fail to exhibit the dynamics of living systems, they are not self-regulating, autonomous agents, and their sense-making is not physically grounded in autopoietic and metabolic processes. Thus, there is good reason to think that they cannot have emotions like ours and are incapable of making sense of their surroundings via affective framing. (239)

Neurologically speaking, Damasio proposes that *convergence-divergence zones* (CDZs) are formed through a subject's neuronal *dispositions* to respond to certain stimuli in characteristic ways based upon the strengthening or weakening of synaptic connections over time. More specifically a "CDZ is an ensemble of neurons within which many feedforward-feedback loops make contact" (2010, 101). CDZs receive "feedforward" connections from "earlier" sensory signal-processing networks and also sends reciprocal feedback signals to those areas. Additionally, CDZs project "feedforward" signals to regions located in the next connectional level of the chain and receives signals from them.

Damasio holds that the activity at the end of the processing chains occurs within what he calls a "dispositional space", made up of microscopic CDZs *and* macroscopic convergent divergent regions (CDRs) in the *association cortices.* These regions are supposedly not engaged in image-making but provide necessary assistance for this process. Though the dispositional space (e.g. in *anterior medial temporal cortices*) contains what are called *grandmother neurons* "whose activity correlates with the presence of a specific object", Damasio holds that this activity does not in itself create recognizable mental images of objects and events without successive *retroactivation*: "To recognize or remember our grandmother, we must reinstate a substantial part of the collection of explicit maps that, in their entirety, represent her meaning. Like mirror neurons, so-called grandmother neurons are CDZs" (107). Thus, for Harris, AE, and Damasio, experiences such as "a grandmother", cannot be isolated from the affective and neurological contexts within which they arise, nor can they be reduced to individual neurological states.

Establishing the second stage of Damasio's argument, he proposes the *core-self*: "a pulse of core self is generated when the proto-self is modified by an interaction between the organism and an object and when, as a result, the images of the object are also modified [...] the modified images of object and organism are momentarily linked in a coherent pattern" (127). This is to say that a momentary *sense of self* is possible when a subject's bodily sensations are influenced by an object such that both CDZs and CDRs are altered by the subject's association and identification of the object-in-relation to oneself. Again, this appears as a more detailed account of what may be considered a *constraint* instantiated by *attentional judgement* (Sect. 7.2, arguments 2 and 4).[9] Thus, Harris appears justified in claiming "corroboration" of what he considers Spinoza's position, in that "perception is not simply the reception of a 'datum' but is an active interpretation of the sensory input requiring at least implicit judgement" (2006, 74).

Following Merleau-Ponty, Brender (2013) relates sensorimotor theory to his conception of symmetry-breaking in a manner that is further supportive of Harris's and Damasio's conceptions of embodiment. Cognition, he argues, is contingent upon the asymmetry of an environment, upon which a body may act, and without which, the organism will receive no perceptual answer to the questions posed by its motor movements. Bodily movement is thus "the original 'transformation' [...] in the body's perceptual field", one that brings a meaningful world into being:

> The particular asymmetries a body perceives will depend on its particular way of moving, the unique motor habits it has developed over the course of its life. As our movements become more complex and asymmetrical, so too does the world we perceive. Thus the organism and its world grow together dialectically, each driving the other to become more articulated and determinate through its own increasing determinacy. This is the growth of sense: the self-articulating field of differences that make a difference to the organism. (271)

Connecting Brender's comments to the above discussion, I suggest that for AE it is precisely the primitive feelings of the proto-self, *together* with their sensorimotor couplings to the environment that undergoes symmetry breaking, thus giving rise to a core-self characterized by particular affective and cognitive capacities ($\mathfrak{C}$). Indeed, a growing body of experimental evidence and theoretical works are revealing that conceptual content is often constituted by sensorimotor and affective systems, rather than disembodied abstract symbols.[10] As Robbins (2013) argues:

> [A]ffect serves what were traditionally believed to be purely cognitive functions. These functions include the consolidation of memory, anticipation of the future, evaluation of the other people and things in the world, and, especially, through the integration and modulation of the process of the transduction of sensation into complex perceptions, the production of a personally meaningful world—a world articulated explicitly by phenomenology but which is lived primarily at a preverbal, implicit level of awareness. (6)

In this vein, Gallagher (2014) contends that enactivist accounts based upon sensorimotor contingencies (e.g. O'Regan and Noë 2001; Noë 2004) are too narrow due to their "neglect of the relevance of the affective aspects including proprioceptive and kinaesthetic aspects—factors that should be of high interest since they derive from movement and contribute to one's practical grasp of sensorimotor contingencies" (234). Gallagher argues that while our affective states are explained by and dependent upon operations of the rest of our body, an affective state such as hunger significantly influences cognitive judgements and sensorimotor abilities. While the subject's environment affords various possibilities for action, each has an affective price tag: "One thus not only has a practical (sensorimotor) understanding of accessibility, but an affective take on that same accessibility [...] a perceptual sense of the ease or difficulty of making something present" (236).

As Thompson (2015) summarizes, the nervous system enacts a self/not-self distinction through a relational process established between *efferent motor signals* for action and the *reafferent sensory signals arising from action*:

> On one side lies a unique sensorimotor perspective that constitutes the subject of perception and the agent of action. On the other side lies the environment as the meaningful locale of perception and action. In this way, sensorimotor I-making and sensorimotor sense-making arise together and are inseparable; they're dependently co-arisen. (334)

At this point it is clear that the Spinozian embodiment to which Harris appeals can be more technically established by Damasio's neurological conception of the emergence of core-self, which can in turn be further philosophically articulated by AE. On this view, feelings, affect, and emotions are not dissociable kinds of process from those capacities commonly considered cognitive, such as reasoning. Moreover, higher-level cognitive capacities reliant upon recursive thought (*idea ideae*) may be both dependent upon and constituted by affective sensorimotor associations. With these sentiments, Harris's place within the AE camp is increasingly evident. However, positing cognition in these terms requires that Harris and others must address to what extent the world is *constructed* by our participation—an issue to which I return in Chap. 8. What remains to be clarified in the following section is how Harris relates to the embedded thesis. It is through this consideration that his proposed relationship between self-maintenance and consciousness may be further elucidated.

7.3 *Conatus*: The Self-preservation of Consciousness

According to Harris, Spinoza's contention *conatus in suo esse perseverandi* denotes the constant effort inherent in all things, to persevere in their own being. This is an effort to increase one's power of action, which provides the incentive for self-improvement. "Accordingly, the effort to persist in one's own being is *ipso facto* the endeavour to act solely from one's own essence or nature" (1995, 58). At the human level of organization Harris contends, "*conative* efficacy" occurs as an enfoldment of simpler feelings manifesting as "emotional states" that are "consequent upon the disturbance of equilibrium" (1965, 324). So, for Harris, the very same self-preservation identifiable in living systems occurs in and through the self-organization of emotions, which (purportedly) permits the complexity of human function. "If this is remembered" he finds, "much misleading talk confusing organic and psychological tensions with mechanical pushes and pulls may be avoided and many mis-conceptions entertained in machine-theorizing may be corrected" (1965, 325).

Anticipating later embodied and embedded conceptions of active meaning-making (such as *affective allure* mentioned above), Harris makes two particularly important points that exemplify the *conatus* of human consciousness: (i) In perception "[l]ong-standing sets may be established which will determine definite perceptual dispositions, as when, for instance, a distant dust cloud is seen by a general as the movement of troops, by a herdsman as a movement of cattle" (1965, 402); (ii) "empathic elements may be contributed by past motor responses (e.g. the appearance of a hill as 'rising,' or a road as 'winding'). Thus, the general set or adjustment of the organism determines the way things appear and emphasis is placed upon the active integrative and organizing process involved" (403).

So, for Harris there is "no such thing as 'pure observation'", and perception is a process by which the subject non-consciously *projects* previously constructed assumptions and generates hypotheses that are either confirmed or disconfirmed so as to revise their original assumptions (1965, 416). As a means of philosophically interpreting this gestalt conception, Harris appeals to Spinoza's argument that causation of any kind between the attributes and their respective modes is denied; "Ideas cannot cause motions nor motions ideas" (1973, 85). However, he argues ideas are "not epiphenomena" since they can become "genuine actions of the mind and are identical in substance with bodily acts involving motion conductive to the health and preservation of the bodily organism" (ibid.). Hence, on this account, the *constraint* of emotions on the basis of *self-maintenance* leads to the emergence of a conscious *self*. To connect his theory to contemporary neuroscience, Harris appeals to Damasio's "autobiographical-self".

In the third stage of Damasio's argument, the autobiographical-self develops as an aggregate of the core-self and is necessary for all the most characteristic qualities associated with consciousness such as identity and self-awareness through time. Accordingly, this cognitive function (responsible for our use of knowledge recorded in innumerable external mediums) is generated when identifiable "objects in one's biography" stimulate the core self to produce a "large-scale coherent pattern" (2010, 127). He claims that this occurs in two stages. (1) A significant set of "biographical memories must be grouped together so that each can be readily treated as an individual object" (150). These networks are then capable of modifying the protoself and re-producing the core-self. (2) Due to their complexity, the brain needs devices capable of coordinating the relevant recall of memories with respect to identifiable objects, situations, and so on:

> From a neural standpoint the coordinating process is especially complicated by the fact that the images that constitute an autobiography are largely implemented in the image workspaces of the cerebral cortex, based on recall from dispositional cortices, and yet, in order to be made conscious, those same images need to interact with the protoself machinery, which, as we have seen, is largely located at brain-stem level. (Ibid.)

The autobiographical self, Damasio contends, involves capacities of memory and reflection that depend upon networks across numerous neurological and bodily strata. However, he further notes that biographical memory "depends on the brain's ability to produce not only mental representations that imitate reality slavishly and mimetically but also representations that symbolize actions and objects and individuals" (2010, 206).

At this point Harris diverges from Damasio's conception of the autobiographical self for its reliance upon representations but follows Spinoza in what he claims is a *nominalist* position (1973, 95). In a later work Harris notes, "it is to the neurophysiologist, who is conscious of the neural activity and its consistent relation to something, and to us who perceive the image, that the relation is represented. There is no implication that the function of the brain is representation" (2006, 65). He goes on to point out that even if such representations were permitted, we would remain puzzled by "how such codes could be decoded to present an image to the conscious subject" (66). Harris complains that Damasio holds the unacceptable position that not only is the brain representational but "all thought consists of images" and "all thinking and acquired knowledge, for him, is conscious" (66–67). For his part however, Damasio does appear to avoid criticisms concerning representation and the non-conscious in his later work: "Nuclei are filled with 'dispositional know-how,' the sort of knowledge that does not require detailed mapped representations" (2010, 62).

Harris's concern is that the brain cannot be responsible for management of the (autobiographical) self because this involves "assessment of circumstances and decision", neither of which seems explicable in terms of representations by neural states (68). In an earlier work, he argues the subject must make some kind of distinction between the concept as remembered and the object it represents, between the present remembering and the past remembered—what is essentially an act of *judgement*: "To do all this, my remembering cannot be confined either to the present event or to the past but must transcend both, as well as the relation between them" (1988, 88). With these contentions, Harris denies that self-awareness is to be either *spatially* located in neurological states or *temporally* located in phenomenal representations of a given moment.

Harris's solution to this problem of locating the idea of the self may be considered in a couple of steps. First, following Spinoza, Harris maintains "sensations in themselves are neither true nor false" (1973, 91). Error, he thinks, consists in our omitting something without our awareness that is pertinent to the idea that we affirm. "It is in virtue of this unrecognized omission that the idea we affirm is confused. Error, for Spinoza, therefore, is mere privation or abstraction—the omission of essential fact or implication" (92). Consequently, no idea or feeling represents something in the world since "what is felt is not just the original activity by itself [...] it is in intimate organic relation to the whole environing universe" (2000, 180). Second, Harris claims perceptions (somatic and kinaesthetic) are "interpretations of impressions in the light of their context and past history. This is a perpetual process of judging [...] measured by its coherence and self-consistency" (2000, 192). I discuss this resulting coherence theory in Sect. 8.3.1, but for now it is only important to orient Harris's rejection of "representations".

In consideration of Steiner's (2014) thorough analysis of enactivist arguments against representationalism, Harris defends the most extreme position against three representationalist theses:

(1) The *role* of representation in cognition and bodily function, that is, against mind independence.
(2) The *formats* of representation, meaning functions like language, meaning memory retrieval and problem solving are not abstract.
(3) The *existence* of representations in the world (44–46).

In contrast to classical cognitive theory, nothing can represent anything else in dialectical holism because such signification depends upon the knowledge of an entire system within which the ***particular relation*** may be *sufficiently* individuated. In other words, nothing can be representational, nor correspond to anything ***in and of itself***, unless it were constructed to do so at the level of the system-as-a-whole (e.g. software created by ***observers***). In the latter case, claiming something is ***representational*** is to provide no more than a mere tautology.[11] As proponents of AE recognize however, taking this position does not conflict with ***methodological*** representationalism.

Harris concludes that self-awareness is not only achieved through "brain activity" but also "inseparable from social activity" (2006, 69). He goes on to argue that Damasio's conclusion that self-awareness depends upon sociality is not strong enough because it still permits us to say that judgement is represented by neural states:

> A moral point of view is possible only for a self-conscious mind, which is aware of the self in its social relations and can make judgments of right and wrong that may induce it to oppose its own instinctual inclinations. How and where patterns of neural transmissions in the brain bring about such self-conscious reflection is puzzling enough [...] but that such patterns could themselves amount to moral judgments [...] hardly makes sense. (2006, 69)

Bearing in mind the discussions from Sects. 7.2 and 7.3.1, the basis of such judgement must be the feeling of self-organization and preservation of *the system as a whole*. At the level of human consciousness, judgements of the *whole* (being within a social context) fall under the heading of morality. Though Harris does not make all of them explicit, three contentions follow: (1) sociality can obtain iff subjects establish "a moral point of view"; (2) if self-awareness depends upon sociality, a moral point of view is necessary for self-awareness; and (3) if moral judgements are *constituted* by social relations beyond the brain, self-awareness is irreducible to neural states.[12] In the next section I consider this line of reasoning in greater detail to clarify Harris's conditions for self-awareness.

7.3.1 Embedded Persons and Responsible Minds

For Harris, elucidating the *conatus* of consciousness involves an analysis of *responsible action*, which he calls the activity of reason that becomes possible only for reflective, self-aware beings, or *persons*:

> Personality and self-consciousness, however, develop only in a social setting and are elicited only through communication with other people, recognized equally as rational agents. Responsible action is, therefore, social action, and its performance presupposes social relations and social structures. (1988, 107)

He goes on to argue "history is the self-knowledge of mind [...] on the grounds that to know oneself is necessarily to be aware of oneself as a social being, and that requires the *a priori* concept of the historical past" (1988, 133). Following Spinoza and consistent with many pragmatists, Harris elaborates what he considers an implied coherence theory:

> [I]nadequate ideas, if properly supplemented and set in proper context, approach the truth [...] There they are fully adequate, completely filled out, and fitted into the web of their actual relations to all others. The nearer ideas approach to this completeness (as is equally the case with their *ideata*) the more reality, and the more perfection, they possess. (1995, 59)

Taking these points together, proper self-consciousness can obtain (and be *maintained*) only on a *moral* and *historical* basis: that is, networks in which a given agent faces pressures to establish judgements concerning the actions and dispositions of other agents with cognizance of the *generative* context (coincident with Peirce's *thirdness*). Achieving *adequate ideas* further implies the agent is capable of re-cognizing their *identity* (self-relation) with respect to other agents and alternative points of view with increasing systematicity. Perhaps another way to understand this proposition (Sect. 7.2, argument 4) is to say consciousness obtains iff the ideatum establishes a relation to (becomes about) some socially constructed metaphysical selection effect (MSE) and cognizes *judgements* about this relation. Here I claim Harris has moved into the domain of *situated* cognition.

Providing general support for Harris's position, Damasio and others have argued that in the absence of emotionally charged somatic markers (as in subjects with damage to the vmPFC), the subject is incapable of effectively determining moral value of outcomes. This emotional impairment leads to moral malfunction, which supports the position that rationality is insufficient for moral reasoning (Strejcek and Zhong 2014). Concerning morality and cognition, after reviewing a range of empirical studies on social isolation after (relatively) normal development, Doris and Nichols (2012) have concluded that "[r]ationality is not only socially *developed*, it is socially *sustained*; in many domains, optimal reasoning occurs when the reasoning process occurs as part of a social process" (434).

Following Husserl's *phenomenology of* culture, Havelange (2010) contends that empathy reverses the *natural attitude*. Comprehension of the "Other" he maintains, comes not from apprehension of an objective body. Rather, empathy provides a "substrate of immediate meaning" that permits the communicative intention of the Other to engender a corresponding intent in the subject. The shared bodily and linguistic expressions he maintains, enables the subject "to intuitively understand the experience of the Other by 'putting myself in his place' and to predict his behaviours" (344).

Similarly, Thompson (2015) argues that social cognition, perspective taking, and empathy are closely related to the ability to cognize the self/not-self distinction. He finds that "being able to think of oneself as a self seems inseparable from being able mentally to grasp an outside view of oneself, that is, from the vantage point of the other" (345). In defending this stance, he appeals to studies from developmental psychology, arguing

> these two mental abilities arise together and build on the capacity for shared or joint attention. Joint attention emerges around nine months of age and comprises the threefold structure of a child, an adult, and something to which they know they are both giving attention. It includes activities such as gaze following (reliably following where one or the other person is looking), acting together with shared objects (such as toys), and imitative learning (the child's acting on or using things the way adults do). (345–46)

He concludes that "self-experience emerges from I-making processes that are fundamentally social and intersubjective" (347).[13,14]

In contrast to *theory of mind* (ToM) approaches that emphasize mind reading, Gallagher's (2014) phenomenological argument holds that the *mirror neuron* system is activated not merely as imitation of another, but *anticipation* of another agent's behaviour. He defends what he calls *the predictive coding model*:

> Intersubjective interaction is not about mindreading the mental states of others, but about directly perceiving their intentions and emotions in their postures, movements, gestures, facial expressions, vocal intonations, etc., as well as in their highly contextualized (by physical environment, social roles, culture, etc.) actions. (238)

These sentiments are in line with AE (and dialectical) evolution as described above, in that cognitive capacities do not arise through intrinsic or genetic alterations alone, but in and through behaviour, which in the human context is necessarily social, moral, and emotional. Concerning the importance of historical cognition, in Bayesian terms, Gallagher further claims that neurological dynamics of anticipation are hierarchical, involving synaptic inhibition "based on an empirical prior—something that depends on the organism's previous experience and context-sensitive learning" (241). Neural states are thus to be understood not as holding intrinsic content, but as being meaningful only in and through the agent's history (see Clark 2016 for further details concerning prediction theory).

Moreover, it is the historical environmental interaction (beyond the brain) that provides the means for and determines ongoing processes of "top-down synaptic inhibition":

> Such inhibitory patterns constitute a prediction which is then matched against ongoing sensory input. If there is a mismatch, i.e., if the new stimulus generates a different firing pattern than the one anticipated, prediction errors are sent back up the line and the system adjusts dynamically back and forth until there is a relatively good fit. (241)

This dovetails with Harris's view that modification of connections sufficient for anticipating future processing relies upon both prior processing and constraints across numerous spatial scales. Harris recognizes that the bi-polarity between self and not-self is not black and white because on occasion, "the body, or part of it, is excluded from the self; at others the self may come to include what, in different circumstances, would be regarded as parts of the not-self" (1965, 338).[15] In agreement with both Bohm (1980) and Silberstein and Chemero's (2015) ENM (Sect. 2.5), Gallagher maintains the "explanatory unit" of perception, cognition, or action, should not be considered the brain, nor some number of brains in social interaction,

> but a dynamic relation between organisms, which include brains, but also their own structural embodied features that enable specific perception-action loops involving social and physical environments, which in turn effect statistical regularities that shape the structure and function of the nervous system. (242)

How then, does this line of reasoning conceive of *self-awareness*? Harris claims that the capacity of a cognitive system to "extricate itself" so as to grasp a structure as a whole means that mind is self-transcendent. The problem is that the *self* remains ephemeral because it must be capable of grasping and he maintains, *going beyond* a differentiated whole of experience, and thus cannot be for itself an *object*: "no such judgements and no awareness of relations could occur without a subject who transcends the single terms of the relationship" (1991, 142). For Spinoza, Harris writes:

> Ideas, so far as they are the awareness of finite extended things, involve the awareness of the limits which define their objects, and such awareness *ipso facto* surpasses those limits; for if it did not it could not be an awareness of them. Thus the self-awareness or idea, of the body, by its very ideal nature, transcends the limits of that body both spatially and temporally. (1995, 66)

Importantly, Harris does not mean to claim that self-awareness is *ideal* in a Platonic sense. Rather, the conscious subject cannot be part of its gestalt, nor the sum of bodily parts at any one time because for Harris (and AE), *it does not even exist at any one time or place*! Harris concludes that while the self is dependent upon and partially constituted by relations with its world, its transcendence of these relations is required to afford knowledge of the world.

Damasio says, "[t]he oddest thing about the upper reaches of a consciousness performance is the conspicuous absence of a conductor before the performance begins, although, as the performance unfolds, a conductor comes into being" (2010, 16). This implies that "No single mechanism explains consciousness in the brain, no single device, no single region, or feature, or trick, any more than a symphony can be played by one musician or even a few" (17). In agreement with Harris and AE, the *explanatory units* of mind are relations across both spatial and temporal scales: "Human experience is always continuous, not discrete; embedded, not isolated; temporal, not eternal" (Gallagher et al. 2015, 115).

Harris's concluding remarks regarding the *conatus* reveal important insight into how he believed the process of mind could be depicted as continuous with that of evolution:

> The *conatus* produces reactions to [the passions] in the endeavour to preserve the self, which take the form of appetites and desires accompanied by the appropriate emotions. But the *conatus* is the endeavour of the thing (*res*)—in this case the human being—to maintain itself in its own essence [...] The *conatus*, therefore, is, in its proper and most adequate expression, the endeavour to increase the power of action of the human self. Ideas corresponding to action (as opposed to passion) are always adequate ideas, and the power of action is the power of reason. (1995, 102)

In summary, for Harris the conative effort of self-preservation that characterizes consciousness is one of historical sensitivity and social relatedness; it is an effort to increase freedom via self-knowledge. Consciousness then involves the effort by an agent to become *about* one's own aboutness by becoming increasingly sensitive to the (endogenous and exogenous) *implicit* dynamics that instantiate their gestalt. For AE and Harris, awareness of this kind provides a means to regulate one's passions and vicissitudinous environments. What Harris attempted to convey, I maintain, is that the *self-preservation of consciousness* is the same process of *internalization* or *broadening of phenomenological horizons* posited in dialectical evolution, the only difference being a matter of scale. I submit that the arguments from Harris and recent proponents of AE can reach the common conclusion that for humans, the *conatus* is to be identified with the capacity to re-cognize one's momentary ideatum *as socially situated.* This process results in increased *freedom* and stands as a necessary condition of consciousness: *Conscientia est ego-transcendentia ad infinitum.*

7.4 Conclusion

Towards establishing a connection between self-maintenance and consciousness Harris has argued that sensations "are differentiations of feeling or sentience, which supervenes on a certain high degree of integration of physiological functioning involving specially integrated patterns of activity at lower levels" (2000, 178). He notes that the "lower levels" consist not only of neural discharges "but probably all, or most of, the metabolic and physiological functions of the body" (ibid.). While he recognizes the importance of the nervous system and brain, he claims, "the whole system of the organism is involved" (2000, 179). Moreover, as discussed above, the process of consciousness is considered inseparable from social networks and historical context. Accordingly, self-awareness emerges not from some isolated neural process, but from one's living, historical, and social *lifeworld* as a whole.

In Sect. 7.3, I have highlighted Harris's anticipation of Damasio's neurological model of mind and sensorimotor enactivism. With Harris, Damasio has recognized the relevance of many of Spinoza's concepts for contemporary neuropsychology (such as *ideatum* and the centrality of emotion). However, I have argued that Damasio's appeal to representations at the level of the biographical-self is incompatible with both Harris and AE. In clear agreement with Harris's conception of the self, enactivists give emotional affect a central role in their theory of sensorimotor cognition. This convergence of ideas was further exemplified in Sect. 7.4, concerning Harris's and AE's arguments that social, moral, and historical relations are partially constitutive of self-awareness.

With this convergence of ideas having been established, I am now in a position to evaluate Harris's revision and naturalization of the *crucial problem*: "It is not a matter of bridging the gap between first-person revelation and third-person description [...] What needs to be overcome is the chasm between third-person description of brain functioning and third person description of conscious awareness" (2006, 60). What remains to be seen is the extent to which the above *descriptive model of mind* (neuropsychological-heuristic) can be expanded into an *ontological explanation*. Towards this end, in accordance with his above (Sect. 7.2) conception of (3) *the scale of forms*, Harris appeals to DST; hence in the following chapter I examine the possible fruits of this approach.

Notes

1. Harris (2006) cites but does not discuss Chalmers's (1996) *The Conscious Mind*, within which the "hard problem" was famously posited.
2. Note that this point does not contradict Thompson's (2007) proposal of neurophenomenology as cited in Sect. 2.3.2 because—as will become increasingly evident—Harris's "isomorphism" is not a reductive identity of mind to behaviour or vice versa, nor does he imply some set of facts that can be read in one way or the other. Rather, in line with Bohm and ENM, both refer to one and the same phenomenon that is *not captured by either mind or* behaviour *alone*.
3. Though they do not cite AE, Hollis et al. (2009) argue, "[i]f metabolism is the primary function of a nervous system, then an elegant theory would be one in which cognitive activity emerges out of metabolism. Such a theory would begin to bridge the chasm between laws of physical processes and cognition" (212).
4. Harris (1965) describes the emergence of a psychical field with the case of awakening from a head injury in a way reminiscent of symmetry breaking: "The state of consciousness described is one of fogginess in which vague inhomogeneities gradually emerge, as the field of awareness becomes progressively organized. The first distinguishable elements are sensory, vaguely unpleasant (in this case) and emotionally coloured. Then the field becomes bipolar with a self distinguished from a not-self [...] From the first there is some inhomogeneity and vague differences in sensory modalities are felt before the self is distinguished as the subject of the experience. The awakening consciousness is from the start an activity of organizing and distinguishing" (314). In a later chapter Harris clarifies that in these lowest stages of awareness "is the nearest approach to homogeneity in the psychical field that is ever experienced" (331). This again suggests that Harris recognized (on some level) despite his later writings, that to speak of a homogenous sensory field is contrary to his system.
5. My reasons for excluding God from the discussions that follow are (i) I maintain that insofar as Harris's logic relies upon Spinoza's reasoning, it can be reduced to a NM without any loss to the force of his resulting system; (ii) much more argument is needed to conclude that the neutral Substance is *Godly* than either Spinoza or Harris provided; and (iii) space prohibits my sufficient consideration of this issue.
6. This interpretation of Spinoza is corroborated by van Bunge et al. (2011), who explain that for Spinoza, "[a]ll knowledge of the human body (E2p19), the human mind (E3p23) and the external bodies (E2p26) begins with affections of the body" (145). What Spinoza called our confused and fragmentary knowledge of the first kind (E2 29).

7. These assertions are particularly interesting considering that Damasio makes no mention of AE in his (2010, 2003, or 1999) works, and offers only brief agreement with Varela et al.'s (1991) *The Embodied Mind* (2003, 308; and 1994, 234).
8. Note the connection between what is here considered the inadequacy of ideas and what has been discussed above as the MSE-PSE fallacy (MPF). See van Bunge et al. (282–83) regarding Spinoza's conception of the *passions*.
9. I fully appreciate how nuanced is the philosophical terrain concerning the nature of "self" and thus how drastic a leap I am making by moving forward in the discussion. Innumerable texts have been devoted to defining the concept and Harris certainly took pains to elucidate what he meant by *self* (e.g. 1965, 315–32; 1987, 83, 117, 238, 259; 2000 ff. 250; 2006, ff. 111). Unfortunately, this topic requires too deep a divergence into phenomenology and neurophenomenology than I currently have space to consider. Thus, the following section will have to suffice for my present purposes.
10. For example, Niedenthal et al. (2014) have found "the body's reproduction of parts of an emotional experience constitute conceptual content for emotion. The words "disgust" and "interest" are not mentally grounded by disembodied symbols but are grounded by parts of the bodily state that are re-enacted to support perception and thought" (247). See Colombetti (2013) for further details concerning affectivity, phenomenology, and AE; see Pessoa (2015) for an extensive discussion on the *cognitive-emotional brain*.
11. In the context of memory, Bechtel (2009) contends that "information is often encoded in a very distributed fashion in which there is no single locus for the engram" (25). He goes on to explain that one consequence of this distributed representation scheme is that the representations or memories are "subject to catastrophic interference as the learning of new information alters the connections that maintained the previously learned information" (ibid.). Hence, even if we admit the term *representation* in philosophical discussion, it may not provide sufficient grounds for a correspondence theory, given how dynamic brains and worlds actually are.
12. Unfortunately, I do not have space to elaborate the fascinating empirical evidence to this end, but the following summarizes two relevant lines of reasoning. In evolutionary neuroscience, Striedter (2005) has argued that the last significant burst of brain growth for Homo sapiens 100,000 years ago was not due to a change of diet or genetics alone, but "altercations with fellow humans" that spurred language and culture (ff. 318). In evolutionary psychology, Corballis (2007) maintains that there is a link between the FOXP2 gene mutation and the mirror system (Broca's area), proposing that this mutation has provided a means to incorporate vocal control into (and thus depended upon) our abilities to mirror physical

actions. "It provides further evidence that the FOXP2 mutation was the final stage in a series of adaptations that allowed speech to become autonomous" (585).

13. Famous examples abound in social psychology of mal development resulting from severe neglect in children that has been shown to arrest moral and reasoning capacities associated with human consciousness (van der Horst and van der Veer 2008).
14. Harvey et al. (2016) have maintained that interactivity reveals a shortcoming to the enactivist conception of operational closure: "where enactivism describes agent-environment relations in terms of in-the-moment coupling, we assume these relations play out on multiple heterogeneous timescales, such that in many cases—perhaps most—agency is fundamentally distributed" (235). Their worry appears to be satiated by considering Thompson's (2007) appeal to generative *phenomenology*, which concerns the *historical, social, and cultural becoming of human experience*: "To investigate the life-world as horizon and ground of all experience therefore requires investigating none other than generativity—the process of becoming, of making and remaking, that occurs over the generations and within which any individual genesis is always already situated" (36). Evidently Harris's conception of historical cognition is in deep agreement with generative phenomenology (1988, ff 79).
15. This and other lines of reasoning concerning the *noösphere* (1991, 139, ff. 149, 173; 1992, 3–4; 1988, 149, 151) and *collective mind* (1995, 90–91) warrant further discussion with contemporary arguments in *extended mind* and *collective intentionality.*

Works Cited

Bechtel, William. "Molecules, Systems, and Behavior: Another view of Memory Consolidation." In John Bickle, ed. *The Oxford Handbook of Philosophy and Neuroscience.* 13–40. New York NY: Oxford University Press, 2009.

Bohm, David. "The Enfolding-Unfolding Universe and Consciousness." In Lee Nichol, ed. *The Essential David Bohm.* 78–120. New York, NY: Routledge,1980; 2003.

Brender, Noah M. "Sense-Making and Symmetry-Breaking: Merleau-Ponty, Cognitive Science, and Dynamic Systems Theory." In *The Canadian Journal of Continental Philosophy.* 246–270. 17 (2) (Fall/Autumn), 2013.

Chalmers, D. J. (1996). *The Conscious Mind: In Search of a Fundamental Theory.* NY New York, Oxford University Press.

Clark, A. (2016). *Surfing Uncertainty: Prediction, Action, and the Embodied Mind.* Oxford University Press.

Colombetti, Giovanna. *The Feeling Body: Affective Science Meets the Enactive Mind*. Cambridge MA: MIT Press, 2013.

Corballis, Michael. "The Evolution of Consciousness." In Philip D. Zelazo, Morris Moscovitch, and Evan Thompson, eds. *The Cambridge Handbook of Consciousness*. 571–595. Cambridge, England: Cambridge University Press, 2007.

Damasio, Antonio R. *Descartes' Error: Emotion, Reason, and the Human Brain*. New Avon Books, 1994.

———. *The Feeling of What Happens: Body, Emotion and the Making of Consciousness*. London Heinemann, 1999.

———. *Looking for Spinoza: Joy, Sorrow and the Feeling Brian*. Orlando, FL: Harcourt Books Inc, 2003.

———. *Self Comes to Mind: Constructing the Conscious Brain*. New York, NY: Random House Inc, 2010.

Doris, John M., and Nichols, Shaun. "Broad-Minded: Sociality and the Cognitive Science of Morality." In Eric Margolis., Richard Samuels., Stephen P. Stich. *The Oxford Handbook of Philosophy of Cognitive Science*. 425–453. New York, Oxford University Press, 2012.

Gallagher, Shawn, and Dan Zahavi. *The Phenomenological Mind*. 2nd Ed. New York, NY: Routledge Press, 2012.

Gallagher, Shawn. "Making Enactivism Even More Embodied." In Tomasz Komendziński, Przemyslaw Nowakowski, and Witold M. Wachowski, eds. *Enactivism: Arguments and Applications*. 232–247. Avant Vol. 2, 2014.

Gallagher, Shaun, Lauren Reinerman-Jones, Bruce Janz, Patricia Bockelman, and Jörg Trempler. *A Neurophenomenology of Awe and Wonder: Towards a Non-Reductionist Cognitive Science*. New York, NY: Palgrave Macmillan, 2015.

Harris, Errol E. *The foundations of Metaphysics in Science*. London: Allen and Unwin Pub, 1965.

———. *Salvation from Despair. A Reappraisal of Spinoza's Philosophy*. Martinus Nijhoff, The Hague, 1973.

———. *Formal, Transcendental and Dialectical Thinking: Logic and Reality*. State University of New York Press, 1987.

———. *The Reality of Time*. Albany NY: State University of New York Press, 1988.

———. *Cosmos and Anthropos: A philosophical Interpretation of the Anthropic Cosmological Principle*. Atlantic Highlands, NJ: Humanities Press International Inc, 1991.

———. *Cosmos and Theos: Ethical and Theological Implications of the Anthropic Cosmological Principle*. Amherst New York: Humanity Books, an imprint of Prometheus Books, 1992.

———. *The Substance of Spinoza*. Amherst New York: Humanity Books, an imprint of Prometheus Books, 1995.

———. *The Restitution of Metaphysics*. Amherst New York: Humanity Books, an imprint of Prometheus Books, 2000.

———. *Reflections on the Problem of Consciousness*. Dordrecht, Netherlands: Springer Pub, 2006.

Harvey, Matthew I, Gahrn-Andersen, Rasmus, and Steffensen, Sune V. "Interactivity and Enaction in Human Cognition." In Tom Froese, Vörös, Sebastjan, and Alexander Riegler, eds. *Exploring the Diversity within Enactivism and Neurophenomenology*. 234–245. *Constructivist Foundations*. Vol 11, 2, 2016.

Havelange, V. (2010). *The Ontological Constitution of Cognition and the Epistemological Constitution of Cognitive Science: Phenomenology, Enaction, and Technology*. (pp. 335–59) In: Stewart, J., Gapenne, O., & Di Paolo, E. A. (Eds). *Enaction: Toward a New Paradigm for Cognitive Science*. MIT Press.

Hollis, Geoff, Kloos, Heidi, Van Orden, Guy C. "Origins of Order in Cognitive Activity." In Stephen J. Guastello, Matthijs Koopmans, David Pincus, eds., *Chaos and Complexity in Psychology: The Theory of Nonlinear Dynamical Systems*. 206–241. Cambridge University Press, 2009.

Maiese, Michelle. "Body and Emotion." In Shapiro, L. A. ed. *The Routledge Handbook of Embodied Cognition*. 231–239. New York, NY. Routledge press, 2014.

Niedenthal, Paula, Wood, Alison, and Rychlowska, Magdalena. "Embodied Emotion Concepts." In Lawrence Shapiro, ed. *The Routledge Handbook of Embodied Cognition*. 240–249. New York, NY. Routledge Press, 2014.

Noë, A. (2004). *Action in Perception*. Cambridge, MA: MIT Press.

O'Regan, K. and Noë, A. (2001). *A Sensorimotor Account of Vision and Visual Consciousness*. Behavioral and Brain Sciences 23, 939–973.

Pessoa, Luiz. "Précis on The Cognitive-Emotional Brain." In *Behavioral and Brain Sciences*. 1–66. Cambridge University Press, 2015.

Robbins, Brent D. "Enactive Cognition and the Neurophenomenology of Emotion." In Susan Gordon, ed. *Neurophenomenology and Its Applications to Psychology*. 1–24. New York: Springer, 2013.

Silberstein, Michael, and Chemero, Anthony. "Extending Neutral Monism to the Hard Problem." In *Journal of Consciousness Studies*. 181–94. Vol 22, No. 3–4, 2015.

Steiner, Pierre. "Enacting Anti-representationalism. The Scope and The Limits of Enactive Critiques of Representationalism." In *Avant: Trends in Interdisciplinary Studies*. (2):43–86, 2014.

Striedter, Georg F. *Principles of Brain Evolution*. Sunderland, Mass: Sinauer Associates, 2005.

Strejcek, Brendan, and Zhong, Chen-Bo. "Morality in the Body." In Lawrence A. Shapiro, ed. *The Routledge Handbook of Embodied Cognition*. 220–230. New York, NY. Routledge Press, 2014.

Thompson, Evan E. *Mind in life: Biology, Phenomenology, and the Sciences of Mind*. Cambridge, MA: Belknap Press of Harvard University Press, 2007.

——. *Waking, Dreaming, Being: New Light on the Self and Consciousness from Neuroscience, Meditation, and Philosophy.* New York: Colombia University Press, 2015.

van Bunge, Wiep, Krop, Henri, Piet, Steenbakkers, van de Ven, Jeroen M.M. eds. *The Continuum Companion to Spinoza.* Continuum International Pub, 2011.

van der Horst, F. C. P., and van der Veer, R. *Loneliness in Infancy: Harry Harlow, John Bowlby and Issues of Separation.* In Integr Psych Behav 42: 325–335. Springer: DOI 10.1007/s12124-0089071-x, 2008.

Varela, Francisco., Thompson, Evan., and Rosch, Elenore. (1991). *The Embodied Mind: Cognitive Science and Human Experience.* Cambridge MA: MIT Press.

CHAPTER 8

Bridging Philosophies of Consciousness and Cosmology

8.1 Consciousness as a Phase of Matter

In the previous chapter I elucidated Harris's cognitive-metaphysical model for the emergence of consciousness. Harris took pains to show that the findings of contemporary sciences were best systematized by appealing to metaphysics of process, dialectical relations, and a series of unifying principles. With these conceptual guides he argued that it is possible to discuss both "subjective" and "objective" phenomena without being hindered by the Cartesian Chasm. Additionally, he argued consciousness can be considered continuous with evolution (as discussed in Sect. 6.3.1). What remains to receive a head-on treatment in this chapter is Harris's ontological depictions of mind and knowledge within the holist metaphysics considered heretofore.

In his later work Harris reiterates his earlier (Sect. 7.2) contention (3), that if mind and neural activity are the same phenomenon, merely described with different languages (as the dual aspect theorist claims), "there must be some similarity or common character in their 'logics.' They must be mutually convertible in some way, as mathematically isomorphic calculi are mutually convertible" (2000, 171). What is this common language? As previously discussed, Harris posits an ontology of self-organizing scales, each dialectically related and inseparable from the Concrete Universal that embodies the forms of Nature. Specifically, the process by

J. Schofield, *A Phenomenological Revision of E. E. Harris's Dialectical Holism*,
https://doi.org/10.1007/978-3-030-65029-2_8

which this scale of forms undergoes self-differentiation is considered the very same process that instantiates consciousness. By linking phenomenology and neuroscience to 𝕰, Harris believed we could establish a common language that resolved the crucial problem of relating mind to matter.

Consciousness, Harris argues, occurs through a *graded* scale, in which earlier grades of presumably lesser complexity "can exist without the later, but not *vice versa*". Hence, "thought and intellect" are unnecessary for earlier phases such as locomotion and appetite, but these capacities cannot develop without the "lower vital functions" (2006, 98). As previously discussed, the lower and higher functions taken together establish Harris's *conditions of consciousness* (2006, ff. 150):

1. Self-organization (entropy management)
2. Embodiment (sensorimotor affectivity and implicit judgement)
3. *Interrelation with other organisms* (normativity-selection)
4. Organization *of emotions* (the explicit judgement of one's gestalt)
5. *Modification of one's gestalt worldview* ("adaptive thought" of a *transcendent ego*).

Following from Harris's original conception of auturgy towards consciousness, phenomenal awareness is by this reasoning understood as emerging through a graded scale. The complexity of cognizance is considered a range of "self-specification" from greater to lesser homogeneity (i.e. symmetry-breaking) of the "psychical field" (1991, 109). Attention, he claims, is established through the construction of "a figure-and-ground datum" that differentiates the lifeworld or gestalt of attention in the same manner as that found in the physical world. Each involve "the same organizing activity as that operative throughout nature [...] It is the same organizing principle that integrates the physical cosmos and unites the biosphere, which unifies conscious experience and ensures the integrity of the experienced world" (1991, 132).

He goes on to claim that "all are mutually continuous dialectical phases or specific forms in the necessary self-differentiation of the universal totality" (ibid.). Here Harris arrives at his TAP: if consciousness is a scale of nature, then it is considered a necessary phenomenon within the history of the Universe. Hence, what remains to be established is an argument that depicts the function of consciousness (as depicted heretofore) as one such scale.

Contrary to the brain/hardware, mind/software arguments of the 80s and 90s that attempted to paint a "grandmother neuron" picture of neuronal activity, Harris maintains that contemporary science has depicted brain function as holistic. "The sensory cortex was found to be hierarchically responsive, successive layers reacting to more detailed and refined characters of objects presented" (2006, 83). Neurons, he maintains respond dynamically under varying condition and are in a continuous flux of death and regeneration. This he believed "undermined the pixel interpretation of neural performance" (84). Population coding suggests that the average firing strength of neural populations, rather than individual neurons, were correlated with consciousness. This implies, "neural reactions were not so much digital as global. They were integrated and inseparably conjoined, so that groups acted as wholes to produce their output" (ibid.). Harris argues that a more accurate approach to the science of mind going forward is *dynamic systems theory* (DST).

Harris finds that in order to represent neural activity computationally, due to the incredible sensitivity of the brain, adjustments have to be made in the measurements relating input to output, which seemed so small as to make no intelligible difference to the outcome of calculations. Chaos theory becomes relevant in this context because it has "discovered that the minutest difference in initial conditions could, in the course of non-linear processes, result in enormous deviations in the eventual outcome" (2006, 85). He goes on to say that what appeared on the surface to be chaotic was in fact a complex order emerging from the total process guided by a strange attractor. "It was impossible to explain the consequent effect by analysing the process and reducing it to disparate items. The organizing principle of the whole is what determined the detail of the total structure" (ibid.). Citing the non-linearity of "re-entrant" feedback loops and their subsequent cascade affects, Harris posits that "the reaction of the brain to any input is never predictable as is the response of a computer" (86). Thus, Harris finds that chaos provides viable methodological approach for neuropsychology.

In this vein, Harris appeals to Edelman and Tononi's (2000) work, which argued the synchronicity of neural firing patterns, rather than specific neurons or nuclei, explains for example how object recognition may be achieved.[1] Harris finds that memory is most adequately conceptualized by the dynamic core because it has been found to arise through the collective efforts of numerous "scattered nuclei" throughout the whole brain and changes in accordance with the subject's experiences. "It is dynamic, not static; that is, it has not only developed over the time of maturation, but is continually changing with varying experiences and learning" (2006, 88). While the dynamic core is meant to model the contents of consciousness, Harris believed it to "occur as one undissectable operation", in which conscious acts both constrain and are dependent upon peripheral non-conscious goals, emotional cues (i.e. value systems), learned motor functions, etc. (89). Following Greenfield (2000) and McCrone (1999), Harris (2006) claims that imaging technology of contemporary neuroscience has provided support for his argument that there is no one-to-one correlation between phenomenal experience and brain region activation. Rather, experiences must be understood as global processes wherein the brain regions correlated with experience execute their functions through an interplay with one another and the body as a whole.

For Harris, the emergence of self-consciousness in and through this *dynamic core* is considered the "actualization of the potentialities of all the earlier stages, each successively matter to the next as form" (2006, 98). Harris maintains this relationship is "analogous to that between a magnetic field and the molecular structure of the body from which it emanates" (2006, 146; see also 1991, 105–6). As with earlier scales,

> the human mind is the form (or idea—*eidos*) of the human body, a highly organized complex phase of matter, incorporating all prior stages in the scale of forms. Yet as a composite of matter and form the living human being is one substance only, not two, as the wax and the impressed pattern are one and the same thing. (2006, 99)

Again, he holds there is no causal "transition" from neuronal function to consciousness, but that awareness is an "enfoldment" of the organism as a whole, "a phase transition occurring when a critical threshold is reached of complex integral unification" (161). In line with my above analysis, Harris thus proposes that consciousness is a field, phase of matter, or further instance of ϕ, whose mode of formal governance is *for the sake of* the intellect and *adaptive thought*. In the following sections I evaluate this position with respect to more recent works.

8.1.1 Neurodynamic Field Theory

In recent years, *dynamic systems theory* (DST) has become increasingly useful for numerous sub-disciplines of psychology. This method provides an overarching framework for investigating the dynamical activity associated with mental activity across multiple space-time scales. For example, the spectrum of electrocortical activity measured by EEG is usually divided into delta, 0–4 Hz; theta, 4–8 Hz; alpha, 8–12 Hz; Beta, 12–30 Hz; and gamma, 30–80 Hz (Minelli 2009, 76). Specifically, *dynamical neural fields* (DNFs) have been useful for studying processes of individual behaviour, development, learning, and environmental constraints. The goal for those applying dynamic field theory (DFT) to cognitive neuroscience is to understand the high dimensional state space of tasks and elementary behaviours as projections from lower dimensional neural fields (Schneegans and Schöner 2008, 269).

What the mathematics of dynamical systems describes is the temporal evolution of state variables "governed by a field of forces, the vector field, which specifies for every possible state of the system the direction and rate of change in the immediate future" (Schöner and Reimann 2009, 454). Besides claiming that nervous systems can be modelled by differential equations, dynamic systems theory (DST) reveals that the functional states resulting from nervous, sensory, and motor system couplings are best captured by attractors and thus fall into a special class of dynamical systems, called "dissipative" systems (456–57). In what follows, I review evidence that dynamic systems theory (DST) provide support for Harris's (Sect. 7.2) contentions (3, 4, 5) and with respect to the previous section, permits consciousness to be treated as a *phase of matter* whose function is the *preservation of adaptive thought*.

In accord with Harris's appeal to the irreducibility of a *dynamic core*, DNFs are holistic in that it is usually not "known exactly where the processes modelled in the DNF take place in the brain, and it may even be doubtful whether any single neural population exists that behaves exactly as the dynamic field does" (Schneegans and Schöner 2008, 252). Consequently, the "functionally significant states of such complete systems emerge as attractor solutions from these dynamics under the appropriate circumstances [...] These functions are not fixed and they do not 'sit somewhere' until activated. They are simply emergent properties of the dynamical system" (268). Accordingly, brain connectivity reveals a "hierarchy of scales" from local neuronal circuits to functional topological networks, the dynamics of each "determined not just by processes at that scale, but by processes at other smaller and larger scales as well" (Richardson and Chemero 2014, 46). DNFs thus arise from multiple distributed areas whose net evolution of stable activation patterns reveals an "interaction dominant system", the components of which cannot be treated in modular terms (48).

DST is therefore complementary with situated cognition, for it depicts the organism as obtaining in and through numerous (spatiotemporally) coupled levels including the nervous system and environmental task context. The brain is thus conceived as "a very high dimensional, complex-dynamical system, built neuronally, but potentially coupled so closely to the environment through its own effector and sensor systems that these become part of the dynamical system" (Schneegans and Schöner 2008, 268). Because interaction-dominant systems are irreducible, some authors have maintained "the cognitive system in question is not encapsulated within an individual brain or even an individual body [...] When human-tool or social cognitive systems are complex dynamical systems with interaction-dominant dynamics, they are extended cognitive systems" (Richardson and Chemero 2014, 48). This idea is clearly consistent with many of the earlier contentions (Sects. 2.5 and 7.4.1) for extended and situated minds.

In accord with Harris's earlier (Sect. 7.2) contention (5), a growing number of theorists in recent years have argued DST provides a method applicable to both physical and phenomenological dynamics. In their introduction to *Chaos and* complexity *in psychology* (2009), Guastello and Liebovitch explain "synergetics" in a way that recapitulates *formal governance* (Sect. 5.3): "The emergence of a hierarchical structure follows the same mathematical and physical dynamics as a phase transition, such as the transition of a solid to a liquid" (24–25). Moreover, supporting a possible extension of ϕ to consciousness, Hollis et al. later contend "the brain and body" may be considered "excitable media" stimulated to create autocatalytic "*traveling waves* or *solitons*"—"temporarily invariant structures that exist as coherent, ordered entities within the space and energy of an excitable medium" (212). They explain such waves do not reduce to components of the nervous system, but are "emergent phenomena" within a hierarchy of constraints: "Constraints are accrued through idiosyncratic experience, and constraints are implicit in the immediate context. Constraints are aspects of biology, culture, history, context, or current states that narrow down the possibility for cognitive activity, prior to its occurrence" (214). It is unsurprising then that Renaud et al. (2009) have found agent-environment coupling can be characterized by fractal patterns, in which stable forms can be represented by specific attractors and repellers that in turn guide an agent's behaviour (182).

In a clinical setting, Van Geert (2009) agrees that the levels of childhood development are best understood as highly dynamic attractors represented by a fuzzy logic (i.e. a function of continuous rather than discrete membership). As attractors, he holds, stages of development are most comparable "to the phases of physical matter (gaseous, liquid, solid)" (262). These stages are "defined by the developmental system's major order parameter", which is captured by brain maturation and habitual skills developed within the subject's niche. He finds the difference between phase transitions of a substance and the emergent self-organized criticality of complex systems such as embodied-embedded brains, is that for the latter, the phases are *possibilities* rather than necessities—the stages are thus *metastable*.

As Hollis et al. (2009) explain, whereas self-organization, by itself ends in an attractor state (e.g. fixed point, limit cycle, Taurus, or strange), humans by contrast, appear to be drawn to *metastable states* of self-organized criticality, "that dance in the neighbourhood of attractors without fully realizing them" (218). Such states have more recently been identified with what Van Orden et al. (2011) call *pink noise*, which reveals *fractal patterns* between random (white noise) and over-regular (brown noise). "Pink noise is a fundamentally complex phenomenon that reflects an optimal coordination among the components of person and task environment" (629). They explain that deviations from this optimum occur when learning a new skill, in advanced aging, and disease.

Schöner and Reimann (2009) find that DST accounts for the capacities of openness to learning, adaptation, the differentiation of skills, patterns of behaviour, and the emergence of novel functions that take place across numerous time scales: "DST makes it possible to talk about the continuous temporal evolution of processes, their mutual or unidirectional coupling and decoupling, and their coupling to sensory or motor processes" (453). Importantly, this is possible without appealing to abstract, symbolic and or computational representations of goals and actions. The system needs only to couple its neural control system driving its motor behaviours to the dynamical neurons that receive sensory inputs about environmental elements of attention. Localized activation peaks of DNFs thus replace the units of representation by acting as "the attractor solutions of dynamical systems that describe the temporal evolution of activation fields" (462). As opposed to information processing as discussed in connectivism, learning is not downloading a function but "a process that eases the constraints on the environmental and task conditions under which a function may emerge" (Schneegans and Schöner 2008, 268). Human consciousness thus involves a process of coupling to an environment in order to generate constraints that offload workload, while simultaneously developing a sensitivity to how those constraints may be altered for increasingly efficient couplings in the future.

In this vein, Dixon et al. (2014) have argued that the classical approach to cognition has proven incapable of accounting for the generation of new cognitive processes, including novel means of problem solving. Towards a more fruitful approach they suggest that "[e]ffects in physical systems, including those within biology, must be understood in terms of the flow of energy and matter" (163). They argue that a truly embodied cognition requires that interactions are understood as a result of "energy consumption", and so should the quantities we measure, which provides what they call a "common currency across all scales of the system and environment, as well as explicit connections to thermodynamic laws" (ibid.).[2] They conclude that the generation of new macroscopic cognitive functions is most adequately defined as *phase transitions*:

> Phase transitions are well understood theoretically, and have been broadly demonstrated empirically in a wide variety of systems [...] Our work shows that the formation of new cognitive structures shows the same signatures as in other embodied, physical systems. That is, an increase in fluctuations as the transition point approaches. (168)

In an effort to *live away the problem of consciousness*, neurophenomenology utilizes DST as a means of modelling neurological and phenomenological dynamics in terms of trajectories within correlated but irreducible phase spaces (Gordon 2013). Methodologically, neurophenomenology combines trained phenomenological reports with third-person observations from the cognitive neurosciences. Both sides are considered irreducible to the other, but each side *constrains* and clarifies the parameters of the other. As Zahavi (2010) explains, an important consequence of this approach is that it

> wouldn't merely be a question of employing phenomenological insights in the empirical investigation of the mind. Rather, the idea would be that the influence goes both ways, that is, it would also be a question of letting phenomenology profit from—and be challenged by—empirical findings. This is why it is entirely appropriate to speak of a *mutual enlightenment*. (9)

In neurophenomenology, experimental subjects are trained to gain greater sensitivity to and subsequently provide detailed description of their experiences using an open-question format without the imposition of predetermined theoretical categories. The resulting neurophenomenological account refutes the *neural correlates of consciousness* model of one–one mapping between brain regions and mental functions. As Fazelpour and Thompson (2015) argue, because "brain dynamics are context-dependent and task-dependent, the same anatomical region may subserve different mental functions in different contexts, and the same mental function may be realized across a functionally connected network of distributed anatomical regions" (224). Following their agenda of irreducibility and co-emergence, this means that rendering the decomposition and localization of functions is no longer viable.

Treated as a complex system, the global state (for either phenomenal or neurodynamic domains) is captured by a single point in some number (*n*) of independent variables that together compose an *n-dimensional* state space:

> A sequence of such states followed in time defines a trajectory, also known as the system flow. The shape of the flow is determined by the system's intrinsic dynamics—the forces that push the system state in one direction or another, depending on where the current state is located. They can be thought of as constituting a kind of landscape over which the behaviour of the system moves. An "attractor" is a trajectory in phase space to which the system will converge from any set of initial conditions. (Van Quyen 2010, 256)

By this methodology, *phase transitions*, attractor *dynamics*, and symmetry breaking are applicable to both phenomenological and neurological domains. The result is that these domains are conceived as "homologous", mutually informing state spaces characterized by a common evolutionary trajectory or "landscape".[3]

For autonomous dynamical systems, the global shape and evolution of the landscape are determined by order parameters (*constraints*) created by collective interactions among the individual parts of the system, which in turn govern the behaviour of the individual parts. Intentional constraints however must be added to those already present in the environment and linked to boundary conditions (like those prescribed in a laboratory). "Once again, in a circular causal manner, the intentions modulate the control parameter, which in turn maintain (or modify) the intentions" (Renaud et al. 2009, 183). DST entails that intentional constraints permit perceptual-motor coordination *on a time* scale *larger than that of* body *movements* and thereby "modulate the intrinsic dynamics of the order parameters linked both to the flow of information and the flow of motor control. By so doing, they define an attractor capable of guiding motor behaviour toward the desired pattern" (183).

Tschacher and Junghan (2009) recognize that this circular causal structure (synergetics) bears important implications for the nature of intentionality. They argue "emergent mental properties *do indeed matter*" and cannot be "epiphenomenal" because this conception of synergetics "is applicable to both sides of the mind-body gap, and may even put forward ways to bridge this gap" (329). Likewise, Van Orden et al. (2011) have argued (continuous with *formal governance*) that DST may naturalize intentionality:

> We have proposed that intentions affect behaviour as constraints, not causes. Intentions as constraints are temporary dynamical structures, soft assembled from interdependent components to function in control parameters to create critical states [...] Constraints are therefore no less natural or no more magical than causes, or the convectional cells of the atmosphere that change the weather. Thus the complexity account makes progress toward naturalizing intentionality. (657)

This section has shown how the tools and methods of DST, when applied to the study of mind reflects Harris's (Sect. 7.2) central ontological claims (3 and 4) and support his contention that the mind is an instance of ϕ. For DST, mind is principally characterized by constraints arising from the organization of a series of irreducible forms, indicated by *phase transitions*, attractors, and *fractal dynamics* across numerous spatiotemporal scales (for further details see Steyn-Ross and Steyn-Ross 2010; Di leva 2016). The capacity to become aware of these constraints thus characterizes adaptive thought. Concerning Harris's methodological contention (5), neurophenomenology utilizes multi-dimensional phase models as a common language for bridging first, second, and third person accounts of consciousness. While the above is merely a crash course in the application of DST to neuroscience, the strongest argument for mind as a phase of matter is in many ways opposed to the enactivist paradigm. In what follows I consider this proposition and articulate a dialectical holist response.

8.1.2 The "Φ", Phase, and Scale of Consciousness

Following Tononi's *integrated information theory* as a means of naturalizing consciousness, Tegmark (2014) contends that "consciousness can be understood as a state of matter, "perceptronium" with distinctive information processing abilities" (1). According to Tononi (2012), integrated information theory can be essentially captured with four axioms:

1. Information *axiom*—every experience is specific; each is what it is by differing in a particular way from a wider repertoire of alternatives. Additionally, a mechanism in a state must "specify either selective causes or selective effects" (e.g. inputs and outputs), so as to generate *cause-effect* information within the system (297). Tononi explains that in ontological terms, the information postulate means that "from the intrinsic perspective of a system, only differences that make a difference within the system exist" (ibid.).
2. *Integration axiom*—each experience is unified and is irreducible to independent components. As an ontological postulate, this means that only information that cannot be partitioned into independent mechanisms or components (irreducible in the past and the future) *exist in and of themselves*: "Integrated information (Φ) can be captured by measuring to what extent the information generated by the whole differs from the information generated by its components (minimum information partition MIP)" (297).
3. *Exclusion axiom*—every experience is *definite* by being limited to particular things and not others and flows at a particular speed and resolution, that is, only experience that have definite borders, temporal, and spatial grain, exist intrinsically. From this Tononi claims a mechanism that specifies only one *maximally irreducible* set of past causes and future effects is considered a *concept*. A *complex* is a maximally irreducible set of concepts. Its concepts signify a maximally integrated conceptual information structure. As an ontological postulate this means that only entities that are "maximally" irreducible exist in themselves (297).
4. Identity *theory*—when in a particular information integration state, the *complex* becomes necessary and sufficient for a *quale* of experience. In this way there is an "identity" posited between the phenomenological and causal-informational aspects of a system:

> [A]n experience is a maximally integrated conceptual information structure. Said otherwise, an experience is a "shape" or maximally irreducible constellation of concepts in qualia space (a quale), where qualia space is a space spanned by all possible past and future states of a complex. In this space, concepts are points in the space whose coordinates are the probabilities of past and future states corresponding to maximally irreducible cause-effect repertoires specified by various subsets of elements. (298)

Although he does not address Tegmark's work, Thompson (2015) takes issue with Tononi's ***integrated information theory*** because to his mind, it incorrectly renders consciousness "an abstract informational property" when consciousness must be understood as "a concrete bioelectrical phenomenon" (343). Thompson argues further that

> consciousness can't be instantiated in an artificial system simply by giving the system the right kind of computer program, for consciousness depends fundamentally on specific kinds of electrochemical processes, that is, on a specific kind of biological hardware. This view predicts that only artificial systems having the right kind of electrochemical constitution would be able to be conscious. (Ibid.)

Given the previous discussions, Harris's theories of mind are in clear agreement with Thompson's contention. Hence, despite Harris's appeal to Tononi's work, current integrated information theory only dovetails with Harris's position to the extent that it identifies some "irreducible" forms as correlates of mental processes. Tononi's claim that qualia are the ***units*** of mind is something neither Harris nor AE can endorse. Tegmark's more recent work however does partially satisfy Thompson's worry by emphasizing that consciousness is fundamentally dynamic and requires the right kind of substrate.

Analogous to "computronium" (the most general substance capable of computation), Tegmark suggests "perceptronium" as the most general substrate capable of supporting perception. He argues that if consciousness is a state of matter, it will be possible to identify, quantify, and model its characteristic dynamics despite the likelihood of there being numerous types (just as there are different types of ***liquid***). In a functionalist spirit, he claims consciousness (just like wave dynamics) can obtain in a range of supportive media. Towards this end, Tegmark argues that such a perceptive state of matter must satisfy at least six necessary conditions: (i) substantial information storage capacity; (ii) substantial information processing capacity; (iii) substantial independence from its world; (iv) it must be integrated, meaning it cannot consist of nearly independent parts; (v) the system is unified such that it specifically records information useful for it; and (vi) the system is *autonomous* such that it has substantial *dynamics* and *independence*, which means "information can be processed with relative freedom from external influence" (2).

Tegmark claims there is a parallel between conceptions of integration in non-living systems with that found in biology. Following Damasio's (2010) work, Tegmark argues:

> [T]o be in homeostasis, a number of physical parameters of our brain need to be kept within a narrow range of values [...] this is precisely what is required of any condensed matter system to be near-critical, exhibiting correlations that are long-range (providing integration) but not so strong that the whole system becomes correlated like [...] a brain experiencing an epileptic seizure. (5)

Tegmark claims that Φ goes to zero bits as the temperature T $\rightarrow\infty$ on one hand and to one bit as T $\rightarrow$ 0 on the other. However, Φ assumes a maximum (>1) at an intermediate temperature that is also near the critical point of phase transition (6). Integration thus requires that memory is relatively efficient (in terms of energy consumption) in both reading *and* writing information. So, the system/substance must not be too ridged, like the ordered molecules of gold, meaning it must be capable of complex dynamics, but it must also be more ordered than a liquid or gas. These remarks agree with the findings of Sect. 6.2.1 and so appear to satisfy Thompson's above concern that consciousness depends upon a *living* substrate.

While Tegmark's contention that consciousness occupies a unique region of phase space is certainly shared by AE and Harris, another point of disagreement is brought to light with respect to the appropriate *scale* of consciousness. Tegmark argues, "consciousness is related to information processing, not mere information storage", which means Φ "applies to classical neural networks rather than general quantum systems" (12). Unlike Harris and AE, Tegmark and Tononi believe that the *function* of neurons obtaining at the classical scale should provide a sufficient identification of consciousness. For example, Tononi reasons:

> [I]t should be possible to establish if in the brain consciousness is generated by neurons or groups of neurons. In this case the exclusion postulate would also mandate that the spatial scale at which ΦMIP is maximal, be it neurons or neuronal groups, excludes finer or coarser groupings: there cannot be any superposition of (conscious) entities at different spatio-temporal scales if they share informational/ causal interactions. (2012, 305)

"Integration" is thus appreciated by the dialectical holist insofar as it agrees with formal governance and third-order emergence, but for Harris (Sect. 7.2), it is a category mistake to identify *measured* information ("mechanism") *at a particular scale with consciousness.*

For the dialectical holist, it would be possible to use Tononi's "Φ" for *predictive* purposes only if the integrated information is considered an *abstracted signature* taken from a contextually embedded system-as-a-whole. In this case, the degree of "integration" would be *indicative* of formal governance arising in and through the particular history of an embodied, embedded, and living system, that is, *as an organizational* constraint *across spatiotemporal scales, not of a particular scale.* It should be noted that this puts an additional demand on the "black box" conception of functionalism: if mind obtains through a particular organization of scales, each can be treated as a black box, but each will also bear dialectical relations to those of its neighbours—relations that are considered necessary for the occurrence of consciousness. It is the process of this organization that appeared to Harris to be indicative of the *phase of consciousness.* While it is expected that living bodies and brains will be analysed as unique phases of matter, the *properties* of which must not be *identified* with perception because doing so abstracts away from the total dynamical system (the total set of material scales) that instantiates consciousness. This is an interpretation that appears congruent with AE appeals to DFT as discussed above, because the attractor dynamics of behaviour, learning, biological adaptation, and environmental alteration are here considered *partially constitutive of consciousness.*

In sum, although Tegmark and Tononi disagree with Harris and AE concerning the metaphysical relations between cosmos and mind, all parties can share the common conception of mind at least associated with a particular *phase of matter.* Additionally, insofar as the integrated information (Φ) of consciousness counts as a *necessitated and irreducible mathematical structure* in Tegmark's mathematical universe hypothesis (MUH), he endorses a much stronger anthropic principle than he has been able or willing to recognize (in line with my conclusion of Sect. 4.5.2). These revelations warrant much further empirical collaboration and philosophical discussion between integrated information theory and AE paradigms in future works.

8.2 Ontological Individuation: *Chasing the Slippery Serpent*

In the following sections, I further unpack Harris's conception of consciousness as a phase or scale with respect to his wider epistemic and cosmological theories. In this section I consider how Harris (or a subsequent dialectical holist) can justifiably refer to the scale of forms as ontological, rather than merely epistemic (i.e. as *conventional* categories). By extension, this means addressing the extent to which Harris's metaphysical arguments provide a means of bridging phenomenal and noumenal domains, thus avoiding Kant's transcendental idealist conclusions. To do so, I first consider two problems resulting from correspondence-driven approaches to a ToE.

Following Rees (1999, 12–13), numerous cosmologists have begun modelling complexity within the cosmos in a way that appears to mirror Harris's scale of forms. Most notably, Carr (2007) maintains that "physics has revealed a unity about the Universe which makes it clear that everything is connected in a way which would have seemed inconceivable a few decades ago. This unity is succinctly encapsulated in the image of the uroborus" (11). He has elaborated this view as follows (Fig. 8.1):

> The physical evolution of the universe from the big bang (at the top) through the Pyramid of Complexity to humans (at the bottom) is just the start of a phase of intellectual evolution, in which mind—through scientific progress—works its way up both sides to the top again [...] interpreted historically as representing how humans have systematically expanded the outermost and innermost limits of their awareness. (2006, 150)

Here, Carr's (phase 1) top-down process is akin to the "natural" aspect of Harris's 𝕮, while the bottom-up intellectual progress (phase 2) mirrors his conception of cognition as increasingly *internalizing the environment* (Sect. 6.3.1). Harris's metaphysical task has been to establish a phenomenological ontology that bridges these two phases.[4]

Towards a clarification of Harris's approach to nature and knowledge, it is instructive to consider Ellis's (2015) argument for a "holistic

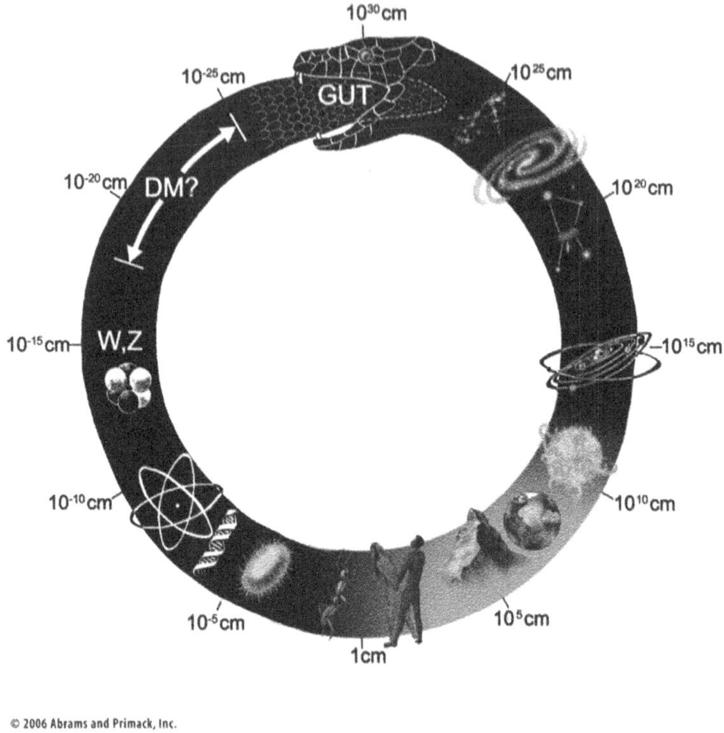

Fig. 8.1 Uroborus as analogue to Harris's scale of forms and explicative process. Figure recreated with permission from Yale University Press (Primack and Abrams 2011)

ontology". Ellis proposes that nature may be understood as a *hierarchy of possibility spaces*, each influenced by both bottom-up and top-down causes, all of which considered equally real (ff. 6). These include: (i) the matter and forces of physics (237), (ii) biological phenomena (ff. 140), (iii) minds and meaning (ff. 291), and (iv) computer software (ff. 35). Importantly, Ellis contends that all these realms exist independently of our minds and each correspond to a super-space of Platonic possibility:

> This Platonic world is being progressively discovered by humans and represented by our mathematical theories. That representation is a cultural construct, but the underlying mathematical features they represent are not, they

are timeless eternal realities whose existence does not in any way depend on the existence of human beings. (365–66)

Although Harris's scale of forms is similar to Ellis's possibility hierarchy, there are a number of important discrepancies. Building upon the discussion of Sect. 5.3, ϕ are to be understood as general classes of spatiotemporal dynamics obtaining through self-organization, phase transition, and symmetry breaking. Much like the *identifiable* scales within a fractal curve however, the regularities (or physical laws) that characterize each are *abstractions* from a continuous higher-dimensional whole. The emergent laws that obtain in each are possible because of *both intra- and inter-relations of instantiated* ϕ. Ellis's and Harris's ontologies are fundamentally at odds because the former appeals to Platonic counterfactuals—that is, to *possible* domains that do not materially exists—whereas for the latter, the (concrete) evolutionary trajectory of differentiating ϕ is all that exist. This means that for Ellis the identifiable domains *exist in themselves*, but for Harris, they are nothing apart from a concrete movement, our knowledge of which is ever incomplete.

The principle conflict then concerns the *individuation* of such domains. As anticipated by the discussions of previous chapters, Tegmark (2007) admits that unless we maintain a Platonic position that mathematical structures are real, "discovered" rather than created, there can never be a ToE due to an unavoidable infinite regress: "one is ultimately just explaining certain verbal statements by other verbal statements" (114). Despite his Platonic sympathies, Ellis has recognized two problems concerning a potential ToE. First, the question of whether a given domain (e.g. humanity) is *determined* will not be solved by a ToE, but only "reinforced" because "such a theory would in essence have the image of humanity built into it—and why that should be so is far from obvious, indeed it would border on the miraculous if a logically unique theory of fundamental physics were also a 'theory of everything'" (2004, 634). Second, Ellis (2006) finds that the inability of a ToE to *predict* future theories and experiments results in "the self-referential incompleteness of physics" (756). Indeed, Ellis's worries should be concerning for anyone sympathetic for his ontology.

Ellis has argued that because we cannot reduce or avoid positing domains of various kinds, we should conclude that those defined by our present capacities of observation (i.e. those "discovered" by respective scientific fields) *successfully individuate* corresponding Levels of Nature. This move not only conflates epistemology and ontology Metaphysical-Physical selection effect fallacy (MPF), but also results in our transcendental problematic. Hence, Ellis's first worry arises because our ability to "discover" Platonic *domains* assumes the very *bridge* between knowledge and nature (via individuation) that is at issue concerning metaphysical arguments about how (if at all) *mind figures into* nature. In other words, it would not be surprising to find humanity built into a ToE because this is inevitable for all metaphysics that purport *completeness* without justifying the means of individuation that generates such a system by some metric beyond the system it creates.

Further, this circularity results in a contradiction, which remains even if we deny a Platonic realm but maintain a correspondence (or representational) theory of truth and reality. In this case, it is believed science should "decrease the degree of discrepancy between objective and subjective" until we acquire a complete picture of nature (Jordan and Day 2015, 3). By this reasoning however, "science is metaphysical" in that it "reveals how reality really is" (ibid.). This means that the methodological examination of nature is somehow *beyond the physical* in order to be capable of representing it, but in so doing, science escapes its own net—thereby leading to Ellis's second worry. Hence, by presupposing the capacities of (empirical) individuation we tacitly presuppose what will be taken as *natural*, which separates observer from observed and so undermines our efforts to *discover* a consistent theory of nature-as-a-whole.

In an attempt to turn reductionism on its head in support of his TAP, Harris contends that if physics is successful in reducing the laws of nature to a mathematical *formula*, we will then exemplify Hegel's articulation of *dialectical reason*: Nature will have become conscious of itself as "pure thought" (2000, 37). Although a prospective ToE is neither necessary nor sufficient vindication of Harris's TAP, the above remarks should sufficiently highlight the problem confronting both contemporary physical theory and metaphysics: attempts to systematize the whole of nature appear to result in an awkward circularity such that the form of our epistemology (formalism) justifies our methods of individuation, which provides an ontology (what we believe exists), and thereby justifies our preconceived relation to the world. Consequently, the relationship between Nature (phase 1) and knowledge (phase 2) is tacitly assumed in our individuation practices, and so it cannot be justifiably established by these practices: *as a mere circle, our conception of* nature *is inherently false*. Despite his rejection of a Platonic reality and representations, it remains to be seen how Harris (and a proponent of dialectical holism) can escape this circularity.

8.2.1 *Pragmatism, Coherence Theory, and Transdisciplinarity: Towards a Unity-in-Diversity of Nature*

Here I consider some efforts to establish an epistemology that avoids the above circularity and remains consistent with Harris's system. In contrast to the thoroughgoing commitment to a correspondence theory of reality and truth, Jordan and Day have developed an alternative *coherence* approach, under the guise of *wild systems theory*, which I maintain is consistent with Harris's epistemology and implied by AE. I contend this epistemology can be fruitfully combined with Clayton's proposal for a plurality of "complexity producing mechanisms", to exemplify the relation between phase 1 and phase 2 within dialectical holism.

According to Jordan and Day (2015), appealing to coherence means refusing to begin a discussion of reality by assuming "its independence of observers" (1). In accord with my conclusion of the previous section, they go on to claim, the "correspondence relation is validated by placing it within an assumed, larger-scale reality [...] the evolved physical world" (6). They argue that this thesis depends upon one making a distinction between intrinsic and relational properties, with the aim of science being to uncover the intrinsic properties of reality. Clearly retracing Harris's reasoning concerning internal relations, wild systems theory is crucially based upon the following move:

> If, however, one assumes that relata are themselves constituted of relational properties [...] then there can be no intrinsic properties. This is because the constitution of all properties, by definition, would be re-lational. In short, reality would constitute a unity in which all things were constituted of all things. (7–8)

This notion of reality being infinitely relational is at odds with the correspondence theory because, they contend, all objects and subjects are here contextually grounded, so what at first appear to be intrinsic properties are re-cognized as relational.

In agreement with Harris, they argue that if we reject the assumptions of subject-object independence and objectivity being of paramount *value* then,

> truth can no longer be measured by assessing the degree of difference between reality and an impression, idea, or representation we have of it, or by investigating an assumed relation we share with it. There exists nothing

> "as it is" to which anything else can accurately correspond. The final, ontological description of what something is must include reality as a whole. (9)

For wild systems theory, scientific endeavours should be understood as "modes of experience", or "abstraction" from reality, which is an "internally related unity in which we are embedded" (11). By "abstraction" they mean science can never be outside of, nor look at reality itself, but is *partially constitutive of the reality it is meant to describe*. Again, building upon Dewey's "situation", scientific practice is "conceptualized as a recursion on reality—an abstraction about that from which it emerged and within which it is entailed" (ibid.). Hence, there is no effort to capture reality as a whole, nor to establish our relation to it from the outside, but only to systematize identifiable relations towards *increasing clarification*.

Consistent with Harris's 𝕮 and neurophenomenology, such clarification in wild systems theory focuses on "multi-scale, self-sustaining homologies [...] that which is common across the internal and external contexts of an organism; namely, energy transformation" (13). In this way, an organism is conceptualized as "the very energy-transformation hierarchy in which it sustains itself; it is an embodiment of the reality (i.e., context) within which it emerged. In short, it is reality within reality" (14). wild systems theory applies the coherence criterion to experience and beliefs by conceptualizing organisms as *embodiments of context* (i.e. as necessarily being *about* their environment in a meaningful way), which avoids the correspondence relation. Hence, by giving up a presumed *completeness* of any given individuation (which is central to correspondence), the notion of "truth" is reconceived. Truth they claim, is a matter of coherence "entailed within one's moment-to-moment embodied context (i.e., phenomenology) and across the beliefs one derives from the moment-to-moment flows of embodied context" (16). Coherence is in this vein no less than a criterion for *sustained sensorimotor enaction* across ever widening environmental domains.[5] wild systems theory thus considers our phenomenal and larger-scale reality as a single recursive hierarchy of energy transformation that is sustained in and through coherent self-reference—a conclusion that is itself maintained only by its coherence.

In his discussions of emergence, Clayton provides a paradigmatic depiction of science, which is extensive of wild systems theory and further clarifies Harris's epistemology. Clayton contends that the theory of emergence is *meta-scientific*, necessary for what he calls "integrative ordering" and predicts "that each relation between neighbouring disciplines will exhibit certain general features recognizable across disciplines" (2004, 603). In a later (2013) text, he makes considerable efforts towards developing this idea. Here, Clayton is careful to voice criticism of the Unity Approach, which holds that complexity may unify all natural sciences. Clayton contends that we should "resist the Unity Approach to complexity [...] *because it obscures a deeper insight that* complexity *theorists offer into the way that cosmic evolution works*" (334).

Clayton argues that we currently do not have a single coherent science of complexity, but a variety of *complexity-producing mechanisms* (CPM) operative at different time scales and observed across numerous disciplines, which provide a "meta-level theory of complexity" (338). Hence, CPM "is meant to produce common ground so that conversations can continue even while theorists deeply disagree about fundamental features of complexity" (340). Importantly, Clayton notes that the common features of CPM can only be identified via comparative studies of their functions in respective empirical fields (i.e. *transdisciplinarity*):

> [N]on-symmetrical relations between disciplines of study produce the "ladder" of the sciences. This ladder is temporally indexed and corresponds to the order of cosmic evolution [...] Later fields of study in the process contain all the complexities of earlier fields while adding new forms of complexity of their own [...] To pluralize complexity studies, as the CPM approach does, does not block the natural scientific study of the world but enhances it. (2013, 342–43)

Indeed, Clayton's pluralistic CPM strategy has been anticipated by both Harris and Bohm. CPM captures Harris's message that individuating the respective wholes of nature requires ongoing interrelation and synthesis between disciplines to resolve conflicts and generate increasingly coherent paradigms. Additionally, Harris recognized that our method should be focused upon not merely understanding the "underlying mechanisms" of natural phenomena, but (from his metaphysics), elucidating the principles of the "highest" scales of organization depicting "different orders of reality" is of equal value (1970, 369). Consistent with Harris and pragmatic epistemology, Bohm and Hiley (1993) has argued that "we do not expect to come to the end of this process of discovery" (e.g. via the ToE) rather, he claims "nature in its total reality is unlimited, not merely quantitatively, but also qualitatively in its depth and subtlety of laws and processes [...] Our knowledge at any stage is an abstraction from this total reality and therefore cannot be expected to hold indefinitely when extended into new domains" (1993, 321). Bohm (1980) further echoes CPM and Harris when he maintained that comprehension of an ultimate law of the whole cannot be the "final goal of scientific research", but rather our goal is to clarify "a movement in which 'new wholes' are continually emerging" (137).

Both Bohm and Harris thus agree that Nature cannot be reduced it to a single law or equation, but should be understood as an emerging process that generates a multitude of internally related levels. This implies our epistemic goal is thus to understand the *process as a whole, in virtue (rather than despite) of its diversity*. As a process of ongoing clarification and revelation of wholes, Clayton is apt to state, "[s]cience's most powerful ally [...] is emergence" (2004, 599). Moreover, CPM provides a way of establishing metaphysical *unity-in-diversity* across the natural sciences that embraces interdisciplinary collaboration and may itself generate further disciplines in the process of clarifying any given phenomenon. The resulting conception of knowing reality (both epistemically and ontologically) is thus an ongoing *becoming of scales* (e.g. a fractal), rather than a means of carving nature at its joints (as though said joints existed *prior* to the carving!).

Clayton's CPM appears consistent with and supportive of wild systems theory because theories of emergence are meta-scientific in *organizing domains of observation*, though scientific observation itself is not posited to capture any *part of reality*. For wild systems theory, this is because there can be no correspondence test of reality, no intrinsic properties, and science cannot be metaphysical: "if reality is an internally related unity, then theories are constitutive of that reality and can never 'point to' reality as if to do so outside of it. They are, by definition, 'in it' just as we are. Thus, they are, by definition, incomplete" (18). They claim wild systems theory provides a way for cognitive scientists in particular to frame their theories and models not as metaphysical tests, but as *pragmatic* tools. Hence, the increased coherence that science affords us provides "more influence over our context; that is, it affords us the ability to more effectively sustain ourselves" (ibid.). This, they conclude, avoids "distracting arguments" concerning the problem of how the meaning of our concepts, symbols, etc., are to be *grounded*.[6]

wild systems theory's appeal to coherence is clearly congruent with Harris's own support of this theory. For Harris, respective fields and paradigms of science, just as sense modalities of individual perspectives, progress by differentiation and synthesis—what Harris considered 𝕮 of social development (1992, ff. 1). Harris's conception of "adaptive" or "reflective thought" as ongoing, implies that one's gestalt is never complete, each "represents a stage in the development of knowledge and none can be finally and totally satisfactory short of omniscience" (1970, 352). Nevertheless, wild systems theory's pragmatic conclusion is *prima facie* at odds with Harris's much stronger attempt to secure a direct-realism about scales:

> Coherence is order and organization. We perceive and come to know the surrounding world by organizing the contents of the sensuous field into an orderly and coherent system. The activity that does this is judgment, and the faculty of reason. It is the same dynamic principle of wholeness that is active at lower levels in the gamut of natural forms, which, in the course of self-specification, has brought itself to consciousness in the human mind, it reveals itself as reason and as dialectically continuous with the organizing principle operative in the physical and the biological phases of reality. (2000, 248)

Like proponents of wild systems theory, Harris proposes that the development of nature (phase 1) and knowledge (phase 2) can be considered a single continuous process. However, Harris argues that if Reason is the fruition of the scale of forms, then we are justified in maintaining a *self*-observation of Nature.

Towards a reconciliation of Harris's realism and wild systems theory's pragmatism, it may be possible to appeal to Brender's ontology of asymmetry. Brender proposes an *endogenous sense* to nature that implicitly supports the PAP as articulated above:

> The autoproduction of sense in nature takes place through symmetry-breaking, in which natural wholes articulate themselves into parts or regions, creating differences out of indifference and form out of uniformity. These differences are neither things nor ideas, neither atoms nor artifacts. They cannot be known by a disembodied mind, but only perceived by a living body. (272)

Brender goes on to dismiss the prospect of obtaining *a view from nowhere*, arguing that ironically nothing could be more anthropocentric than the efforts of mechanistic science to "strip nature of all anthropic predicates in order to arrive at an account of reality as it exists 'in itself'"

(272). In line with wild systems theory, CPMs, and my conclusions of the previous section, Brender finds that the mechanistic reductive project has thus failed for deciding a priori "how nature is to be divided" (273).

Here again, symmetry breaking provides further illumination of Harris's 𝕰 by serving as a *neutral base* of evolution (phase 1) and knowledge (phase 2), but insofar as it serves as an ontic conclusion, it nevertheless remains *pragmatic*. This is because the goal of science, Brender holds, is to permit a natural phenomenon to reveal "which difference make a difference to it", and this involves continually learning to recognize differences that were previously unrecognized while at the same time discounting some that were previously considered relevant (273). Accordingly, *coherence* may be understood as yet another instance of the same recurrent process of self-differentiation and formal governance that has been fundamental to my exposition of dialectical holism heretofore. In the following three sections I consider how this conclusion may be formalized with respect to Harris's TAP.[7]

8.2.2 Towards a Topo-logic of Phenomenology

Towards an elucidation of the rationale by which Harris believed he had overcome the problem of consciousness, the recent development of *Klein bottle logic* (KBL) appears invaluable. I argue that Harris's theory of consciousness as an instance of ϕ anticipates Rosen's and Rapoport's independent arguments that KBL provides an onto-epistemology for understanding mind, life, and (cosmological) evolution. Additionally, though the metaphysics of ENM has yet to be sufficiently specified, I maintain KBL makes significant strides towards this end.

Rosen (2008) contends that what is needed to surmount the present difficulties in theoretical physics is a completely new "philosophical base" that provides an altogether different "way of intuiting the world" (xi). By recognizing, rather than suppressing philosophical assumptions he aims to embrace subjectivity and surpass the subject-object split. Taking a "dialectical approach", Rosen claims "neither mind nor body, neither subject nor object nor space" is primary to the other, but each arises from a common "inchoate flux", which he calls *apeiron* (i.e. without measure) (7).[8] Following Rosen, Rapoport holds that the dominant worldview today, *Aristotelian Boolean logic* (ABL) is responsible for reinforcing rifts of philosophy-science, ontology-epistemology, mind-matter, etc. Moreover, Rapoport contends, our reliance upon ABL dualisms has led us to speak about the manifestations of things and processes as contained in space and time, which has resulted in the *myth of finality in science* (2012, 7). In accord with the conclusions of the previous two sections, Rapoport claims the central problem is that ABL has "failed to describe its own generation and its association with becoming" (14).

Rapoport and Rosen propose a paradigm shift based on the topology of the Klein bottle as a principle that surmounts the *Cartesian cut* (CC). Appealing to cybernetics and multi-valued logic they challenge the fundamentality of unity/symmetry in physics, which Rosen claims, permits a self-transforming function for cosmogony—"a unity-in-diversity of nature." Rapoport writes that the resulting model is essentially a "second-order cybernetics" in which, "the steerer has to be included in the system that is steered, and thus self-reference is placed as the lifeworld of the system in which the observer cannot be excluded [...] subject and object become fused into a single being" (2011, 35).[9] Both authors agree that the analytic assumption of *object-in-space-before-subject* stands as their primary obstacle.

Our task is to signify Being in some way that avoids linguistic representation, says Rosen, because such methods render "Being as an *other*, a free-floating semantic object cast before the reading subject" (2014, 253). This, he claims, would be an act of abstraction, which misses the concrete nature we aim to signify precisely because Being cannot be understood from a purely *objective* standpoint. Towards this end, he argues that for some relationships between identifiable variables to remain invariant, we must posit a container that is also invariant. However, "space is not an object to be known but is the means by which the subject does the knowing" (2008, 18). Consequently, spatiotemporal dimensionality can no longer be considered a changeless backdrop: "Dimensionality itself is seemingly thrown into the arena of concrete change, thereby mitigating the absolute distinction that had been drawn between the spatial container and the dynamic processes it contains" (20). This implies, there are no invariant variables to be identified and even the "laws of nature" are "viewed as evolving" (ibid.).

Towards a phenomenological signification of Being, Rosen first illustrates a solid cube, noting that under normal circumstances we cannot view our object from all sides simultaneously. In the case of the Necker cube however, two perspectives can be superimposed, which demonstrates an internal relation between opposing perspectives. In the case of touching one's own hand, we find a similar situation: "while what was subject can be known as object a moment later, I cannot know subjectivity as such. The subject is still the one who does the knowing, while the object remains that which is known" (2014, 257). Here, he claims there is likewise a dialectical relation between subject and object and a continuous switching from one to the other, but the subject cannot be completely transformed into the object. Following Merleau-Ponty, he claims the resulting gap is epistemic, in that we cannot see (or objectify) our own seeing despite our ability to infer its possibility. Rosen considers the Necker cube to be limited as an iconic sign-vehicle of Being because it appears as "an object cast before our detached gaze" (259). What we require is a solid semiotic body that "could be read in such a way that while standing before us, it would also stand *within us*" (260).

Rosen thus exhibits the Moebius strip, which is paradoxical for being simultaneously one- and two-sided. Paradoxical because there are differences between inside and an outside, yet they are also one and the same depending upon vantage point. Nevertheless, the Moebius strip is merely a two-dimensional model and thereby objectifies the subject-object paradox of Being. If left- and right-oriented Moebius strips were welded together however, Rosen contends, the resulting Klein bottle might satisfy our signification of Being. In classical mathematics the resulting Klein bottle is conceived as being embedded in a higher dimension, which preserves its continuity but renders the structure abstract. Rosen suggests that thinking of the Klein bottle in this way avoids contending with its concrete discontinuity, that is, its hole. Following Merleau-Ponty's conception of "depth" (1964, 180), Rosen suggests that the discontinuity of the Klein bottle provides an intuitive phenomenological means of signifying a first, self-contained dimension, from which all others can be derived, and in which subject and object are fused (2014, 269). Rosen thereby takes the Klein bottle to be at once the "signifying" and "signified object" (270) (Fig. 8.2).

According to Rosen, the point at which the uncontained intersects the contained, a discontinuity seems to be created, but the true Klein bottle

> enacts a dialectic of continuity and discontinuity [...] Depicted here is the process by which the three-dimensional object of the lifeworld, in the act of containing itself, is transformed into the subject. This blueprint for phenomenological interrelatedness gives us a graphic indication of how the mutually exclusive categories of classical thought are surpassed by a threefold relation of mutual inclusion. It is this relation that is expressed in the primal dimension of depth. (72)

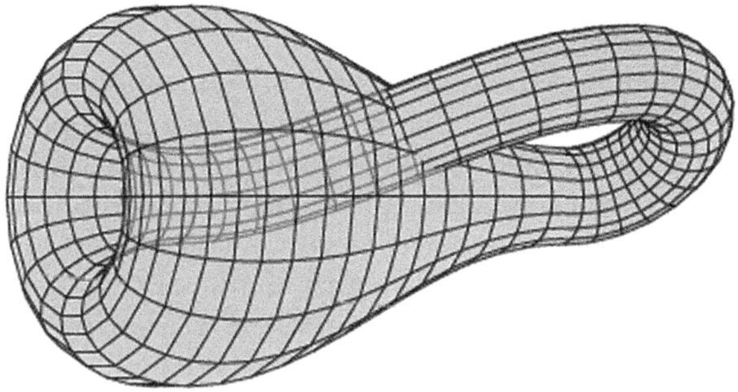

Fig. 8.2 Klein bottle. Recreated with permission by the *Manifold Atlas Project*: http://www.map.mpim-bonn.mpg.de/File:Klein_bottle.png

To make this move however, Rosen argues that the standard mathematical conception must be wrong because as an abstraction, it presupposes an "extensive continuity" that cannot account for the "inherent *dis*continuity of the Klein bottle created by its self-intersection" (75). Rosen thereby rejects the notion that the Klein bottle is a closed object embedded in a hyper-dimensional continuum. Instead, as a phenomenological structure, he maintains the Klein bottle breaks the continuity of space itself. Thus, there is no need for "extra" dimensions within which the form is embedded, rather the hole or point of intersection (D-0 *singularity*) is considered the "dynamic ground", acting as source and goal of the whole.

Following the same phenomenological intuition as Harris and wild systems theory, Rosen argues this model dispenses with the "external relations" of contemporary science", and posits that "all relations are *internal* [...] in the underlying *lifeworld* there is no object with boundaries so sharply defined that it is closed off completely from other objects" (2008, 44). Likewise, Rapoport argues, "if objects are related by mutual containance, no separate container is required to mediate their relations, as would have been [...] the case with externally related objects" (2011, 39). Hence, there is a "dialectical relation" between object and subject in which "space and time are not mere containers but the essential core of this relation" (ibid.). What is meant by "core"? Rosen insists that "the dialectical features of perceptual experience ('orientation, polarity, [and] envelopment') are not merely secondary to a space that itself is devoid of such features" (2008, 47). The "depth dimension" thus "blends subject and object concretely, rather than serving as a static staging platform for the objectifications of a detached subject" (49).

Essentially then, Rosen considers the depth dimension to be an instance of Kant's synthetic a priori, "the immutable organizing principle for all experience", a principle side-tracked by mathematical abstraction employed to fill the void of what he calls "phenomenological intuition" (ff. 66). Rapoport proposes the Klein bottle torsion as a means of surmounting the CC by serving as a logo-physical field that unites our description of subject with object, thus providing an *onto-epistemological principle of self-reference* (2012) (see Kauffman 1987). Rosen finds that the simultaneous open and closed nature of the one-sided Klein bottle provides a paradigmatic model of *depth*, demonstrating "a continuous, self-containing movement that flies in the face of the classical trichotomy of contained, containing, and uncontained—symbolically, of object, space, and subject" (72).[10] Hence, neither subject, Nature, nor the Klein bottle are contained within space, but are self-contained, *embodying both global inside and outside through self-reference*. In this way they claim, this structure-process supplies a bridge between the physical and logical domains that "incorporates" and "produces" dualism without permitting independent ontologies because the Klein bottle remains a unified field. Thus, the Klein bottle captures the self-referential aspect of Harris's "dialectical whole", which is likewise posited *synthetic a priori*.

Both authors likewise reject the standard view of cosmogenesis because in place of a primal symmetry, KBL implies a dialectic of symmetry and asymmetry ("synsymmetry"). Rosen explains, "[a]ssociating wholeness with symmetry and 'holeness' with asymmetry, we can also see the intrinsically synsymmetric nature of Kleinian action. In flowing through itself as it does, the Klein bottle continually breaks and remakes its own symmetry" (119). Cosmogenesis is thus considered a process of dimensional (global) transformation, a "self-transforming symmetry process" (120). Consequentially, dimensions cannot be considered objects of analysis, but as "spatio-subobjective concrete universals" whose evolutionary dynamics are accessible to "phenomenological intuition" (167). It is here maintained that the "first" symmetry break is one of *self-reference* in and through depth dimension:

> If linking the evolution of the universe to transformations of consciousness seems odd, it is because we have been deeply conditioned to regard cosmological events in purely "objective" terms, as physical occurrences "out there

> in space," with subjective awareness playing no role. But phenomenological intuition guides us to a different sensibility. From this standpoint, cosmogony cannot properly be understood as a sequence of happenings devoid of inner life, of lived subjectivity. Instead we must regard it as a process of Individuation that involves the self-transformations of sentient dimensional organisms. (207)

This line of reasoning renders consciousness continuous with cosmology and results in a cyclic model of cosmic evolution, each phase of which permitting a "new dimension, including new forms of matter and a new force of nature" (Rosen, 206). In virtue of its self-containing and self-transforming quality, the Klein bottle eigenstructure εD2/εD3 develops from what Rosen calls "dialectical action upon itself", or "Kleinian autopoiesis" (110). Likewise, Rapoport holds that applying KBL to cosmology entails an eternal return symbolized by the *Ouroboros*. In agreement with Kurakin and Smolin above, Rapoport argues "Mathematically, non-linear systems have an increasing entropy until they suffer a transition, a 'blow-up' [...] in which they go through an infinity (the Klein bottle re-entering) after which the entropy is lowered, instead of the thermodynamic death that the Big-Bang mythology claims for the Universe" (2012, 90). Hence, both Rapoport and Rosen claim to have linked cosmological evolution with phenomenological dynamics. Following Peirce, Rosen proposes a model of evolution that also provides a means of accounting for *individuation* (Fig. 8.3):

This self-signification, Rosen maintains, involves three *contained* dimensions plus a fourth dimension that realizes "concrete self-reference" and reflects our lived subjectivity. While the Klein bottle presents a single surface that is simultaneously "open" and "closed", distinguishing three regions provides a phenomenological means of experiencing how *contained*, *uncontained*, and *containing* spaces can be dialectically related within a "self-containing" whole. Though examining our experience of ordinary objects tends to reinforce an intuitive subject-object split, proprioceiving one's experience of the Klein bottle, he maintains, should facilitate such an embodiment of self-containment: "if the Cartesian gap between subject and object is to be bridged, the property of self-containment is what must be digested" (2014, 273). Rosen notes however that this proprioceptive act is not merely accomplished with one's eyes, but is cognitive, which means that viewing the self-containment of the Klein bottle requires engaging our *whole* being.

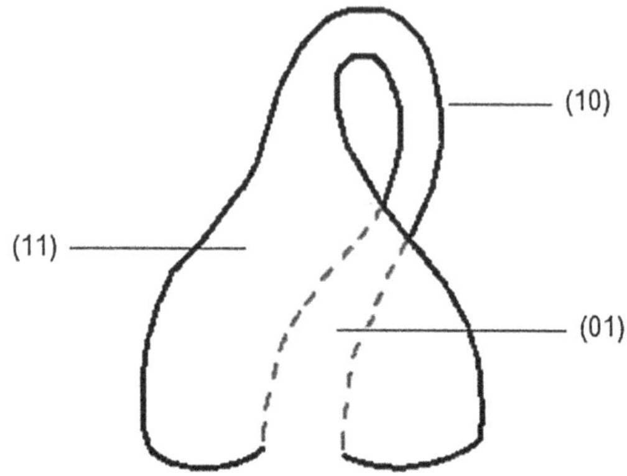

Fig. 8.3 Klein Bottle topology as a *triadic phenomenology*. (01) *Outside-Inside*—the *contained* consists of bounded and finite "facts", structures, and objects. In this primitive form of individuation, subject/object, past/present, inside/outside remain undifferentiated. (10) *Inside-Outside*—the *uncontained* denotes the individuation of agent in relation to their object via an action exclusively from the former to the latter. This reveals an implicit and dialectical *lifeworld*. (11) *Inside-Inside*—the *containing* space serves as the means by which reflection occurs. This represents a return of the lifeworld to itself via "proprioceptive circulation", wherein the identities of subject and object become fused. The *synthesis* of this stage becomes the object upon a return to (01) (Rosen 2008, 77, 127; Rapoport 2012, 72)

Rosen finds psychical and physical domains become so connected that movement from one to the other "is at once a return to the first" (237). This results in what he calls a "dialectical cosmogony", that involves "situating the analysis of cosmogony within cosmogony itself" (242). In the act of cognizing the self-transformation of the cosmos, the theorist herself surpasses the projective construction of oneself as an independent observer and "the analyst figures essentially in the reflexive enactment of this process" (ibid.). Echoing Harris's TAP, Rosen claims phenomenology reflects the very same self-referential dynamics of Nature because the "self that participates in the concrete universality of cosmic transformation must also be universal" (243). In accord with the conclusions of Sect. 2.5, emphasizing a sub-objective and psycho-physical quality of dimensional transformation results in a *self-referential ontology* and a *reflexive epistemology*, that is, an *onto-epistemology*.

In sum, for KBL the "transcendental problematic" is side-stepped by rejecting the assumption of object-in-space-before-subject and embracing an inherent reflexivity of Nature that is directly observable in and through embodied phenomenology. Brender's symmetry breaking and KBL are mutually supportive because in each case object and subject may be reconsidered as a single (dimensionally) differentiated field. In agreement with Clayton's CPM, KBL posits a unity in diversity of nature that also links Jordan and Day's wild systems theory of science as a "recurrence on reality" with the findings of Sects. 5.2 and 5.3, that is, further illuminating an *enactivist cosmology*. As will be discussed in the following sections, I contend this reasoning accommodates the *synthetic act* of observation in the very same manner as Harris's 𝕮.

8.3 Dialectical Holism

Harris (1965) argues that the world revealed by contemporary science is "undoubtedly monistic" (452). On the following page he makes it evident that what he has in mind is *neutral-monism* based upon process ontology: "Physics, if it does not actually reveal, nevertheless gives ample reason to presume a primordial matrix of the world, which though *mater*, is not *materia*, because it is prior to all matter. The matrix is in no way static but is some form of perpetual activity, in essence process" (453). What is this process? Harris later concludes: "The process of evolution thus transpires as one of progressively brining to explicit self-consciousness the principle of organization inherent in the organism from the start, that which galvanizes its auturgy and is itself the immanent principle ordering the cosmos as a whole" (1991, 91). As self-specification, the Concrete Universal is thus identified with processes of both mind and evolution. I examine his reasoning in three steps:

(1) *Epistemological* coherence—invoking coherence theory Harris says, "the diverse components must be mutually adjusted so that they fit together without conflict or friction, and this means that they are organized in accordance with a principle of order that determines how they are reciprocally adapted" (2006, 152). Harris argues that if the world is the result of a dialectical evolutionary process, the "'external world' is not extruded from the mind so as to be inaccessible to it, nor is the proffered theory of knowledge such that, if it were true, it would be impossible to ascertain" (163). Maintaining

realism about the scale of forms does not result in circularity he claims, because as long as chaos is parasitic on order, *modus tollens* shows that "unless the world had the sort of dialectical structure revealed in knowledge, we could not have the sort of knowledge that we enjoy" (1970, 383). This latter point will be critically examined below.

(2) *Ontological* coherence—in order for the Universe to "fulfil its own holistic character" Harris maintains, it must complexify its physical constituents so as to "elaborate its intrinsic differences as a noetic world, in which the physical and the organic are sublated and re-presented in a theoretical system of unified science" (1991, 155). Explication of the world is for Harris achieved in and through explication of the subject: "What is organized and integrated is a relational complex and relations are made explicit only when they are cognized" (2006, 147). The synthetic act of consciousness is thus considered the "universal function of relations" (152). The principle of coherence is here considered the same ϕ identified in previous forms, now manifested as the awareness of one's body-in-relation: "The 'transcendental subject,' therefore, turns out to be the ultimate dynamic principle of organization itself" (2006, 154–55). Individuation of any *phenomenon* is thereby achieved via the differentiation of a *phenomenological space*. The re-cognition of any object within this space is inextricably tied to the enactment or *projection* of a dialectical whole. Cognition of which, though forever incomplete, expands towards coherence with the *whole becoming of* Nature.

(3) *Anticipating KBL*—addressing the *priority problem*, Harris argues that the problem of the transcendental ego has not been resolved or avoided by appeal to Husserlian or Heideggerian phenomenology because in either case,

> relationships are being posited, awareness of which, even though only incipient, requires synthesis of a manifold in the Kantian sense, spontaneous and *a priori*, which can only be attributed to an apperceptive subject logically and ontologically prior to any of the related terms (one of which is Dasein). (1988, 95)

Harris contends the result of this recurrent situation is as follows: "On the one hand, Dasein is in the world, but at the same time, on the other, there is an important sense in which the world is in Dasein" (96). His solution is that the content of consciousness is "the entire scale of forms, dialectically related each to the next, as which the universal principle of organization has specified itself—the very process through which the mind has been generated. What becomes object is itself the autogenesis of the subject" (2006, 163). This is to say the object of the mind is the world in becoming and "the subject is no less than the world come to consciousness of itself" (1991, 115).

Harris hereby identifies the *concept* maintained by consciousness as a partial reflection of the very same self-specifying Concrete Universal considered heretofore (ℭ) (1987, 200; 1988, 104). He anticipates the idea that understanding the dynamics of respective kinds of complexity or scales ought to require disparate cognitive abilities of corresponding *predictive* capacities, development of which is the ongoing aim of consciousness (1970, ff. 352; 1991, ff. 121; ff. 135).[11] By implication, if the material world and phenomenology are best conceived as a self-specifying and self-transcending scale of forms, then it is justifiable to maintain that consciousness is itself a re-entrant scale of nature, thus vindicating Harris's TAP.[12]

8.3.1 On the Topology of Self-transcendence

Towards a development of dialectical holism, it may be noted that many of the above-mentioned authors who appear sympathetic to the system have made a common case that consciousness must be included within any sufficient cosmology. On numerous occasions, Harris and his would-be proponents have articulated this as some kind of "emergent hierarchy". I propose that such a conception would render the resulting metaphysics incoherent and therefore requires an alternative pragmatic conclusion, which repudiates Harris's grand Hegelian conclusion. Such a pragmatic conclusion, I maintain, is not only consistent with AE, but provides a sufficient metaphysical framework that extends the enactivist paradigm into a broader arena.

As Rapoport has maintained, due to its cyclic dynamics, appealing to the Klein bottle entails a *heterarchical* paradigm captured by *hyper Klein bottle logic* (HKBL): "For the HKBL the 'outmost' system in the hierarchy, re-enters in all systems which also exteriorize as the outermost

Outside-Outside state as well as in a myriad of intercommunicating systems" (2012, 86). Rapoport claims that the archetypical form of all systems in the world is the *hyper Klein bottle* (HKB), all of which are "interrelated in hierarchies with re-entrances of all elements on themselves and other selves and the whole re-entering in the elements as well" (97). In other worlds, the self-recurrent structure of Being entails an ongoing process of ontopoiesis that involves constraints across all scales—as Bohm has maintained, none of which being at the *bottom* or *top*. Rapoport argues that if conscious observation requires such a contextualization (i.e. if it must be embedded in culture), then consciousness must be described with HKBL, meaning "reality is a joint construction of the dialectical relation of object with subject" (24). For Rapoport, HKBL implies "that the world indicates itSelf through ourselves, which by doing this we indicate ourselves (and the Socius) and the Self of Nature" (96) (Fig. 8.4).

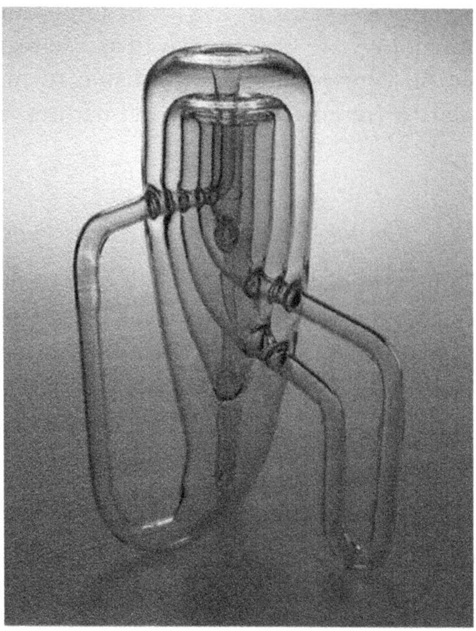

Fig. 8.4 *Nested glass* Klein bottle as a topological depiction of self-transcendence in and through cultural development. Created by Alen Bennett, shown at the London Science Museum. http://www.sciencemuseum.org.uk/images/i046/10314758.aspx. Image reproduced under creative commons license

Interestingly, this heterarchical conception is in line with Thompson's ontology:

> In the process view, "up" and "down" are contextual-relative terms used to describe phenomena of various scales and complexity. There is no base level of elementary entities to serve as the ultimate "emergence base" on which to ground everything. Phenomena at all scales are not entities or substances but relatively stable processes, and since processes achieve stability at different levels of complexity, while still interacting with processes at other levels, all are equally real and none has absolute ontological primacy. (2007, 441)

KBL is further reflected in Thompson's blog post, in which he builds on his earlier notion of "groundlessness" (1991) by maintaining "Nature, by way of the brain, beholds itself [...] You might think that you contain your mind, but it's actually the other way round: your mind contains you—your sense of self—and your mind is too vast for you to comprehend" (2015a).

Thompson (2015) further argues that the scientific method provides us no access to consciousness that is independent of consciousness. This means that the intersubjective confirmation of perceptual experience that "necessarily presupposes empathy or the recognition of others as having the same kinds of experiences as oneself, are the bedrock of experimental science" (99). So, scientific methods (including asking questions, formulating hypotheses, doing background research, analysing data, and communicating results) is possible and intelligible as a human activity only by presupposing consciousness (99–100):

> The upshot is that there's no way to stand outside consciousness and look at it, in order to see how it fits into the rest of reality. Science always moves within the field of what consciousness reveals; it can enlarge this field and open up new vistas, but it can never get beyond the horizon set by consciousness. In this way, direct experience is primary and science secondary. (100)

Thompson argues we can never acquire an objective measure of consciousness by something outside of consciousness, and consciousness never occurs without a context of embodiment, which likewise only occurs within the field of consciousness. The primacy of consciousness is thus considered methodological, epistemological, and now existential: "Consciousness is something we live, not something we have. [...] Consciousness is our way of being, and it cannot be objectified, that is, treated as just another kind of object out there in the world, because it is that by which any object shows up for us at all" (100).

This means that while the horizon (Lebenswelt) of phenomenology is situated within and conditioned by Nature, any conception of matter proposed by science is conditioned by and situated within our phenomenological horizon. Crucially, although unrecognized, Thompson's phenomenological approach has depicted mind as a *torsion of* Nature and has here (in only one page!) reiterated the epistemic side of Harris's TAP. Dialectical holism goes further. The argument (from Fig. 8.5) is that any attempt to explain *phenomena* (01), by investigating empirical *matter*—that which is apparently *beyond* self (10), or appealing to a priori first principles (11), results in a return and reliance upon the opposite. This is because the self-reference of Being not only distinguishes 01, 10, and 11, but also couples each to the rest in a *continuous self-differentiating movement*. Moreover, the trajectory (or *bias*) of this movement remains transparent (i.e. *unconscious*) to the subject at each stage. Consequently, the subject is engaged in an ongoing self-transcendence in and through their embodied engagement with their *Other* (i.e. self-projection).

An examination of cultural development serves to exemplify the above points. Harris and AE share the common appeal to coherence that requires reflexive attendance to a *finite set of relations deemed relevant within a given context*. For Harris, by *going beyond* this structure we fulfil the final cause of Nature in our philosophical and scientific practices. Gunn (2011) indirectly engages with Harris's original teleological proposal: the "goal" of Nature, he claims, "cannot involve the attainment or bringing into existence of some specific physical state [...] but only the continuous unfolding or objectification of the activity itself" (416). What is meant by "objectification"? Gunn proposes that human activity is a "special instance" of physical or "pneumatical force", whose goal is the progressive and "never-ending development of culture" (ibid.).[13] This implies a simultaneously *circular* and *linear* development of Nature:

> [O]bjectification of anthropic force at a given time alters the environment in which this force will subsequently objectify itself, and this in turn alters the subsequent objectification itself [...] Therefore, the correct geometrical image for cultural development and the history of thought is neither the circle nor the line but the spiral. (Ibid.)

More specifically, for dialectical holism it appears that re-cognizing our practices of objectification (i.e. our *selection effects*) within a given paradigm results in HKB *ascension*: effectively taking an *orthogonal* vantage point of said practices. While Harris maintained an *ontological realism* about a culmination of this process (i.e. *objective* idealism), I contend contemporary dialectical holism can only remain consistent (with an appeal to coherence) if its ontology is endorsed on *pragmatic* grounds, *with no final end in sight*. While the ends towards which we aspire (e.g. ethically, scientifically, and in terms of our ontological models) serve to guide us towards increasing coherence, there is no evidence, nor argument in Harris's system that supports a sufficient justification of any individuation—least of all an *optimal* state of consciousness. What remains is a continuous clarification or *explicative process* that is increasingly homoeomorphic across all epistemic and ontological domains, but one which also leads to the enactment of new domains.[14]

For Thompson, classical emergentism is inadequate for maintaining "that physical nature, in itself, is fundamentally non-mental, yet when it's organized in the right way, consciousness emerges" (2015b). He goes on to say that as long as we maintain this kind of conceptual dichotomy the emergentist view won't work. He claims this means developing a theory of "neutral monism—which says that the 'physical' and the 'mental' are concepts that capture different aspects of something that's more fundamental and neutral between them, being neither 'physical' nor 'mental'" (ibid.). Following Merleau-Ponty, Brender (2013) has proposed that our "new ontology", or "milieu" common to philosophy and the natural sciences that serves to unite "the phenomena of form and morphogenesis", thereby closing the CC is "asymmetry and symmetry-breaking" (272). Providing a *primal process of transformation*, symmetry breaking is epitomized by $\mathfrak{E}$: I propose that the common conclusion for Harris, Bohmian ontology, and AE is that the subjective and material domains are considered co-arising and mutually constraining scales; each of which incomplete abstraction from a *higher dimensional* Concrete Universal that achieves HKB self-reference via self-differentiation (i.e. symmetry breaking). On this *pragmatic* view, consciousness is to be *treated* as a higher-dimensional transformation of Nature (or *order of self-reference* in HKB).

Here, Bohm evidently mirrors Harris's scale of forms:

> [T]here is a universal flux that cannot be defined explicitly but which can be known only implicitly, as indicated by the explicitly definable forms and shapes, some stable and some unstable, that can be abstracted from the universal flux. In this flow, mind and matter are not separate substances. Rather, they are different aspects of one whole and unbroken movement. (1980, 14)

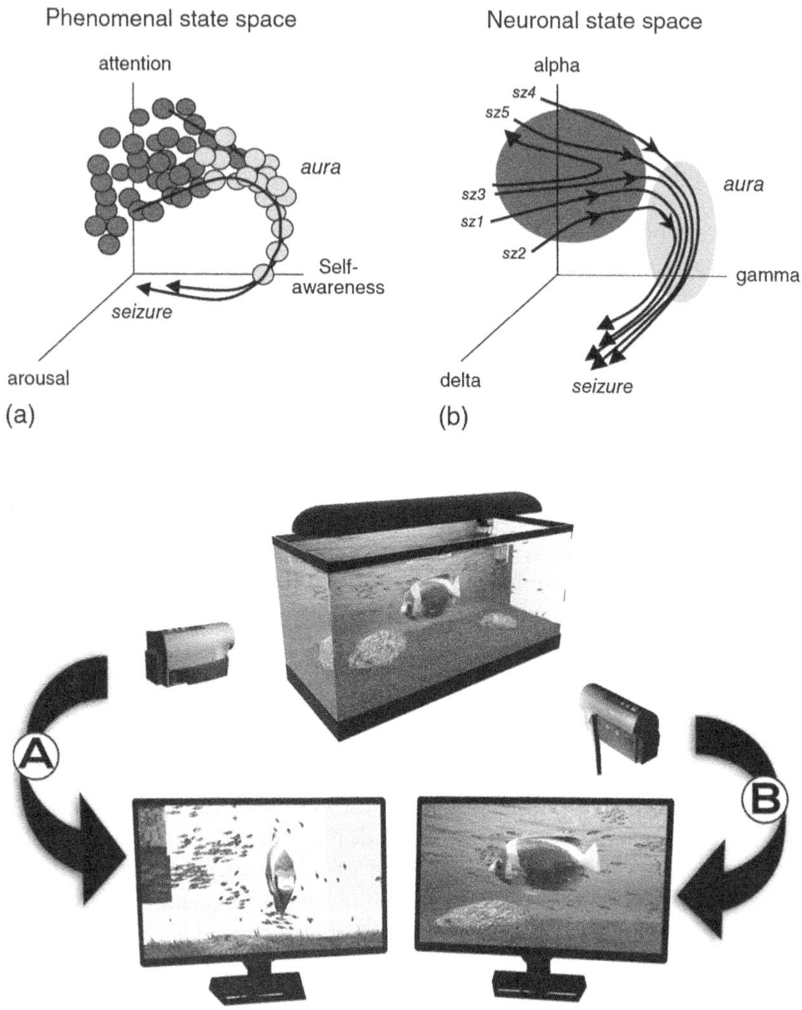

Fig. 8.5 *Relating* neurophenomenology *and the* implicate *order*. Top: a "homeomorphic" relationship between (a) phenomenological and (b) neuronal dynamics used in neurophenomenological seizure prediction (Van Quyen 2010, 259). Recreated with permission from the MIT Press. Bottom: based upon Bohm's (1980, 237) idea that the correlations of phenomenal and neurodynamics are analogous to respective 2D recordings of a single 3D object. Bohm holds that the reality *behind* respective "aspects" will be of a "higher dimensionality than are the separate images on the screens" (238)

Juxtaposing these diagrams is philosophically illuminating of the neurophenomenological methodology because in both cases "mind" is not to be identified with or constituted by the data from either of the abstracted phase spaces; it is rather to be identified with a *phase of matter* extended across spatiotemporal scales. Thus, mind must be treated as a *higher dimensional whole*, one that is constituted by the dynamics of neuronal, bodily, ecological, historical, and social scales, of which *we have incomplete access regardless of what methodology* (or language) is employed.

The dialectical holist response to the hard problem of consciousness is thus to back away from discussions presupposing ABL in analysing phenomena of the world. The dialectical holist refuses to take properties, T/F atomic *facts*, and the *object-in-space-before-subject* as given. Appealing to HKBL, consciousness may be framed as a natural function of the Universe that is ultimately neither mind nor matter but *contains* both. This topologic provides a sympathetic response to Harris's TAP, a revision of emergentism, and a common ground for mind and matter. Taking the lessons from KBL and wild systems theory together, Nature's self-referentiality is posited on pragmatic grounds with the expectation of theoretical *unity in diversity* in and through agential participation. This participation is believed to facilitate an ongoing clarification of our individuation practices directly correlated with an ongoing differentiation of scientific disciplines and philosophical domains (e.g. Clayton's CPM).

As formal governance is considered the torsion of *system*, so *phenomenology* is considered *a recurrence of* $\mathfrak{C}$. To my mind, a model of this kind fulfils Varela's conception of "walking the razor's edge" between scientific realism and subjective idealism. Thus, the resulting onto-epistemology encapsulating "self-reference", and "symmetry breaking" of the "Concrete Universal", appears to metaphysically ground the neurophenomenological programme and elucidate a cosmological dimension of the enactivist paradigm.

8.3.2 Ethical Implications: The Responsibility of Being a "Philosopher"

I believe that a particularly appropriate testament to Harris's character is eloquently exemplified in the preface of one of the last books he published, *The Restitution of Metaphysics* (2000). Even at the age of 90, still hoping to inspire new minds to the tremendous value of philosophy, he explains that a metaphysic and moral philosophy premised on the conception of wholeness is our best chance of counteracting the environmental and social challenges that have resulted from our modern technological capacities and empirical outlook. He explains that this work

> presents a world view that I think I can claim is genuinely post modern [...] If there is little sign at the present time of widespread adoption of a fresh philosophical outlook, I can but hope that what I have written may catch the attention of some of the younger generation, in whose hands the direction of the future lies. (2000, xiv)

While Harris's works on ethics, value, and politics are too extensive for even the briefest of summary here, I intend to bring attention to a characteristic feature of Harris's approach that fuses the practices of knowing with an analysis of normativity.

Harris argues that viewing the development of knowledge in the terms outlined heretofore produced a number of ethical implications in general and provided criteria for the practice of philosophy in particular:

> For the Ancients, philosophy meant what it said: the love of wisdom. Wisdom was held to be the fruit of intelligence and thoughtful reflection upon as wide a range of experience as possible. It was held to bring insight into every aspect of human life and to give enlightenment in the face of every problem, both theoretical and practical [...] Philosophy, as the love and practice of wisdom—for to love it must be to use it and not just to affirm its theoretical teachings—was not merely an intellectual pursuit but a way of life. (2000, 69)

To this end, while a review of the past and analysis of scientific claims are necessary, for Harris the aim of philosophy is to reveal connections between scientific disciplines thus unifying our understanding of nature to solve contemporary problems. As Harris states above, the problems we face are not just theoretical, but involve the interrelations of actual people. Ultimately Harris's metaphysics implies a deep concern for the evolutionary trajectory of humanity; an imminent responsibility to avert global disaster and aspire towards enlightenment. What is meant by "enlightenment"?

Harris laments that logic "is simply a matter of convention" in that the logician is free to adopt whatever logic validates their chosen theory (2000, 74). Consequently, "the philosophy of analysis, in its eagerness to clear away rubbish, has been so zealous that it has removed everything of any positive value and left a vacuum, into which the forces of extremism may move without effective resistance" (76). Interestingly, Harris found that over-emphasizing analysis was to blame for many ecological and socio-political problems of our day. He claims that when applied on its own, such intellectual dissection could not provide morals, cannot facilitate human development, nor account for one's fellow beings as *ends in themselves.* For Harris, the solution is to recognize "the organizing and integrating activity of the mind in all its functioning", believing that the "proper activities of reason" consists in a process both "analytic and synthetic in mutually complementary phases. Thus, it is the principle of insight and rational enlightenment" (77). Only the professional philosopher, he claimed, with an appreciation of such holism, is equipped to assess the clash of ideologies and their resultant social conflicts in the real world (80). Hence, enlightenment will at least *involve* an appreciation of the evolutionary trajectory of nature and a sense of how human thoughts and behaviours can either fit or conflict with this *explicative* process.

Harris argues that we cannot but adopt some philosophical theories about the world and that these theories unavoidably become *value judgements.* In line with the pragmatists, these judgements he claims, cannot but affect our conduct: "Philosophy always changes the world, and only if this is clearly recognized and acknowledged by philosophers will they understand, and can they fulfil their proper function in society conscientiously and responsibly" (81). As discussed above, he argues that for both common sense and scientific practice, observation is "embedded in and inseparable from interpretation" (88). Any thought or observation requires the projection of a paradigm, which means any "metaphysical explanation is the same in character as, and is continuous with, scientific explanation" (92). He goes on to remind us that, "[t]he aim of metaphysics is a comprehensive conceptual scheme in the light of which the whole of experience can be organized and become intelligible" (ibid.). It is the philosopher's responsibility, therefore, to not only critically assess the form of a paradigm, but to also synthesize such worldviews with ever increasing coherence.

In applying this dialectical logic, Harris (2005) advocates for a Global Federation, a unification of humanity and dissolution of sovereign states. Each nation, he maintains, shares the earth and must recognize their interdependence with those of their neighbours. As a species, Harris claims, our very survival depends upon our ability to facilitate the emergence of a "noösphere" of "communal minds and wills" (1988, 161). To do so he argues, human efforts across disciplines and nations must be unified in and through our differences, and there is only one way by which this can be achieved:

> [T]he pervading and consolidating sentiment that alone can bind together the members of any society [...] is love—not a sensuous or sentimental love, but a genuine felt concern for the true welfare of its object. In any community whatsoever that can maintain itself as a society, there must be some degree of mutual reliance and trust. (162)

Harris argued that the only way any sort of social contract or intellectually constructed social organization can remain a non-contradictory and self-preserving whole is through love. Upon "self-reflection" he claims, love of one's neighbour and of God become coincident: "It is genuine concern for the welfare of each and every individual, it is the universal respect for persons, that treats each as an end and none merely as a means" (163).

In relation to the whole of Nature, mankind is understood as a single community, the *concern* of which is the maintenance of oneself, world, and society. "That maintenance is a responsibility upon man, so that his relation to nature is ethical, rather than simply biological or technical" (1987, 262). To act upon this moral concern is considered a "responsibility to the ultimate totality", one fulfilled only "in a spirit of unreserved self-giving", towards what he calls a "a spiritual *Heimat*" (263). As with previous unifying principles, the individual "is not solely constituted by his or her subjective character, but by the social functions and capacities exercised and the consequent web of relations woven by interaction with other people" (1992, 15). *Evil* is thus considered whatever *part* "denies or contradicts the universal principle of order, or obstructs or counteracts its dynamic urge towards wholeness" (42). Harris certainly took pains to extend his theory of coherence to an analysis of social organization (e.g. 1992, 1993, 3, 10, 106), where he makes it clear that the interdependence and interconnectivity of diverse social systems coupled with an efficient means of energy use is not only necessary for human survival but teleologically driven by nature itself. In this way globalization, suppression of war, and the increasing development of renewable energy are a result of the same explicative process considered heretofore. In this vein, dialectical holism produces a form of *natural law theory*, in which the "law" is $\mathfrak{E}$.

Considering Harris's ethics with respect for the above discussions reveals a number of interesting conclusions. All conscious beings (especially philosophers) have the responsibility to pursue enlightenment. This involves relating a given theory, concept, etc., to an ever more inclusive system in order to establish a clearer understanding not only of the object, but also of *oneself* as the subject(s) *responsible* for bringing the whole into being. In line with contemporary phenomenology, *truth* thus requires an embrace, rather than a suppression of our subjective disposition, which stands as a prerequisite to *accounting for the world we enact and transcending our momentary means of individuation.* Accordingly, *Reality* is not only revealed to us, but comes into being as the observer and community can enact a greater range of relations. Whereas *evil* is instantiated by an *individual* "value" assuming dominance over, and thus contradicting the whole to which it belongs, love is considered the only way by which a (social) system can achieve coherent unity in and through difference. *Hence, the coherence of* love *and the* coherence *of truth are thus coincident; both enacted through the ongoing empathic* identity *with and synthesis of perspectives beyond one's own.*

8.4 Conclusion

Despite his entire career being focused upon a study of consciousness, when Harris finally finished his one and only text devoted to the subject, he merely remarked, "I have been able to get hold of several useful books on this subject and have managed to complete a short book of my own, 'Reflections on the Problem of Consciousness'" (2015, 259–60). Ever so humble, this is all we are told regarding the culmination, the conclusion of his lifelong intrigue. However, in the preface of this work we gain a somewhat deeper insight that rings with well-warranted condescension:

> From whatever angle one approaches philosophy one cannot avoid this problem [...] The problem has been central to my own thinking throughout my philosophical career and has figured in almost everything I have written. The solution I have reached is not new and I claim for it no originality; it is derived from Aristotle, Spinoza, Hegel and Collingwood [...] I know of none that is better founded. But I can claim that it has been sorely neglected and universally overlooked by contemporary writers whether from ignorance or from deliberate oversight; and I have felt the necessity repeatedly to remind them of it. (2006, ix)

I think that at this point Harris's humility had clouded his judgement. I hope it is clear from the contents of this text that the extensive efforts he made to *derive* his conclusion from earlier philosophers and read the resulting system through such a wide range of contemporary and consistent scientific theories assures the originality of his solution.

For my own part, I have shown that Harris's conception of the dialectical whole in his theories of cosmos, life, and mind, anticipates various philosophical and scientific works of recent years. I maintain that although Harris did not frame his work in this way, the metaphysics of dialectical holism is broadly a *phenomenological* response to both the problem of mind and anthropic reasoning. By way of making the value of this contribution clear, it will be useful to mention a number of avenues for future research within the resulting paradigm:

1. *Interdisciplinarity*—identifying the unifying principle with *formal governance* (e.g. solitons) provides a metaphysical framework for understanding natural phenomena from the formation of galaxies, to autopoietic systems of varying spatiotemporal scale and complexity. In line with CPM, dialectical holism predicts that principles from disparate disciplines will be mutually informative yet remain irreducible to one other. Ongoing collaboration should provide illumination of such principles and thereby lead to the development of further scientific disciplines. Specifically, as discussed in Chap. 5, to the extent that phase transitions and symmetry breaking provide support for $\mathfrak{E}$, the resulting theories of evolving laws, second- and third-order emergent systems will demand further scientific and philosophical development. Towards this end, it is hypothesized that phases of matter should be *innumerable* upon further investigation and give rise to corresponding disciplines of study.
2. *Theoretical applications*—depicting the whole of nature as a self-governing, self-differentiating open thermodynamic system (Concrete Universal) appears to have provided a viable teleological response to the anthropic principle. In support for this conclusion, Harris's theory of *environmental internalization* that was originally conceived as a biological telos is now logically extended to the evolution of the cosmos in and through conscious observation. If dialectical holism is to be maintained, proponents should critically develop and apply the HKBL that supports this approach as broadly

as possible, for example, to phenomenology, ethics, anthropology, cognitive neuroscience and of course cosmology. It is hypothesized that the emphasis on self-reference and second-order science will be invaluable in the above-mentioned transdisciplinary project.

3. *Bridging research programmes*—for Harris, the telos of nature in general but of humanity in particular, is the re-cognition of the universe as a whole. Broadening enactivist sensorimotor-affectivity and ENM, Harris has maintained that at the level of human (i.e. social) cognition, the same universal process of auto-genesis and self-reference appears as empathy and self-transcendence. Continued neurophenomenological research should in this vein be combined with projects in anthropology, complexity, and cosmology, to further clarify the purported scale of consciousness. Specifically, the philosophical interpretations of physics that underpin neurophenomenology must be further developed. Towards this end, research in DFT may find significant fruits by drawing upon Bohmian quantum mechanics and research into ***dimensionality***.

Harris died on 21 June 2009 at the age of 101, leaving behind over 25 books that he had authored or co-authored throughout his 70-plus-year career. In this text, I have shown that Harris's metaphysics is alive and well, though until now, maintained within disparate camps and research programmes, their proponents each unaware of one another. I have set out what I believe to be the strengths of the dialectical holist naturalization of mind and the challenges it ought to face in future research. The principal value of this work has been to pose a challenge to philosophers of mind to recognize the wider cosmological and underlying metaphysical implications of their position. Specifically, I hope this work stirs those sympathetic for AE to respond with either detailed refutation or support of the respective camps I argued they implicitly endorse. It warrants further mention that Harris system was extensive, developing tributaries into issues of free will, religion, political philosophy, and ethics, none of which have I been able to adequately address above. To fully grasp the import of the system born from Harris's efforts and the present text, the dialectical holist development of all these topics demands attention in future works.

Notes

1. Harris may have gone too far in positing "that theories of artificial intelligence can throw little or no light on how the brain operates, any more than they can on what I have said is the crucial problem" (2006, 87). He maintained that without a programmer there is no program, without an informed observer there is no behaviour, read out, etc. To judge an algorithm as intelligent or not, meant for Harris, presupposing an intelligent creator. Following Penrose's (1989) appeal to the Gödel theorem, he maintains that we must be able to determine whether or not the system is capable of non-computable *insight*, short of this no test will suffice (1991, 118).
2. Concerning the relationship between neuroanatomy and cosmology, Cherniak (2009) has found certain biological structures have originated "directly from simple physical processes, without need of DNA involvement" (371). An example of this is the way in which nature solves the *Steiner tree problem.* According to Chemiak's research, the brain solves this problem "via a lucky if counterintuitive factoid of computational geometry, that embedding dominates over topology; topology can in effect be ignored at comparatively little cost [...] Neuron arbors solve their minimization problem to within a few present of optimality, about as well as non-living arbors, such as river drainage networks" (372–73). Interestingly, this is to say that brains achieve volume minimization "via the same basic fluid dynamic processes as do water networks" (ibid.). Indeed, this should come as no surprise given the findings from Sects. 5.2 and 6.2.1.
3. Concerning its practical applications, Van Ouyen explains that in the case of seizure research, neurophenomenology can provide a significant advance in our understanding of the disease. What must be remembered however, is that "the efficiency of this approach needs a continuous circulation between the field of phenomena revealed by the patient's experience and the correlated field of phenomena established by the neurodynamics [...] this circulation can be based on a state space strategy, revealing homeomorphisms linking the topologies of phenomenal and neuronal states" (261). The research, he maintains, reveals a "co-determination" between respective domains.
4. Interestingly, Rosen (2008) argues that we cannot distinguish between uncertainties of the "large scale" (singularities in general relativity) and "small scale" (quantum measurement problem) because they "constitute the very same limitation" (31–32). In both cases he claims, the problem is that this uncertainty "entails the collapse of scale" (33).
5. In agreement with Harris's appeal to the ideatum, this implies that mental "models" are not representing a reality external to us, but "would have to

be stretched to such a point that the organism itself constitutes a model of reality" (Jordan and Day 2015, 16). In other words, avoiding an epistemic gap created by representations means conceptualizing the organism "as a self-sustaining prediction" (17).

6. As opposed to the "hard naturalism" that resulted from Dewey's system, which eschewed metaphysics and the reality of the subject, following Nagel (2015), Jordan and Day (2015) argue meaning is "ubiquitous throughout reality" (3) and later, "meaning is constitutive of reality" (4). Additionally, they contend that we need not appeal to the *a priori*, the transcendental, objective reality, or the absolute (3). It is difficult to imagine how the ubiq uity of meaning in Nature can be maintained and idealism avoided, while the transcendental, *a priori*, and absolute are rejected. These moves appear inconsistent but are clarified no further. Harris's holism depends upon the *synthetic a priori* and at least some implications for the absolute. To my mind, any dialectical holist would be obliged to maintain Nature is inherently meaningful because we are inherent within it, but "meaning" remains an abstraction from the whole.
7. Petit (2014) has also provided arguments concerning the relationship between autopoiesis, coherence theory and neurophenomenology that largely agree with the conclusion of this section. He proposes coherence as a "multi-scale concept" that may be "applicable to different levels of organization of the living being" (219). However, Petit gives up the neurophenomenological project because he finds it an impossible task to "identify" neurodyamics with mind since there are too many relevant parameters both within and beyond the brain to be considered. I address this issue in the final sections.
8. On numerous occasions Rosen (2008) recognizes Bohm's anticipation of this line of reasoning with his arguments for the holomovement (2–3, 26, 29–30, 46, 80, 83, 108, 126, 201, 211).
9. For a rigorous discussion concerning the present status and viability of *second-order cybernetic*, see Riegler and Müller (2016).
10. ABL is a two-valued system of T/F (0, 1), which are inter-transformable through negation. In Matrix Logic, on the other hand, Rapoport contends T and F are non-dual operators in which the negation operator does not transform the values. "True and False are no longer in specular reflexive relation; there is a folding in the cognitive plane, which is embodied in the twist of the Mobius Band and the Klein Bottle" (2012, 19). Here the torsion defines a cognitive operator, M, given by the difference of True and False, that is, M = True – False, which it not equal to the null operator. In Matrix Logic, ABL is a projection within the superposed topological states of true and false, nor true nor false, etc. (2012, 20).

11. To my mind KBL and symmetry breaking are really two sides of the same coin. This is because the very completion of self-reference by a Klein bottle introduces a fundamental kind of limitation, an *exclusion* of what is not included in the referential process, that is, what is *beyond* the system. It is this inclusion/exclusion that (on the dialectical holist account) necessitates an open-ended process for nature and mind. In the following section I shed some further light upon this process.
12. Additionally, the closest Harris comes to an analogous and succinct HKBL (as I depict in what follows) can be found in his (1987) discussion of objectivity (ff. 227).
13. Rosen claims his conception of "proprioception" is similar to Fox-Keller's "empathy", in her scheme of *dynamic objectivism*. In line with sensorimotor enactivism, Rosen emphasizes feeling, embodied empathy, and love that constitutes the lifeworld: "It is a world in which the dialectic of difference and identity is enacted proprioceptively, through an intimate knowledge of other that requires and is inseparable from the knowledge of self (a 'consciousness of self')" (241). Though not discussed by Gunn, empathy is a key aspect of AE and Harris's conception of cultural development—the latter view will be summarized in Sect. 8.4.
14. *Endnote to Plato*—this HKBL line of reasoning hereby implies that we are never free of the "cave", and never have an adequate grasp of the "forms", but must always seek the limitations of any given *shadow schema*, thereby perpetually liberating ourselves from one cave to another via the genesis of orthogonal perspectives: *conscientia est ego transcendentia ad infinitum*.

Works Cited

Bohm, David. *Wholeness and the Implicate Order*. New York NY: Routledge, 1980.

Bohm, David and Hiley, Basil J. *The Undivided Universe: an Ontological Interpretation of Quantum Theory*. New York NY: Routledge, 1993.

Brender, Noah M. "Sense-Making and Symmetry-Breaking: Merleau-Ponty, Cognitive Science, and Dynamic Systems Theory." In *The Canadian Journal of Continental Philosophy*. 246–270. 17 (2) (Fall/Autumn), 2013.

Carr, Bernard. "Cosmology and Religion." In Philip Clayton, ed. *The Oxford Handbook of Religion and Science*. 139–155. New York NY: Oxford University Press, 2006.

Carr, Bernard. "Introduction and Overview." In Bernard Carr, ed. *Universe or Multiverse?* New York: Cambridge University Press, 2007.

Cherniak, C. (2009). *Neuroanatomy and Cosmology*. (pp. 370–378) In: Bickle, J. (Ed). *The Oxford Handbook of Philosophy and Neuroscience*. New York NY: Oxford University Press.

Clayton, Philip. "Emergence: Us from It." In John D. Barrow, Paul C.W. Davies, and Charles L. Harper, Jr., eds. *Science and Ultimate Reality: Quantum Theory, Cosmology and Complexity*. 577–606. Cambridge University Press, 2004.

Clayton, Philip. "On the Plurality of Complexity Producing Mechanisms." In Charles H. Lineweaver, Paul C. W. Davies, and Michael Ruse, eds. *Complexity and the Arrow of Time*. 332–351. New York: Cambridge University Press, 2013.

Damasio, Antonio R., *Self Comes to Mind: Constructing the Conscious Brain*. New York, NY: Random House Inc, 2010.

Di Ieva, Antonio, ed. *The Fractal Geometry of the Brain*. Springer Pub, 2016.

Dixon, James A., Kelty-Stephen, Damian C., and Anastas, Jason. "The Embodied Dynamics of Problem Solving." In Shapiro, Lawerence A. ed. *The Routledge Handbook of Embodied Cognition*. 160–170. New York, NY. Routledge press, 2014.

Edelman, G. M., & Tononi, G. (2000). *Consciousness: How Matter Becomes Imagination*. London, Allen Lane. Penguin Press.

Ellis, George. "True Complexity and its Associated Ontology" In John D. Barrow., Paul C.W. Davies., and Charles L. Harper, Jr., eds. *Science and Ultimate Reality: Quantum Theory, Cosmology and Complexity*. 607–636. Cambridge University Press, 2004.

Ellis, George. "Physics, Complexity, and the Science-Religion Debate." In Philip Clayton, ed. *The Oxford Handbook of Religion and Science*. 751–766. New York NY: Oxford University Press, 2006.

Ellis, George. *How Can Physics Underlie the Mind? Top Down Causation in the Human Context*. Berlin Heidelberg: Springer Pub, 2015.

Fazelpour Sina, Thompson Evan. "The Kantian brain: brain dynamics from a neurophenomenological perspective." 31: 223–229. In *Curr Opin Neurobiol*, 2015.

Gordon, Susan, ed. *Neurophenomenology and its Applications to Psychology*. New York, NY: Springer, 2013.

Greenfield, Susan. *The Private Life of the Brain*. London, Allen Lane. The Penguin Press, 2000.

Guastello, Stephen J., and Liebovitch, Larry S. "Introduction to Nonlinear Dynamics and Complexity." In Stephen J. Guastello, Matthijs Koopmans, David Pincus, eds. *Chaos and Complexity in Psychology: The Theory of Nonlinear Dynamical Systems*. 1–40. New York, NY: Cambridge University Press.. 2009.

Gunn, David. *Pneumatology of Matter: A Philosophical Inquiry into the Origins and Meaning of Modern Physical Theory*. Winchester UK. Iff Books, 2011.

Harris, Errol E. *The foundations of Metaphysics in Science*. London: Allen and Unwin Pub, 1965.

Harris, Errol E. *Hypothesis and Perception: The Roots of Scientific Method*. London: Allen and Unwin Pub, 1970.

Harris, Errol E. *Formal, Transcendental and Dialectical Thinking: Logic and Reality*. State University of New York Press, 1987.

Harris, Errol E. *The Reality of Time*. Albany NY: State University of New York Press, 1988.

Harris, Errol E. *Cosmos and Anthropos: A philosophical Interpretation of the Anthropic Cosmological Principle*. Atlantic Highlands, NJ: Humanities Press International Inc, 1991.

Harris, Errol E. *Cosmos and Theos: Ethical and Theological Implications of the Anthropic Cosmological Principle*. Amherst New York: Humanity Books, an imprint of Prometheus Books, 1992.

Harris, Errol E. *The Spirit of Hegel*. Atlantic Highlands, NJ: Humanities Press International, Inc, 1993.

Harris, Errol E. *The Restitution of Metaphysics*. Amherst New York: Humanity Books, an imprint of Prometheus Books, 2000.

Harris, Errol E. *Earth Federation Now: Tomorrow is Too Late*. Sun City: AZ. The Institute for Economic Democracy, 2005.

Harris, Errol E. *Reflections on the Problem of Consciousness*. Dordrecht, Netherlands: Springer Pub, 2006.

Harris, Errol E. *Bound in Shallows: Autobiographical Reminiscences*. Marquette University Press, 2015.

Hollis, Geoff, Kloos, Heidi, Van Orden, Guy C. "Origins of Order in Cognitive Activity." In Stephen J. Guastello., Matthijs Koopmans., David Pincus, eds., *Chaos and Complexity in Psychology: The Theory of Nonlinear Dynamical Systems*. 206–241. Cambridge University Press, 2009.

Jordan, Scott, and Day, Brian. "Wild Systems Theory as a 21st Century Coherence Framework for Cognitive Science." In Thomas Metzinger and Jennifer M. Windt, eds. Open MIND. 21(T). Frankfurt am Main: MIND Group, 2015.

Kauffman, Louis H. "Self-Reference and Recursive Forms." In Journal of Social Biology. Struc. 53–72. Vol 10, 1987.

Merleau-Ponty, Maurice. "Eye and Mind." In Edie, James. M. ed. *The Primacy of Perception*. 159–90. Evanston, IL: Northwestern University Press, 1964.

McCrone, John. *Going Inside: A Tour Round a Single Moment of Consciousness*. London, Faber and Faber, 1999.

Minelli, Tullio A. "Neurodynamics and Electrocortical Activity." In Stephen J. Guastello., Matthijs Koopmans., David Pincus, eds., *Chaos and Complexity in Psychology: The Theory of Nonlinear Dynamical Systems*. 73–107. Cambridge University Press, 2009.

Nagel, S. K. (2015). *Thickening Descriptions with Views from Pragmatism and Anthropology: A commentary on Scott Jordan and Brian Day*. In: Metzinger, T. & Windt, J. M. (Eds). Open MIND. 21(T). Frankfurt am Main: MIND Group.

Penrose, R. (1989). *The Emperor's New Mind: Concerning Computers, Minds, and the Laws of Physics*. Penguin Books.

Petit, J. (2014). *Naturalizing Husserlian Phenomenology Along a Leibnizian Pathway.* (pp. 218–231) In: Komendziński, T., Nowakowski, P., & Wachowski, W. (Eds). *Enactivism: Arguments and Applications.* Avant Vol. V, No. 2/2014 ISSN: 2082-6710.

Primack, Joel, and Abrams, Nancy. *The New Universe and the Human Future: How a Shared Cosmology Could Transform the World.* The Terry Lectures Series. Yale University Press, 2011.

Rapoport, Diego L. "Surmounting the Cartesian Cut Through Philosophy, Physics, Logic, Cybernetics, and Geometry: Self-reference, Torsion, the Klein Bottle, the Time Operator, Multivalued Logics and Quantum Mechanics." In: *Foundations of Physics.* 33–76. January, Volume 41, Issue 1, 2011.

Rapoport, Diego L. "Hyper Klein Bottle Logophysics Ontopoiesis of the Cosmos and Life." In *Analecta Husserliana: The Yearbook of Phenomenological Research.* 275–350. Vol CXVII, Book 2. New York: Springer, 2012.

Rees, Martin J. *Just Six Numbers: The Deep Forces that Shape the Universe.* Great Britain: Basic Books, 1999.

Renaud, Patrice, Chartier, Sylvian, and Albert, Guillaume. "Embodied and Embedded: The Dynamics of Extracting Perceptual Visual Invariants." In Stephen J. Guastello., Matthijs Koopmans., David Pincus, eds. *Chaos and Complexity in Psychology: The Theory of Nonlinear Dynamical Systems.* 177–205. Cambridge University Press, 2009.

Richardson, Michael J., and Chemero, Anthony. "Complex Dynamical Systems and Embodiment." In Shapiro, Laurence A. ed. *The Routledge Handbook of Embodied Cognition.* 39–50. New York, NY: Routledge Press, 2014.

Riegler, A., & Müller, K. H. (2016). *Varieties of Contemporary Second-Order Cybernetics.* Constructivist Foundations. Vol. 11, 3.

Rosen, S. M. (2014). *How Can we Signify Being? Semiotics and Topological Self-Signification.* In: *Cosmos and History: The Journal of Natural and Social Philosophy.* vol. 10, no. 2. (pp. 250–277).

Rosen, Stephen M. *The Self-Evolving Cosmos: A Phenomenological Approach to Nature's Unity-in-Diversity.* Singapore: World Scientific Publishing Co, 2008.

Schneegans, Sebastian., and Schöner, Gregor. "Dynamic Field Theory as a Framework for Understanding Embodied Cognition." In Calvo, Paco, and Gomila, Antoni, eds. *Handbook of Cognitive Science: An Embodied Approach.* 241–271. San Diego. CA: Elsevier, 2008.

Schöner, Gregor., and Reimann, H. "Understanding Embodied Cognition Through Dynamic Systems Thinking." In Symons, John., and Calvo, Paco. eds. *The Routledge Companion of Philosophy of Psychology.* 450–473. New York, NY: Routledge Pub, 2009.

Steyn-Ross, Alistair D., Steyn-Ross, Moira, eds. *Modeling Phase Transitions in the Brain.* New York, NY. Springer pub, 2010.

Tegmark, Max. "The Multiverse Hierarchy." In Bernard Carr. ed. *Universe or Multiverse?* 99–125. New York: Cambridge University Press, 2007.

Tegmark, Max. "Consciousness as a State of Matter." Massachusetts Institute of Technology, Cambridge, MA, 2014.

Thompson, Evan. *Mind in life: Biology, Phenomenology, and the Sciences of Mind.* Cambridge, MA: Belknap Press of Harvard University Press, 2007.

Thompson, Evan. *Waking, Dreaming, Being: New Light on the Self and Consciousness from Neuroscience, Meditation, and Philosophy.* New York: Colombia University Press, 2015.

Thompson, Evan. "The Brain Is Wider Than the Sky: What Emily Dickinson Can Teach Neuroscience." https://www.psychologytoday.com/blog/waking-dreaming being/201504/the-brain-is-wider-the-sky, 2015a.

Thompson, Evan. "Unified Thought: 'Philosophical Ruminations,' Interview with Evan Thompson." In OldPiano.org, 2015b.

Tononi, Giulio. "Integrated Information Theory of Consciousness: An Updated Account." In *Archives Italiennes de Biologie.* 290–326. Vol 150, 2012.

Tschacher, Wolfgang, and Junghan, Uli. "Psychopathology." In Stephen J. Guastello., Matthijs Koopmans., David Pincus, eds. *Chaos and Complexity in Psychology: The Theory of Nonlinear Dynamical Systems.* 307–334. Cambridge University Press, 2009.

Van Geert, Paul. "Nonlinear Complex Dynamical Systems in Developmental Psychology." In Stephen J. Guastello., Matthijs Koopmans., David Pincus, eds. *Chaos and Complexity in Psychology: The Theory of Nonlinear Dynamical Systems.* 242–281. Cambridge University Press, 2009.

Van Orden, Guy C., Kloos, Heidi, and Wallot, Sebastian. "Living in the Pink: Intentionality, Wellbeing, and Complexity." In Cliff Hooker. ed. *Philosophy of Complex Systems.* 629–672. The Netherlands, Amsterdam: Elsevier Pub, 2011.

Van Quyen, Michel Le. "Neurodynamics and Phenomenology in Mutual Enlightenment: The Example of the Epileptic Aura." In John Stewart, Olivier Gapenne, and Ezequiel A. Di Paolo, eds. *Enaction: Toward a New Paradigm for Cognitive Science.* 245–265. MIT Press, 2010.

Zahavi, Dan. "Naturalized Phenomenology." In Shaun Gallagher., and Daniel Schmicking, eds. *Handbook of Phenomenology and Cognitive Science.* 3–19. New York, NY: Springer Pub, 2010.

Index[1]

[1] Note: Page numbers followed by 'n' refer to notes.

J. Schofield, *A Phenomenological Revision of E. E. Harris's Dialectical Holism*,
https://doi.org/10.1007/978-3-030-65029-2

L

M

The manufacturer's authorised representative in the EU is Springer Nature Customer Service Centre GmbH, Europaplatz 3, 69115 Heidelberg, Germany. If you have any concerns regarding our products, please contact ProductSafety@springernature.com

Printed and bound by CPI Group (UK) Ltd, Croydon, CR0 4YY
30/09/2026
02225326-0018